Issues in
Psychotherapy
Research

APPLIED CLINICAL PSYCHOLOGY

Series Editors: Alan S. Bellack, *Medical College of Pennsylvania at EPPI,
Philadelphia, Pennsylvania,* and Michel Hersen, *University of Pittsburgh, Pittsburgh,
Pennsylvania*

FUTURE PERSPECTIVES IN BEHAVIOR THERAPY
Edited by Larry Michelson, Michel Hersen, and Samuel M. Turner

CLINICAL BEHAVIOR THERAPY WITH CHILDREN
Thomas Ollendick and Jerome A. Cerny

OVERCOMING DEFICITS OF AGING: A Behavioral Approach
Roger L. Patterson

TREATMENT ISSUES AND INNOVATIONS IN MENTAL RETARDATION
Edited by Johnny L. Matson and Frank Andrasik

REHABILITATION OF THE BRAIN-DAMAGED ADULT
Gerald Goldstein and Leslie Ruthven

SOCIAL SKILLS ASSESSMENT AND TRAINING WITH CHILDREN
An Empirically Based Handbook
Larry Michelson, Don P. Sugai, Randy P. Wood, and Alan E. Kazdin

BEHAVIORAL ASSESSMENT AND REHABILITATION OF THE
TRAUMATICALLY BRAIN DAMAGED
Edited by Barry A. Edelstein and Eugene T. Couture

COGNITIVE BEHAVIOR THERAPY WITH CHILDREN
Edited by Andrew W. Meyers and W. Edward Craighead

TREATING CHILD-ABUSIVE FAMILIES
Intervention Based on Skills Training Principles
Jeffrey A. Kelly

ISSUES IN PSYCHOTHERAPY RESEARCH
Edited by Michel Hersen, Larry Michelson, and Alan S. Bellack

---

Issues in Psychotherapy Research

Edited by

Michel Hersen

Larry Michelson

Western Psychiatric Institute and Clinic
University of Pittsburgh School of Medicine
Pittsburgh, Pennsylvania

and

Alan S. Bellack

Medical College of Pennsylvania at EPPI
Philadelphia, Pennsylvania

Plenum Press • New York and London

Library of Congress Cataloging in Publication Data

Main entry under title:

Issues in psychotherapy research.

(Applied clinical psychology)
Includes bibliographical references and index.
1. Psychotherapy—Research. I. Hersen, Michel. II. Michelson, Larry, 1952–
III. Bellack, Alan S. IV. Series. [DNLM: 1. Psychotherapy. 2. Research—Methods.
WM 420 I866]
RC337.I87   1984   616.89'14'072   83-22938
ISBN 0-306-41401-5

©1984 Plenum Press, New York
A Division of Plenum Publishing Corporation
233 Spring Street, New York, N.Y. 10013

Printed in the United States of America

To
Victoria, Jonathan, and Nathaniel
Ellin Blumenthal
Barbara, Jonathan, and Adam

Contributors

Geary S. Alford, *Department of Psychiatry and Human Behavior, University of Mississippi Medical Center, Jackson, Mississippi*

Ted P. Asay, *Department of Psychology, Brigham Young University, Provo, Utah*

Curtis L. Barrett, *University of Louisville School of Medicine, Louisville, Kentucky*

Robert E. Becker, *Department of Psychiatry, Albany Medical College, Albany, New York*

Alan S. Bellack, *The Medical College of Pennsylvania at EPPI, 3300 Henry Avenue, Philadelphia, Pennsylvania*

D. Cartwright, *Department of Psychology, University of Colorado, Boulder, Colorado*

Anthony J. Conger, *Department of Psychology, Purdue University, West Lafayette, Indiana*

Norman I. Harway, *Department of Psychology, University of Pittsburgh, Pittsburgh, Pennsylvania*

Richard G. Heimberg, *Department of Psychology, State University of New York at Albany, Albany, New York*

Michel Hersen, *Department of Psychiatry, Western Psychiatric Institute and Clinic, University of Pittsburgh School of Medicine, Pittsburgh, Pennsylvania*

William G. Johnson, *Department of Psychiatry and Human Behavior, University of Mississippi Medical Center, Jackson, Mississippi*

Alan E. Kazdin, *Department of Psychiatry, Western Psychiatric Institute and Clinic, University of Pittsburgh School of Medicine, Pittsburgh, Pennsylvania*

Thomas R. Kratochwill, *School Psychology Program, Department of Educational Psychology, The University of Wisconsin-Madison, Madison, Wisconsin*

Michael J. Lambert, *Department of Psychology, Brigham Young University, Provo, Utah*

F. Charles Mace, *Department of Human Development, Lehigh University, Bethlehem, Pennsylvania*

Larry Michelson, *Department of Psychiatry, Western Psychiatric Institute and Clinic, University of Pittsburgh School of Medicine, Pittsburgh, Pennsylvania*

Peter E. Nathan, *Graduate School of Applied and Professional Psychology, Rutgers University, Busch Campus, Piscataway, New Jersey*

Richard E. Nay, *Psychology Department, University of South Carolina, Columbia, South Carolina*

W. Robert Nay, *Behavioral Medicine Associates, 6870 Elm Street, Suite 100, McLean, Virginia*

J. R. Wittenborn, *Interdisciplinary Research Center, Rutgers University, New Brunswick, New Jersey*

Gideon Weisz, *Department of Psychology, University of Colorado, Boulder, Colorado*

Jesse H. Wright, *University of Louisville School of Medicine, Louisville, Kentucky*

Preface

Psychotherapy research is undoubtedly one of the most puzzling, diverse, complex, controversial, and multidimensional areas tackled by clinical psychologists, psychiatrists, and psychiatric social workers. The numerous theoretical, methodological, and clinical-research issues dealt with by workers in the field have increased exponentially in the past three decades. To do full justice to the area, monographs in each of the specific subareas would be warranted.

In this volume, we, as editors, have endeavored to present the student and interested professional and practitioner with an understanding of the most salient issues and trends confronted by the psychotherapy researcher. In order to accomplish this task, we asked our colleagues, who are experts in their respective areas, to share their current thinking with us and with you, the readers. Thus, many theoretical viewpoints are represented, with none having a monopoly over the others. This is as it should be, given the data collected by clinical researchers at this time. We have also attempted to capture the excitement that has permeated the field in the past 30 years or so.

The book is divided into four parts. In Part I we have traced the main developments over the past 30 years, with specific focus on the 1950s, 1960s, 1970s, and 1980s. In Part II the important nosological, assessment, and measurement issues are considered in detail. Part III deals with design and statistical considerations in conducting psychotherapy research. Finally, in Part IV, general issues—including patient and therapist variables, ethics, and technical diversity—are given an incisive analysis.

Of course, many individuals have contributed to the development of this book. First and foremost are the contributors, who articulated their most current views. We are extremely grateful for the time they took out from their crowded schedules. Next, we would like to thank Mary Newell and Susan Capozzoli for their excellent secretarial assistance. Finally, we thank our editor at Plenum Press, Eliot Werner, for his confidence in our ability and his patience in the face of delay.

MICHEL HERSEN
LARRY MICHELSON
ALAN S. BELLACK

Contents

PART IV. GENERAL ISSUES

1

Introduction

1

Historical Overview

MICHEL HERSEN, LARRY MICHELSON, and ALAN S. BELLACK

INTRODUCTION

The objective of this introductory chapter is to provide a historical overview of the major theoretical, conceptual, and methodological issues and trends in psychotherapy research. The chapter is intended to illustrate the issues rather than detailing them exhaustively. However, the reader should find in-depth reviews of the substantive issues in the remaining chapters of the text. For purposes of this review, historical trends will be divided into three decades: 1950s, 1960s, and 1970s. Although these demarcations are admittedly artificial, they do provide boundaries that facilitate description and understanding of the issues in psychotherapy research over the past 30 years.

WHAT IS PSYCHOTHERAPY?

Prior to discussing the research issues of the last three decades, it is important that we come to some agreement about the parameters of "psychotherapy." The literal translation of the term means "treatment of the psyche." However, what "treatment" is composed of (e.g., theory, methodology, intervention strategies, etc.) varies widely.

According to Reisman (1966), there are over 31 distinct definitions of the term *psychotherapy*. Probably in no other area in the discipline of psychology

---

MICHEL HERSEN and LARRY MICHELSON ● Department of Psychiatry, Western Psychiatric Institute and Clinic, University of Pittsburgh School of Medicine, Pittsburgh, Pennsylvania 15213.   ALAN S. BELLACK ● The Medical College of Pennsylvania at EPPI, 3300 Henry Avenue, Philadelphia, Pennsylvania 19129.

are there more diverse, and perhaps divisive, opinions regarding definitions of a term. Some definitions are couched in terms of reconstructing personality dynamics by altering basic unconscious conflicts and motivations. Some emphasize changes in the self or self-concept. Others focus on changes in more observable behaviors and the contingencies under which they occur, with emphasis on the socioenvironmental context of behavior rather than personality *per se.*

Although no single definition of the term *psychotherapy* is universally accepted, there is generally agreement that it is (1) an interpersonal process, (2) designed to change feelings, cognitions, attitudes, and behaviors, and (3) conducted by a trained professional.

Historically, psychotherapy has its roots in ancient medicine, religion, faith healing, and—more recently—hypnotism. Indeed, psychotherapy is a relatively new profession with a long history. Previously, religion, philosophy, and theories of supernatural forces were posited to explain human behavior and disorders. Over the past 50 years a wide variety of therapeutic modalities has emerged, each alleging a unique perspective and often superior outcome over alternative interventions. Rather than addressing the numerous therapeutic modalities individually, it may be useful to outline their general commonalities and differences.

1. *An opportunity for relearning.* It is a fundamental assumption of all psychotherapies that human behavior can be improved. This primary assumption has two corrolaries. First, as problems develop from life experiences and learning, so too can they be undone through unlearning or relearning. Second, psychotherapy provides an optimal environment for relearning. The last point is derived from the position that, despite the psychotherapeutic benefits of many natural experiences—including support from family, friends, and so on—psychotherapy implies a *systematic* interaction, guided by theory and applied by a highly skilled practitioner to change an individual's behavior in a desired manner.

2. *Talking and experiencing.* Irrespective of the specific psychotherapeutic modality, interventions seek to engage the patient in either one or both of the activities of talking and experiencing. The "talking cure" is really only one avenue to providing the essential experiences for relearning. Indeed, it would be a gross oversimplification to focus merely on the verbal aspect of treatment. Although intellectual and emotional affective exchanges, as well as clarification, are necessary parts of the therapeutic process, they are not typically considered sufficient in regard to effecting improvements. Indeed, as described below, many therapists have asserted that the crux of effective therapy lies in the qualitative dimensions of the therapeutic "process" between helper and client.

3. *Psychotherapeutic relationship.* The importance of the relationship between therapist and patient has been the focus of innumerable articles, books, debates, and studies, particularly during the 1950s and 1960s. In addition to some form of advanced training, psychotherapists are expected to possess particular characteristics that are felt to be contributory, if not essential, to therapeutic effectiveness. These qualities have been widely cited and include genuineness, empathy, unconditional positive regard, authenticity, congruence, self-disclosure, and warmth. Clearly, the psychotherapist, lest he or she be just another sympathetic ear, must possess several outstanding capacities. He or she must be able to listen with the proverbial "third ear" without responding in terms of his or her own feelings, needs, opinions, or demands of social convention. He or she must be sufficiently endowed intellectually, emotionally, and academically to assess, understand, and remediate the patient's problems. Psychotherapists must exercise patience, optimism, good faith, curiosity, breadth of knowledge, practicality, and creativity—a difficult task at best! Not only do psychotherapists differ widely on many of these dimensions but opinions among experts regarding the therapeutic importance of these qualities range from "absolute necessity" to "neither necessary or sufficient." Thus, although many of these elements are commonly observed in most therapeutic relationships, there appears to be little overt agreement among the diverse theoretical/psychotherapeutic orientations as to their importance. Interestingly, despite the apparent verbalized differences among schools, oft-quoted studies by Fiedler (1950) indicate that psychoanalysts, client-centered therapists, and Adlerians were more alike in the manifestation of these therapeutic behaviors than their theoretical orientations would suggest. Similarly, more recent comparative outcome studies have shown equivalency of these therapeutic qualities across diverse psychotherapeutic modalities (see Chapter 11).

DIFFERENCES AMONG THE PSYCHOTHERAPIES

The preceding discussion has briefly examined the major commonalities among the psychotherapies; we will now highlight the major differences.

Goals and Strategies

The ultimate goal of any psychotherapy depends, in large part, on the underlying theoretical assumptions and philosophy of the psychotherapist. These goals derive from the views of the nature of human beings and typically entail helping people achieve a particular state of well-being. This state of well-being has also been termed self-actualization, mental health, achieving existential freedom, finding personal meaning, achieving liberation, as well as numerous other descriptions. However, the ultimate goal of treatment is not only variously defined but may differ substantially across modalities. Whereas one modality may seek insight into the etiology of a disorder, another may focus on symptomatic relief of the same. Similarly, as therapies may differ markedly in their articulated goals and objectives, so too are they different in regard to their intervention strategies.

Although anything but a cursory discussion of this issue is beyond the scope of the present chapter, it is a truism that there are a multitude of modalities. Recent surveys have estimated that over 250 different therapeutic modalities are presently in use, ranging from conventional to unorthodox, with foci varying from psychic to somatic to environmental. Needless to say, such diversity creates difficulties for researchers in ascertaining similarities, differences, and relative efficacy. Likewise, no two therapists, even of the same orientation, practice identically. Therefore, even within modalities, therapists may show major differences. Recent examples include schisms between behavior therapists of operant and instrumental persuasion versus those more cognitively oriented. Likewise, many contemporary psychoanalytically oriented therapists have divorced themselves from the strict tenets of Freudian doctrine regarding the importance of long-term treatment, the Oedipus and Electra complexes, and related, previously revered analytic notions.

THE DECADE OF THE 1950S

During the early stages of any clinical science there is great reliance on the case study method. Initially, these cases provide supportive evidence for both theory and practice. Indeed, Freud's psychoanalytic hypotheses were almost exclusively supported with evidence derived from relatively few clinical case studies. Case studies can generate hypotheses that then can be subjected to more empirical verification or rejection and can result in the stimulation of more systematic research (Bolgar, 1965). However, they do not allow objective conclusions to be drawn or generalizations to be made. Limitations—including uncontrolled and often unsystematic assessment, diagnosis, treatment, and related methodological parameters—severely attenuate both internal and external validity.

The 1950s saw a proliferation of uncontrolled and anecdotal clinical case studies, typically supporting the particular theoretical orientation of the proponents. However, as a result of several important events and historical trends that occurred during the 1950s, the case study came under increasing disrepute, in favor of more refined and empirical tools of research (namely, clinical surveys and experimental investigation). The major areas of psychotherapy research during the 1950s included examination of patient characteristics, therapist characteristics, process of psychotherapy, and so-called nonspecific treatment factors.

Patient Characteristics

Therapy outcome depends, to a significant degree, on the characteristics of the patient. Patients, even with identical diagnoses, show more differences than similarities. Indeed, patients differ vastly across a wide variety of salient dimensions, including intelligence, age, sex, education, race, social class, religious affiliation, mental status, motivation, insight, and defensiveness, to name but a few. Similarly, patients differ dramatically in regard to developmental experiences, "personality organization," assets and resources, interpersonal relations, social competency, and related properties. Recognizing the potential mediating effects of these patient characteristics, numerous studies were conducted during the 1950s to delineate which, if any, were related to treatment success or failure.

A fair number of studies examined the types of individuals who voluntarily seek out psychotherapy. Garfield and Kurtz (1952) and Rosenthal and Frank (1958) noted that approximately one-third of those judged to be in need of treatment refused when offered the opportunity. Likewise, Weiss and Schaie (1958) reported that 38% of their patients failed to return for treatment or disposition although they had been given a definite appointment.

Rosenthal and Frank's (1958) study found significant correlations between lower socioeconomic status (SES) and failure of clients to keep initial appointments. Conversely, higher social class was found to be positively related to being accepted for treatment (Brill & Storrow, 1960; Cole, Branch, & Allison, 1962). Similar results were reported in a study of a Veterans Administration Mental Hygiene Clinic (Bailey, Warshaw, & Eichler, 1959). It was found that assignment to psychotherapy was positively correlated to high SES, age, expressed desire for psychotherapy, and previous experiences in psychotherapy. Likewise, Hollingshead and Redlich (1958) reported that patients from different social classes received different types of treatment, with middle-to upper-SES patients more frequently being given long-term psychoanalysis. Another SES-related finding of Schaffer and Myers (1954) was the relationship between rank and status of the therapist and the social class of his patients. The study reported that basically the more senior and administratively powerful therapists saw the presumably more desirable patients (e.g., upper-SES).

As previously mentioned, many patients who were offered psychotherapy declined. Moreover, the apparent rejection of psychotherapy by patients in need was both difficult to explain and potentially harmful. Thus, the widespread phenomenon of dropping out and premature termination of treatment served as an impetus for studies designed to examine this problem. Prior to discussing these studies, it should be noted that according to reviews of this problem, 30% to 65% of all patients are dropouts (Eiduson, 1968). Moreover, those who terminate treatment prematurely only rarely seek additional or subsequent treatment (Garfield, 1963; Riess & Brandt, 1965), making this issue even more critical.

The major findings of these studies suggested that education, income, and occupation were positively associated with continuation of treatment (Imber, Nash, & Stone, 1955). Although all investigations did not confirm these results (Garfield & Affleck, 1959), most did report a positive correlation between SES, education, and length of treatment (Bailey *et al.*, 1959; McNair, Lorr, & Callahan, 1963; Rosenthal & Frank, 1958; Rubinstein & Lorr, 1956; Sullivan, Miller, & Smelzer, 1958).

In addition to SES and education, several studies examined the role of sex, age, diagnosis, and race. With regard to sex, no consistent differences emerged. Cartwright's (1955) study found that only a small proportion of the total variance in length of treatment could be accounted for by the patient's sex. Likewise, while some evidence suggested that therapists preferred younger patients (Bailey *et al.* 1959), age was not an important variable in regard to continuation of treatment. Similarly, psychiatric diagnosis was not found to be a contributory factor to early termination (Affleck & Garfield, 1961; Bailey *et al.*, 1959). In regard to race, Rosenthal and Frank (1958) reported that about twice as many white as opposed to black patients remained for six sessions. In conclusion, it appeared that patients with higher SES and more education were less likely to terminate treatment prematurely and were more likely to be offered the preferred treatment of that era, long-term psychoanalysis.

A considerable amount of research also focused on client variables as they related to psychotherapy outcome. First, social class was not found to be significantly related to outcome (Cole *et al.*, 1962; Rosenthal & Frank, (1958). Several studies find evidence suggesting a positive correlation between level of education and outcome (Bloom, 1956; Sullivan *et al.*, 1958), but others reported no significant results (Knapp, Levin, McCarter, Wermer, & Zetzel, 1960; Rosenblaum, Friedlander, & Kaplan, 1956). Likewise, variables such as sex and age did not appear to be clearly related to therapeutic outcome. Furthermore, studies investigating intelligence, expectations for improvement, and severity of illness have generated equivocal and often contradictory results. Thus, this area of research raised more questions than it answered, and many remain unanswered today.

Therapist Variables

The failure to identify client predictors of outcome helped stimulate psychotherapy researchers to begin systematically to examine therapist characteristics and their potential relationship to outcome. The research suggested that therapist psychopathology inhibited effectiveness of treatment (Bandura, Lipsher, & Miller, 1960; Cutler, 1958; Holt & Luborsky, 1958). With regard to sex of therapist, most studies indicated no outcome effect, with a few suggesting a moderate sex effect. Although several pilot studies were undertaken to examine the role of therapist level of experience on outcome, the data base lacked sufficient validity from which to draw any firm conclusions.

In regard to examining therapist qualities, psychotherapy research in the 1950s was more notable for posing questions than providing definitive answers. Typically, these studies (although representing a promising beginning) were flawed, with numerous methodological inadequacies and confounds apparent. Three decades later, we now acknowledge that the evidence for alleged therapeutic conditions is quite modest, suggesting that need to pursue more complex models (Parloff, Waskow, & Wolfe, 1978).

During the 1950s the interaction and combination of therapist and patient characteristics became a new area of interest, leading to numerous subsequent investigations in the 1960s and 1970s. Whitehorn and Betz (1954, 1957, 1960) proposed that therapists might be differentiated on the basis of their ability to work effectively with schizophrenics versus neurotics. A positive "match" between therapist and patient presumably facilitated efficacy, whereas a "mismatch" would allegedly inhibit positive outcome. Although this hypothesis was posited in the 1950s, it was not actively pursued by researchers until the 1960s and 1970s. Unfortunately, more recent reviews such as one by Razin (1977) of A–B research, suggests no "interaction hypothesis" for inpatients and only minimal applicability for outpatients under special circumstances. Thus, as the research became progressively more controlled, the results yielded diminishing support for the A–B hypothesis posited in earlier decades.

One of the largest process–outcome research efforts was undertaken to examine the relationship between the total amount of treatment and outcome. The reader is referred to Luborsky, Chandler, Auerbach, Cohen, and Bachrach (1971) and Meltzoff and Kornreich (1970) for more exhaustive and in-depth reviews. Overall, the evidence suggests a complex relationship: positive in many studies; no relationship in others; and curvilinear, negative, and equivocal findings in the remainder.

Once again, the search in the 1950s for simple linear process–outcome relationships was not successful, particularly in light of subsequent and more methodologically refined studies. Looking retrospectively at this line of research, it is easy to utilize hindsight and remark on the naiveté of merely considering quantitative variations in treatment without controlling for comparable qualitative variations. As discussed by Orlinsky and Howard (1978),

> More of a good thing is better than less of it; more of a bad thing is worse; and there may very well be a point of diminishing returns in any therapeutic relationship beyond which only negligible (or even retrogressive) results are attainable! (p. 313)

However, given the pilot status of these studies, such efforts must be regarded as initial steps that subsequently resulted in more developed and programmatic research during the 1960s and 1970s.

In the mid- to late 1950s, several researchers began examining the role of therapist values in psychotherapy. It was found that therapists do indeed transmit their values to patients and that outcome is positively correlated with the extent to which patients adopt these values (Parloff, Goldstein, & Iflund, 1960; Parloff, Iflund, & Goldstein 1957; Rosenthal, 1955). This pioneering research led to more in-depth studies, stimulating an important and previously dormant area of psychotherapy investigation.

''Milestones'' in the 1950s

Several significant developments occurred which, when viewed retrospectively, greatly facilitated the advancement of the field of psychotherapy research. One was the outstanding and perhaps most widely publicized (and debated) study of Eysenck (1952), which seriously questioned the effectiveness of psychotherapy. Eysenck asserted, based on data provided by Landis (1937) and Denker (1946), that two out of three neurotic patients could be expected to recover within 2 years without the benefit of treatment. Eysenck subsequently buttressed this argument with data from Shepard and Gruenberg (1957).

His major conclusions were as follows:

1. Neurotics who did not receive psychotherapy showed parallel improvements when compared with those who underwent treatment. This finding also held true for soldiers experiencing neurotic breakdowns and children with emotional disorders.

2. Neurotic patients treated with behavioral interventions improved more readily than those receiving psychoanalysis, eclectic therapy, or no treatment. Moreover, neither analytic nor eclectic treatments were more efficacious than no treatment. Therefore, except for the efficacy of learning-theory-derived interventions, alternative psychotherapies appeared to have nominal, limited, or even negative effects.

These conclusions, as well as the procedures employed to arrive at them, have been the subject of three decades of debate, ranging from praise to heated criticism. However, irrespective of Eysenck's conclusions *per se,* the catalytic influence of these pioneering works on the field of psychotherapy has left an indelible and, in most cases, a positive influence in stimulating psychotherapy research. Eysenck's arguments have fostered many productive discussions and reconceptualizations and served as an impetus in regard to promoting greater concern for psychotherapy's effectiveness. He also served to fuel the fire of discontent regarding the presumed effectiveness of psychoanalytic treatment. This dissatisfaction also helped play a role in the eventual development and rapid growth of behavior therapy. This leads to the next area of inquiry in the 1950s: psychoanalysis.

In the preceding 50 years or so, psychoanalytic theories remained largely untested, with most reports limited to case studies, retrospective/anecdotal surveys, and uncontrolled clinical data. Recognizing the paucity of data in regard to the efficacy of psychoanalytic treatment, the American Psychoanalytic Association in 1952 organized a central fact-gathering committee (Hamburg, 1967).

Briefly, the investigators requested and received 10,000 completed questionnaires from approximately 800 practicing psychoanalysts. Unfortunately, further information could be obtained from only 210 of the original 10,000 cases contacted. Of course, this information would have been used to evaluate the effectiveness of psychoanalysis. In addition to the above methodological confound, 60% of all patients were at least college graduates, compared to 6% in the general population. In 94% of the cases, patients received private treatment, with 61% attending analysis four to five times per week. Approximately 7% of the sample was made up of psychiatrists, presumably receiving treatment as part of their training. The committee noted that the survey had major flaws (which prohibited any conclusions regarding outcome) but hoped it would serve to stimulate further research. Therefore, although the intentions of the committee were laudable, the methodology employed to address the important questions was inadequate. On a positive note, the concept of utilizing large-scale clinical surveys to examine the experiences of thousands of patients across hundreds of therapists offered another strategy for monitoring the efficacy of psychotherapy. Clearly, more rigorous and controlled studies need to be performed. However, this pilot effort helped to shed light on potential pitfalls in clinical survey research.

Ellis (1957) compared the effectiveness of conventional psychoanalysis with "rational psychotherapy," with the latter reportedly being superior and requiring a shorter period of time. Unfortunately, the presence of numerous methodological confounds prohibited definitive conclusions from being drawn. However, this research was noteworthy in that it represented the beginning of rational-emotive therapy, which would subsequently receive widespread clinical and research interest among psychotherapists (Ellis, 1962).

Another major development during the 1950s was the emergence of client-centered psychotherapy, which dictated that the necessary and sufficient conditions for therapeutic change were therapist warmth, empathy, unconditional positive regard, and genuineness. Rogers (1957) claimed that the presence of these conditions was highly related to treatment outcome irrespective of the specific modality. In a pioneering outcome study, Rogers and Dymond (1954) reported on the successful outcome of client-centered treatment, leading the way for a new approach to treatment. Rogers made a significant contribution by (1) emphasizing the importance of research in psychotherapy; (2) attempting to assess the dimensions of what constitutes effective treatment; and (3) advancing a new and bold model of psychotherapy which, in many respects, stimulated much of the subsequent thought of nonanalytic models.

Psychotherapy Research Conferences

In April 1958, a conference was held in Washington, D.C., under the auspices of the Division of Clinical Psychology of the American Psychological Association and financed by the National Institute of Mental Health. The primary objectives of the conference were to (1) assess the present status of psychotherapy research and (2) stimulate future research. The list of participants included such luminaries as Lester Luborsky, Jerome Frank, Joseph Matarazzo, Morris Parloff, Carl Rogers, and Hans Strupp.

The proceedings of the conference were published (Rubenstein & Parloff, 1962) and resulted in widespread interest. Indeed, a second conference was held in 1961 and published shortly thereafter (Strupp & Luborsky, 1962). These conferences provided an initial and important opportunity for clinical researchers. The participants represented the full spectrum of contemporary orientations and shared their theoretical, conceptual, methodological, and therapeutic strategies with one another in the hope of refining psychotherapy research.

The conclusions of the conferences are too numerous to be given justice in this brief review. Therefore, the reader is referred to the proceedings for more detail. However, it is safe to state that these efforts served for many years as a reference and standard by which psychotherapy research might be designed, as well as enhancing comparability of independent researchers and facilitating "the growth of a systematic body of knowledge prerequisite to the development of the science of psychology" (p. 292).

With the close of the decade of the 1950s, psychotherapy research had transcended its case study origins, posing and answering complex questions.

The 1950s saw early studies examining simple demographic factors and descriptive parameters of patients and therapists, with progressive emergence of sophisticated methods for assessing interview behavior (Saslow & Matarazzo, 1962) and making psychophysiological evaluations (e.g., Lacey, 1962) of process and outcome phenomena. Similarly, as the technological and methodological sophistication significantly improved over the decade, there was also growing recognition of the possible limitations of psychotherapy, which in the past had often been regarded as a panacea for all patients. Related to the growing skepticism of the efficacy of certain treatments was a concomitant interest in and expansion of Rogerian, behavioral and rational-emotive approaches (modalities which today are both widely practiced and researched). In conclusion, it appears that the 1950s was a period of beginnings, some false starts, and a time marked by a spirited commitment to examining the many facets of psychotherapy.

THE DECADE OF THE 1960S

The decade from 1960 to 1970 was a difficult period for the "traditional" psychotherapies. In contrast to the tacit acceptance accorded in early years, psychotherapists and their favored techniques were beleaguered by friends and foes alike during this period. In contrast to the philosophical, political, and economic critiques forthcoming in the 1970s and 1980s, the primary villains in the 1960s based their attacks on research results (or the lack thereof). The increasing importance of research on psychotherapeutic procedures and outcomes was clearly apparent at the beginning of the decade. In his article in the *Annual Review of Psychology,* Rotter (1960) stressed the need for research on four variables: the patient, the therapist, therapy techniques, and measures. The role of the patient–therapist relationship soon became a fifth topic of study. At the beginning of the decade, all five variables were primarily considered in the context of outcome. In later years, they began to be examined independently.

The decade began with the second volley from Eysenck's formidable cannon (Eysenck, 1960). His earlier blast (Eysenck, 1952) had rattled some windows, but was generally taken as little more than the prattling of a malcontent. But his 1960 article was bolstered by more and somewhat better data. Moreover, in the period since 1952, psychotherapists had failed to provide any solid evidence to refute his arguments that psychotherapy was not effective. Consequently, this second report was more difficult to ignore. It stimulated a tremendous amount of research (designed to substantiate and repudiate Eysenck's claims) over the next 10 years.

One of the most important responses to Eysenck was a careful reexamination of the existing outcome data conducted by Bergin (1963). Among his findings was the (then) startling fact that while some therapists seemed generally to produce positive results, some generally produced negative outcomes (i.e., they made their patients worse). The resulting statistical average across therapists gave the misleading impression that psychotherapy was uniformly not effective. It could be argued that this was a case of "damning with faint praise," or "with friends like that you don't need enemies." In any case, this assuaged those clinicians who (somehow) knew that *their* patients usually got better. It also stimulated a great deal of research on the therapist factors.

Once the rock of therapist inviolability began to crack, it was not long before additional fissures developed. Not only were some therapists (apparently) harmful but there seemed to be a possibility that even the good ones were superfluous. A number of studies were conducted demonstrating that lay counselors (e.g., nonprofessionals given a minimum of training) could function quite well (e.g., R. Matarazzo, 1971). Probably the best-known study was reported by Rioch, Elkes, Flint, Usdansky, Newman, and Silber (1965). Eight homemakers were trained to conduct therapy; indeed, they proved to be about as effective as traditionally trained professionals. For a time, it was even hoped that indigenous "listeners," such as bartenders, hairdressers, and policemen could be trained to provide a therapeutic service to their customers. However, this has proven to be more idealistic than practical (or ethical). Challenges to the traditional therapeutic role of professionals also came from behavior therapy studies on the use of paraprofessionals (e.g., Ayllon & Azrin, 1968; Tharp & Wetzel, 1969). Nurses, psychiatric aides, parents, teachers, and B.A.-level assistants proved highly competent to carry out behavioral programs. It appeared that highly trained doctoral-level clinicians were not needed either to conduct therapy or implement specific behavioral interventions. This is not to say they were viewed as useless. Rather, the implication was that they would be more productively used in evaluation, diagnosis, administration, and treatment planning.

In addition to questions about effectiveness and the role of therapists, research in the 1960s confronted several other cherished notions about psychotherapy as well. To a great extent, psychotherapy had been regarded as a unique and somewhat mysterious process. The content, principles, and research procedures of mainstream psychology were seen as tangential at best. Yet with the increased number of psychologists involved in psychotherapy and the rise of the Boulder model of training, this distinction was increasingly seen as inappropriate. Two broad developments helped to demonstrate that psychotherapy was not inherently different and unresearchable.

First was the utilization of analogue research designs. Rather than being restricted to research on actual therapeutic intervention with real clients, researchers began to look at models or parallels of therapy as well as contrived or simulated therapy situations. Analogues were more economical, permitted much greater experimental control, and avoided many of the ethical constraints associated with field research. They facilitated the shift from survey research to controlled experimental designs. For example, Matarazzo and his colleagues (Matarazzo & Wiens, 1972) conducted an extensive series of studies on interview behavior, which was seen as a parallel of the dyadic interchange between therapist (interviewer) and client (interviewee). Many aspects of interviewee–client behavior (e.g., time spent talking, pace and duration of utterances) were shown to be affected by interviewer–therapist response style. Another perspective on this issue was provided by the tremendous number of studies on verbal conditioning. This research attempted to show how learning theory was applicable to the therapy situation as well as to demonstrate that therapists, consciously or otherwise, shaped client verbalizations by selective reinforcement of specific topics and response styles. Thus, Noblin, Timmons, and Reynard (1963) showed that psychoanalytic interpretations could reinforce client responses regardless of their accuracy. Krasner (1962) went so far as to call the therapist a "verbal reinforcement machine." But perhaps the classic example of this phenomenon was reported by Truax (cited in Truax & Mitchell, 1965), who analyzed transcripts of a case conducted by Carl Rogers. In contrast to the client-centered ethos, Rogers appeared systematically to reinforce client responses through the selective use of empathy, warmth, and directiveness.

The second major development, which also relied heavily on the use of analogue designs, was the growth of behavior therapy. It is difficult to identify the beginning of the "behavioral revolution," but it certainly had not progressed very far by 1960. Important books by Wolpe (1958) and Eysenck (1960) had caused hardly a ripple among traditional therapists, but they had a significant impact on the burgeoning number of young academicians and graduate students who saw clinical psychology as a scientific endeavor. The rapid development of behavior therapy can be traced through the yearly chapters on psychotherapy in *The Annual Review of Psychology* series during this period. Early volumes made no or minimal mention of behavior therapy. By the middle of the decade, Matarazzo (1965) concluded that behavior therapy was established as an "important new force" and "was here to stay." The following year, Dittmann (1966) indicated that "By now the group of workers identified with behavior therapy exhibit all of the characteristics of a school: ideas and people to espouse, others to reject, and a journal" (p. 68). Finally, the 1971 volume contained no chapter on psychotherapy; rather, the treatment chapter was devoted entirely to behavior therapy.

Four behavioral studies stand out as especially important in terms of methodological innovation as well as advancement of behavior therapy *per se.*

Peter Lang and his colleagues (Lang & Lazovik, 1963; Lang, Lazovik, & Reynolds, 1965) stimulated hundreds of studies by developing a clinical analogue of therapy: (1) in lieu of clinic-referred patients, they recruited subjects with a mild clinical problem which was common but would not ordinarily have required treatment (i.e., snake phobia). This strategy allowed researchers on just about every college campus to conduct clinically relevant research. (2) They administered a brief, standardized intervention rather than a longer-term idiosyncratic treatment. Thus, their procedure was highly replicable by other clinicians and researchers. (3) They trained graduate student therapists rather than employing experienced clinicians. This lowered the cost of the research dramatically. (4) Finally, they developed treatment-specific, objective measures (i.e., the Behavioral Avoidance Test) to assess outcome rather than relying on the subjective judgments of client and clinician. This helped ensure the validity of the results. The Lang studies not only demonstrated the efficacy of systematic desensitization but also set the pattern for a decade of research.

Many readers argued that the Lang studies were too removed from the clinical situation to have much generality. Gordon Paul (1966) solved this problem with probably the single most influential treatment study of the decade. There were five groups: systematic desensitization, insight-oriented psychotherapy, attention-placebo, waiting-list control, and a no-contact control. Paul used rigorous criteria to select highly speech-anxious subjects. He employed multiple dependent measures covering the three primary response modalities: self-report, physiological arousal, and motoric behavior. Treatment was conducted by experienced, nonbehavioral clinicians who were committed to insight-oriented psychotherapy; Paul trained them to conduct desensitization. Given these careful and rigorous controls, the clear superiority of desensitization was difficult to challenge.

The fourth study demonstrated how a rigorously defined treatment could be carefully analyzed in order to identify the probable mechanisms of action as well as to determine which elements of the procedure are really critical. Davison (1968) examined the role of muscle relaxation in desensitization, using a research strategy referred to as "systematic dismantling." It involved the critical evaluation of a treatment package, comparing groups containing all the treatment components with groups that paralleled the entire procedure with, however, the omission of a particular component or components. The Davison (1968) study included four conditions: (1) standard systematic desensitization, (2) deep muscle relaxation with irrelevant hierarchy, (3) exposure to a relevant hierarchy without relaxation, and (4) no-treatment control. Group 1 was significantly more improved than the other three groups, which did not differ from one another. Hence, use of the entire package was supported. More importantly, the design strategy illustrated that therapy need not be viewed as an inviolable gestalt. Rather, it could be broken down, analyzed, and modified when necessary.

All psychotherapy research between 1960 and 1970 was not as challenging as the examples cited above. A tremendous amount of supportive work was also conducted. Early in the decade, an important distinction was made between *outcome* research and *process* research (cf. Goldstein & Dean, 1966). As might be apparent, outcome research deals primarily with whether or not therapy works and how different forms compare. This research was and is extremely difficult and expensive to conduct. Also, the results are invariably confounded by problems in the selection of measures and subjects. Process research focuses on within-therapy variables: how therapy operates and what factors affect the operation rather than how well it works. These issues tend to be much more researchable than outcome questions, they are especially suitable for analogue procedures, and they avoid controversy about outcome criteria. According to Goldstein and Dean (1966), there were two broad assumptions that supported interest in process studies: (1) Therapy obviously works, so there is no need to study outcome; one merely needs to examine what therapists do. (2) The best way to understand and improve therapy procedures is first to identify the key elements (e.g., what happens what the participants do and then relate them back to outcome).

Process research can be loosely divided into three general categories: patient variables, therapist variables, and relationship variables. Patient variables include factors such as expectancy for success, intelligence, verbal skill, and social class. One of the most important findings of the decade was the so-called YAVIS pattern, first described by Schofield (1964). Successful patients tended to be young, attractive, verbal, intelligent, and successful (e.g., middle or upper SES). Apparently, therapy worked best for patients most like, and most liked by, their therapists.

We have already considered some examples of therapist and relationship factors above (e.g., verbal conditioning and interview behavior). We will also briefly illustrate some other popular themes. One major area of interest was the therapist–patient match. Data began to accumulate showing that most therapists were more effective with some types of patients than others. For example, some therapists (Type A) tended to be more successful with hospitalized schizophrenics than others (Type B; Whitehorn & Betz, 1960). Conversely, the B therapists were more successful than A's with neurotic patients (McNair, Callahan, & Lorr, 1962). Curiously, the A–B categorization, which is still the subject of considerable research, is made on the basis of a few items on the Strong Vocational Interest Blank.

Another important series of studies, emanating from client-centered therapy, concerned identification of therapist factors that led to success. An excellent summary of this work is provided by Truax and Mitchell (1971). Briefly, three therapist characteristics appeared to be of importance: genuineness, nonpossessive warmth, and accurate empathy. Early client-centered theory held that these attributes were essentially all that was needed for therapy to be effective. They were later thought to be necessary but not sufficient. Even this is no longer a supportable premise. Moreover, problems with measurement, definition, and replicability have raised serious questions about the significance of the early results. Yet, the three conditions are now widely regarded as being desirable in any therapy, regardless of theoretical orientation.

It is difficult to be objective or fair when evaluating research from 15 to 20 years ago. Given "20-20 hindsight," it is easy to be overly critical. Hence, we will conclude this section with two evaluative comments written at the time.

> The King is dead, long live the king. The picture of psychotherapy as a condition in which two people sit privately in an office and talk about the thoughts and feelings of one of them with the expectation that changes in these will automatically produce changes in overt behavior outside that office has been shattered. A new generation is emerging in the field of psychotherapy. A much wider range of procedures is being used by people with a variety of theoretical persuasions. Out of this innovative activity will undoubtedly come major theoretical changes. (Ford & Urban, 1967, p. 366)

> At first, surveying the research studies, we were only dismayed at the rarity of good studies (although in this respect, 1968 is not different from previous years). We were struck by the lack of control groups and properly blind ratings. We were also surprised by the variety of straightforward outcome studies. Does it mean only that research is still poor? It means at least that. (Gendlin & Rychlak, 1970, p. 156)

THE DECADE OF THE 1970S

Even the most casual observation of psychotherapy research in the 1970s clearly reveals many exciting developments in this field. The contributions were not only extremely important but also characterized by extensive diversity. Given the scientific impetus of this decade, bolstered by relatively good funding sources, many new investigators from psychology, psychiatry, and social work entered the burgeoning arena of psychotherapy research. A great diversity of interest corresponded to the varied educational, professional, and theoretical backgrounds of the evaluators. Thus, it should not be surprising that many different questions were posed, matched by an equally diverse armamentarium of strategies to answer them.

In this section we will highlight the contributions, the trends, and even the fads of the 1970s. First, we will consider the relevance of single-case research to the process and outcome of psychotherapy (cf. Hersen & Barlow, 1976). In our discussion we will not delve into the specifics of the individual designs, since this is dealt with in Chapter 6. However, we will discuss the philosophical impetus of this strategy of research. Second is the application of large-scale clinical trials contrasting therapeutic approaches with bona fide psychiatric patients (e.g., Sloane, Staples, Cristol, Yorkston, & Whipple, 1975). Along with implementation of large-scale outcome research is the attempt to standardize psychotherapy change measures (see Waskow & Parloff, 1975). This, of course, was and is critical to an intelligent understanding of interstudy comparisons. Third is the application of specific techniques with particular diagnostic categories (e.g., Bellack, Hersen, & Himmelhoch, 1981), following Gordon Paul's (1967) oft-quoted dictum concerning the right psychotherapy for the right patient. In this connection, very detailed treatment manuals have been developed in order to make public the specific transactions that take place during treatment (e.g., Bellack, Hersen, & Himmelhoch, 1980). Fourth is the concern in the 1970s of looking at the comparative effects of drugs and psychotherapy in addition to their possible complementary effects (e.g., Klerman, DiMascio, Weissman, Prusoff, & Paykel, 1974). Fifth is the possibility that psychotherapy could lead to *negative* as well as positive effects in patients (Strupp, Hadley, Gomes, & Armstrong, 1976). Thus, it became apparent that psychotherapy was a *potent* treatment which, when poorly applied, could be harmful to the recipient. Sixth is the attempt to evaluate, *en masse,* all research efforts in psychotherapy in order to ascertain their general value and utility. This approach has been labeled the *meta-analysis* by its proponents (Smith, Glass, & Miller, 1980). However, its merits have been questioned by some authorities (Garfield, 1981; Wilson & Rachman, 1983), particularly as to the arbitrariness and bias of the strategy. Seventh and last, is the overall greater concern with accountability in the practice of psychotherapy. Given the prospect of diminished funding combined with extensive third-party-payer interest and congressional concerns in a general atmosphere of increased scientific precision, the onus increasingly was on the therapist to demonstrate that what he or she was doing really benefited patients.

Single-Case Research

The single-case approach to evaluating psychotherapeutic techniques gained considerable popularity in the 1970s. Although this approach long has been a legitimate method of inquiry in psychoanalysis, in Pavlovian psychology, and in the laboratories of physiologists, the rigor of the operant laboratories gave it a new lease on life. Indeed, under the aegis of Skinnerian psychology, operantly oriented psychotherapy researchers used a variety of single-case *experimental* designs to document the controlling influences of their treatment methods (e.g., Baer, Wolf, & Risley, 1968). This was done by alternating baseline assessment (with repeated measures) and evaluation during the course of treatment (also with repeated measures) Not only were the controlling effects of treatment evinced in the single case strategies, with the patient as his own control, but use of repeated measures during baseline and treatment phases permitted an analysis of the vicissitudes of the treatment (i.e., the ups and downs of the therapy process). This, of course, is not typical of the larger controlled group-outcome studies, where there may be as few as two assessment points (pre- and posttreatment).

Many of the initial single-case analyses of therapy strategies involved children and were published in the *Journal of Behaviorial Analysis*. Much about the theory of single-case research has appeared in the psychological (Kazdin, 1978; Leitenberg, 1973) and psychiatric literature (e.g., Barlow & Hersen, 1973). In addition, more comprehensive descriptions have appeared in chapters (e.g., Hersen, 1982) and in books (e.g., Hersen & Barlow, 1976).

In the space remaining in this section, we will briefly describe the philosophy underlying employment of this design strategy. One of the basic arguments articulated by single-case researchers on behalf of their approach is its ability to demonstrate unequivocally the controlling effects of the treatment under consideration. This, in combination with the flexibility of the single-case strategy, makes it very attractive in initial hypothesis testing. As noted by Hersen and Barlow (1976):

> The major advantage of the single-case approach in beginning an investigation is the ability to isolate mechanisms of therapeutic action in a global treatment. Isolation of these mechanisms of action then makes it possible to combine various treatment variables in a more powerful treatment "package." There is little question that a single case approach, with its flexibility, can determine individual sources of variability and quickly bring the investigator to the point where he is ready to construct a global treatment package. (p. 63)

However, single-case experimental designs *do have their limitations.* Leitenberg (1973) acknowledges that

> Single-case experimental designs are no panacea. If used appropriately, that is, with provision of unconfounded distinctions between experimental phases and with provision of relevant patient behaviors, they can probably greatly clarify the immediate effects of important aspects of most psychotherapies. The purpose of these designs, however, is somewhat limited. They are primarily addressed to evaluating effects of therapy during the course of therapy. Thus, they can make less direct contributions to outcome research which is concerned with long-term treatment effects, with patient behavior *after* treatment has been completed. (p. 100)

Moreover, to contrast two or more psychotherapies, a controlled group-comparison design generally is needed.

Although single-case design strategies have been very popular with behavioral psychologists, they have been infrequently employed by nonbehavioral psychotherapy researchers. (Of course, there is no good reason why aspects of traditional psychotherapy cannot be evaluated with this strategy.) In addition, single-case research has not had much of an impact in the psychiatric arena. Finally, in recent years funding agencies seem to have given preference to psychotherapy researchers who use traditional group-outcome designs. Nonetheless, the single-case approach has found its deserved place as an respectable method of assessing initial treatment hypotheses. In our opinion, psychotherapy researchers of all theoretical persuasions first ought to assess their treatment notions before proceeding to the large scale designs that are time-consuming and very costly to carry out.

Outcome Research with Psychiatric Patients

Beginning with the 1970s and into the 1980s (e.g., Olson, Ganley, Devine, & Dorsey, 1981), we have witnessed extensive interest in comparing psychotherapeutic techniques with psychiatric patients exhibiting serious *clinical* symptomatology. This is in marked contrast to the plethora of short-term treatment analogues (using volunteer populations with little or no follow-up) that filled the pages of our psychological journals in the 1960s. Although these analogue investigations with subclinical populations are of some value (see Kazdin, 1978), they certainly are no substitute for the conclusions derived from the more difficult treatment problems tackled in *clinical trials*.

In general, these clinical trials involved the contrasting of two or more psychotherapeutic techniques with individuals requiring therapy for their neurotic or characterological problems. Treatment in such studies ranged from 12 to 30 sessions and in the better ones was administered by highly trained specialists (psychiatrists and psychologists) who were adherents of the particular therapeutic position being evaluated. This, of course, is vastly different than the short-term procedures carried out by graduate-level clinicians in analogue work. Furthermore, in clinical trials, much greater attention was accorded to the fate of study patients after treatment was concluded (i.e., follow-up).

The prototype of this study is the one conducted by Sloane, Staples, Cristol, Yorkston, and Whipple, (1975) with anxiety neurotics and personality disorders over a 4-month treatment period. This investigation contrasted behavior therapy, psychotherapy, and no-treatment controls. As noted by Joseph Wolpe in the foreword to the Sloane *et al.* (1975) book:

> The comparison between behavior therapy and brief psychoanalytically-oriented psychotherapy that was the central purpose of the study was statistically inconclusive, but a wealth of information was obtained that will be indispensable to researchers in psychotherapy for years to come. Some interesting trends were found. There were indications that in the more severe disturbed patients behavior therapy was more effective, though in milder cases, both approaches were in equal measure superior to a no-treatment control. There may be a parallel to this in the finding, in recent studies on small animal phobias, that almost any program of psychotherapeutic intervention is effective in overcoming mild neuroses, in contrast to the fact that explicit conditioning techniques are significantly superior in the treatment of major neuroses . . .
>
> Although the foregoing remarks have concentrated on behavior therapy, the results of the study were by no means one-sided, and there is much in the book that will give gratification and comfort to the psychoanalyst as well. (pp. xix–xx)

Parenthetically, we might note that Joseph Wolpe himself was one of the participating therapists in this study. Other notable therapists were recruited for conducting treatment for both groups. Thus, each therapy orientation was given a maximum test of its efficacy.

Despite the fact that Sloane's work is held as a model to emulate, there are some problems with the study. The first is that treatment was applied to a variety of diagnostic groupings. Second is the fact that a multitude of behavioral strategies (e.g., desenitization, assertion training, thought stopping) were carried out. Thus, precise conclusions about the given efficacy of a treatment for a given type of patient (à la Gordon Paul, 1967) were not possible. More specific attempts at delineating particular treatments for given populations are to be described in a subsequent section.

In addition to evaluating the two therapies, Sloane and his colleagues looked at the differences between behavior therapists and psychotherapists (Staples, Sloane, Whipple, Cristol, & Yorkston, 1975), patient characteristics as related to outcome (Sloane, Staples, Cristol, & Yorkston, 1976), and process issues (Staples, Sloane, Cristol, & Yorkston, 1976).

Concurrent with publication of the work of Sloane and his colleagues, the National Institute of Mental Health undertook to propose a series of psychotherapy change measures of acceptable reliability and validity that could be used by researchers in the field. The aim of their recommendations was eventually to facilitate interstudy comparisons (Waskow & Parloff, 1975). As argued by Tuma in the preface to Waskow and Parloff (1975):

> The present volume emerges from a concerted effort to advance the development of systematic information concerning the efficacy of psychosocial therapies and the measurement of psychotherapeutic change. Its immediate goal is to encourage the use of a range of standard procedures for the measurement of change in psychotherapy, in the hope of achieving greater comparability in the assessment of outcome across various studies. . . . It is our position that at the present time the use of a minimal, yet broad-gauged battery of standardized methods would prove helpful to the further development of the field. (pp. v–vi)

At this time it is a bit premature to ascertain whether these recommendations have borne fruit. However, we certainly acknowledge that the range of measures recommended by Waskow and Parloff (1975) is comprehensive in that it encompasses patient, therapist, relevant other, and independent clinical evaluator variables. Further, the chapter by Waskow (1975) concerned with "the process of choosing outcome measures" should be mandatory reading for every psychotherapy researcher. The logic of choosing a given set of measures is carefully articulated by simulating a conversation between the psychotherapy researcher and his or her consultant.

Specific Techniques for Particular Diagnostic Categories

Perhaps the 1970s will be remembered as the decade when specific psychotherapeutic techniques were developed for particular diagnostic categories. Although outcome research frequently did not reveal the superiority of one treatment strategy over another (cf. Luborsky, Singer, & Luborsky, 1975), the importance of clearly delineating and articulating in press the specifics of the therapy can only be applauded. Such documentation certainly ensured that future investigations involving the same treatment would lead to relevant interstudy comparisons. Moreover, for the first time the interactions between therapist and patient were truly made public. This naturally is of great importance in terms of the therapist's accountability.

Several psychotherapeutic approaches (each apparently successful) were developed and evaluated for the treatment of unipolar (nonpsychotic) depression in women. These include interpersonal psychotherapy (Weissman, Klerman, Prusoff, Sholomskas, & Padian, 1981), cognitive therapy (Rush, Beck, Kovacs, & Hollon, 1977), self-control therapy (see Rehm & Kornblith, 1979), and social skills training (Bellack *et al.*, 1981; Zeiss, Lewinsohn, & Munoz, 1979). Proponents of each of the therapies mentioned above have prepared comprehensive treatment manuals that include how to interact with the patient, what to discuss and when, homework assignments, length of treatment, maintenance, follow-up, and problematic issues. Ample case material appears in these manuals. Such material, of course, facilitates teaching the neophyte in addition to providing guidelines and standards. We would anticipate that more of these manuals will be developed and published in the future.

Reviews of the literature in the 1970s began to indicate which of the specific techniques developed were successful with given diagnostic categories. Indeed, limitations of the techniques also were noted. For example, it is quite apparent that systematic desensitization and other exposure techniques (e.g., flooding) are most effective with clinical phobia (Emmelkamp, 1979). However, for agoraphobia, flooding is more efficacious than desensitization. Moreover, treatment of panic attacks in agoraphobia is best accomplished with tricyclic drugs (e.g., imipramine). On the other hand, exposure treatment combined with response prevention seems to be the treatment of choice for obsessive-compulsives (Marks, 1981). But some ritualizers with accompanying depression require the addition of clomipramine, also a tricyclic drug. However, when the drug is discontinued, depressive symptoms tend to return.

As previously noted, a number of treatment regimes are successful in dealing with unipolar depression. None, however, is clearly superior to any of the others. As research continues in the future, we would expect that treatment may reach a prescriptive basis, given the type and subcategory of the disorder in question (Beutler, 1979).

Drugs and Psychotherapy

In actual clinical practice it is not at all uncommon for the patient to receive a "cocktail" combination of drugs and psychotherapy. However, until the 1970s the value of the combined approach was not evaluated empirically. A number of questions quickly came to mind: (1) Is psychotherapy as effective or better than drugs for certain classifications? (2) Are the two approaches complementary? If so, what is the contribution of each? (3) What approach is more cost-effective?

A few of the above questions have received initial assessment in the 1970s. For example, in the work of Klerman *et al.* (1974) with depressed women, it appears that amitriptyline was most effective in reducing depressive symptoms, whereas interpersonal psychotherapy (IPT) led to improved social functioning. However, in a more recent study, Weissman *et al.* (1981) showed that interpersonal psychotherapy and amitriptyline were equally effective in bringing about symptomatic remission. However, a combined condition of interpersonal psychotherapy and amitriptyline was superior to either treatment alone. At the 1-year follow-up,

> There were no differential long-term effects of the initially randomized treatment on clinical symptoms ... since most of the patients were asymptomatic. While most patients were functioning reasonably well, there were some main effects of IPT on social functioning at the one-year follow-up. Patients who received IPT with or without pharmacotherapy were doing significantly better on some measures of social functioning. (Weissman *et al.*, 1981, p. 51)

On the other hand, Rush *et al.* (1977) published a study showing the superiority of cognitive therapy over tricyclic medication for depressives with respect to (1) reduction of symptomatology, (2) maintenance of gains, and (3) dropout rate. More recently, Bellack *et al.* (1981) contrasted amitriptyline, social skills training plus amitriptyline, social skills plus placebo, and psychotherapy plus placebo with female unipolar depressives. Although each of the treatments led to significant improvements, the drug alone condition had the highest dropout rate. Moreover, the social skills plus placebo condition resulted in the highest proportion of "substantially improved" patients. There were no complementary effects of drug and social skills on the depression and skill dimensions evaluated.

The contrasting and complementary effects of drugs and psychotherapy have just begun to receive careful attention with respect to depression. Similar outcome studies have been (see Emmelkamp, 1979; Foa & Steketee, 1979; Marks, 1981; Mavissakalian & Michelson, 1982a, b) and are in the process of being completed with the anxiety disorders. With work continuing in this direction, future therapists will be in a better position to determine which psychotherapeutic and/or chemotherapeutic approach will best fit a given patient of a particular diagnosis or subcategory.

Negative Effects

The 1970s will also be remembered as the decade when psychotherapy researchers became concerned with the possible negative effects of their treatments. The impetus for such concern basically came from three directions. The *first* stems from Bergin's (1971) continued interest in evaluating the apparent minor therapeutic change seen in several treatment outcome studies. The most impressive aspect of Bergin's reanalyses of such data is that he takes into account patients who clearly *deteriorated* as well as those who *improved*.

> This conclusion was drawn from seven (well-designed) psychotherapy outcome studies and was startling in that it directly implied that some treatment cases were improving while others were deteriorating, thus causing a spreading of criterion scores at the conclusion of the therapy period, which did not occur among the control subjects. Evidently there is something unique about psychotherapy which has the power to cause improvement beyond that occurs among controls, but equally evident is a contrary deteriorating impact that makes some cases worse than they were to begin with. (p. 246)

A second impetus for concern with negative effects of psychotherapy can be traced to the precise work of operantly oriented therapists using single case methodology to evaluate their efforts. For example, the negative effects of differential attention were documented for oppositional and hyperactive children in a series of A-B-A type designs (Herbert, Pinkston, Hayden, Sajwaj, Pinkston, Cordua, & Jackson, 1973). Data indicated that behavior worsened under the effects of differential attention but improved when such treatment was withdrawn. This is in contrast to the successful use of differential attention in classroom situations with disruptive but nonoppositional children. Thus, the upper limits of the treatment were discovered using single-case analyses (cf. Hersen & Barlow, 1976).

Third and more recently, Hans Strupp and his colleagues (Strupp *et al.*, 1976) completed a report for NIMH in which the clinical and theoretical issues concerning negative effects in psychotherapy were examined. Based on their own review of the literature, opinions of experts, and a conceptual analysis of the issues, Strupp *et al.*, agree that "psychotherapy may have noxious effects." This, however, only is viewed as very preliminary evidence. The need for more empirical study of the problem obviously was recommended. But simply at the conceptual level, Strupp *et al.*, (1976) argue that existence of negative effects

> is a corollary of the proposition that if psychotherapy is a potent force in effecting positive change in feelings, cognitions, and behavior, it must be capable of producing negative changes as well. To reject this proposition means accepting the alternative that the effects of psychotherapy are essentially trivial, a position which has indeed been taken by some critics. (p. 83)

We might point out that this notion is generally no longer articulated in press (Bergin & Suinn, 1975). On the contrary, the various psychotherapies are seen as leading to behavioral change (Smith *et al.*, 1980).

Meta-Analysis

The 1970s also saw the emergence of meta-analysis, a strategy that purports to integrate treatment research through the statistical analysis of separate investigations (Smith & Glass, 1977). Smith and Glass report the evaluation of some 400 disparate treatment studies using this method. The range of studies was considerable and included analogue, counseling, behavioral, and nonbehavioral treatments. The magnitude of the treatment effect of each study was determined and related to therapist and patient characteristics. Smith and Glass (1977) conclude the following from this exhaustive analysis:

> On the average, the typical therapy client is better off than 75% of untreated individuals. Few important differences in effectiveness could be established among many quite different types of psychotherapy. More generally, virtually no difference in effectiveness was observed between the class of all behavioral therapies (systematic desentization, behavior modification) and the nonbehavioral therapies (Rogerian, psychodynamic, rational-emotive, transactional analysis, etc.). (p. 752)

Garfield (1981), a noted psychotherapy researcher, has questioned the "clinical significance" of findings from the meta-analysis. We must concur that other than documenting that psychotherapy appears to have a general effect, the utility of the meta-analysis is somewhat questionable. The confounding of this vast mélange of therapy studies, in our opinion, amounts to the comparison of "apples and oranges." Therefore, we see this as a passing fad. Others, of course, see this as a possible beginning for more sophisticated quantitative analyses that attempt to account for qualitative differences among studies.

Accountability

Accountability in the 1970s has become a key consideration for psychotherapists and psychotherapy researchers. As noted in an earlier section, such concern has come from a number of sources. One, of course, is the fact that psychotherapy has been widely publicized by the media, thus making its existence more accessible to the general public. As noted by Garfield (1981), "As a result of wider utilization, clients, third party payers, governmental agencies, the press, and others may all have a greater interest in psychotherapy and its efficacy" (p. 298). Books such as Tennov's (1975) *Psychotherapy: The Hazardous Cure* have also placed psychotherapy in the spotlight.

Probably the greatest impetus toward accountability has come from governmental agencies and insurance carriers in the 1970s. Earlier in the 1970s when it looked like a National Health Insurance Program might become a reality, the importance of documentation in this area was underscored. Indeed, congressional interest was sparked by this possibility (cf. Marshall, 1980; Parloff, 1979). Who would be reimbursed for what services and for how long was at issue. Also at issue was the clinical efficacy of the technique(s) to be administered. But of course, with economic conditions deteriorating, the notion of a National Health Insurance Program dissipated.

Currently, with deteriorating economic conditions, the third-party payers (e.g., insurance companies, governmental agencies) have been more concerned with the efficacy of the psychotherapy treatments given to patients. Moreover, there definitely is concern with the inordinate length of some of the therapies being administered.

Yet another example of accountability involves the concern that some of the newer psychotherapies prove effective only in the hands of the originators or their followers. Recently, multicenter treatment outcome studies have been put into operation to evaluate whether such treatments can be taught to other professionals who do not necessarily share the implicit theoretical bias of the originators. Also, the important question is: Will the level of success be as great in this case? This latter question has gained greater momentum in the 1970s. Not only is it important to document the superiority of one treatment over another, but it behooves investigators to document that these differences are of some clinical import. Here, of course, we are referring to the distinction between statistical and clinical significance.

In looking toward the future (the remainder of the decade of the 1980s), we can only anticipate that a greater measure of accountability will be called for both by the consumers of psychotherapy as well as those who pay for the services (i.e., third-party payers). In an atmosphere of economic "belt-tightening," we can easily predict the survival of psychotherapeutic practice increasingly will be determined by efficacy of the approach. Regardless of the economic situation, we are certainly in favor of improving the quality of care given to patients in need of psychotherapy. This should be the goal of psychotherapy research in the 1980s.

SUMMARY

From our analysis, it is clear that psychotherapy research has had an extremely brief history. Indeed, we can truly account for little more than three decades of work in this area. And at that, only the most important empirical findings have appeared since the 1960s. In our survey we have attempted to present the reader with a rather broad overview of trends rather than detailing a whole host of specific studies. In so doing we have tried to capture the thinking and spirit of three distinct (albeit artificially delineated) decades: 1950s, 1960s, 1970s. Over these three decades the trend toward increasing sophistication and methodological rigor in the psychotherapy research enterprise has increased substantially. Now being in the 1980s, we prognosticate that this trend will continue. We trust and hope that as important new findings emerge from controlled study, a parallel improvement will appear in the actual clinical arena.

REFERENCES

Affleck, D. C., & Garfield, S. L. Predictive judgments of therapists and duration of stay in psychotherapy. *Journal of Clinical Psychology,* 1961, *17,* 134–137.

Ayllon, T., & Azrin, H. H. *The token economy: A motivational system for therapy and rehabilitation.* New York: Appleton Century Crofts, 1968.

Baer, D. M., Wolf, M. M., & Risley, T. R. Some current dimensions of applied behavior analysis. *Journal of Applied Behavior Analysis,* 1968, *1,* 81–97.

Bailey, M. A., Warshaw, L., & Eichler, R. M. A study of factors related to length of stay in psychotherapy. *Journal of Clinical Psychology,* 1959, *15,* 442–444.

Bandura, A., Lipsher, D., & Miller, P. E. Psychotherapists' approach–avoidance reactions to patients' expressions of hostility. *Journal of Consulting Psychology,* 1960, *24,* 1–8.

Barlow, D. H., & Hersen, M. Single-case experimental designs: Uses in applied clinical research. *Archives of General Psychiatry,* 1973, *29,* 319–325.

Bellack, A. S., Hersen, M., & Himmelhoch, J. M. Social skills training for depression: A treatment manual. *JSAS Catalog of Selected Documents in Psychology,* 1980, *10,* 92. (Ms. no. 2156).

Bellack, A. S., Hersen, M., & Himmelhoch, J. M. Social skills training compared with pharmacotherapy and psychotherapy in the treatment of unipolar depression. *American Journal of Psychiatry,* 1981, *138,* 1563–1567.

Bergin, A. E. The effects of psychotherapy: Negative results revisited. *Journal of Counseling Psychology,* 1963, *10,* 244–250.

Bergin, A. E. The evaluation of therapeutic outcomes. In A. E. Bergin & S. L. Garfield (Eds.), *Handbook of psychotherapy and behavior change.* New York: Wiley, 1971.

Bergin, A. E., & Suinn, R. M. Individual psychotherapy and behavior therapy. In M. R. Rosenzweig & L. W. Porter (Eds.), *Annual review of psychology* (Vol. 26). Palo Alto, Calif: Annual Reviews, 1975.

Beutler, L. E. Toward specific psychological therapies for specific conditions. *Journal of Consulting and Clinical Psychology,* 1979, *47,* 882–897.

Bloom, B. L. Prognostic significance of the underproductive Rorschach. *Journal of Projective Techniques,* 1956, *20,* 366–371.

Bolgar, H. The case study method. In B. B. Wolman (Ed.) *Handbook of clinical psychology.* New York: McGraw-Hill, 1965.

Brill, N. Q., & Storrow, H. A. Social class and psychiatric treatment. *Archives of General Psychiatry,* 1960, *3,* 340–344.

Cartwright, D. S. Success in psychotherapy as a function of certain actuarial variables. *Journal of Consulting Psychology,* 1955, *19,* 357–363.

Cole, N. J., Branch, C. A., & Allison, R. B. Some relationships between social class and the practice of dynamic psychotherapy. *American Journal of Psychiatry,* 1962, *118,* 1004–1012.

Cutler, R. L. Countertransference effects in psychotherapy. *Journal of Consulting Psychology,* 1958, *22,* 349–356.

Davison, G. C. Systematic desensitization as a counterconditioning process. *Journal of Abnormal Psychology,* 1968, *73,* 91–99.

Denker, P. Results of treatment of psychoneuroses by the G. P. *New York State Journal of Medicine,* 1946, *46,* 2164–2166.

Dittman, A. T. Psychotherapeutic processes. In P. R. Farnsworth, O. McNemar, & Q. McNemar (Eds.) *Annual review of psychology* (Vol. 17). Palo Alto, Calif.: Annual Reviews, 1966.

Eiduson, B. T. A note on patients' reasons for terminating therapy. *Psychological Reports,* 1968, *13,* 38.

Ellis, A. Outcome of employing three techniques of psychotherapy. *Journal of Clinical Psychology,* 1957, *13,* 344–340.

Ellis, A. *Reason and emotion in Psychotherapy.* New York: Lyle Stuart, 1962.

Emmelkamp, P. M. G. The behavioral study of clinical phobias. In M. Hersen, R. M. Eisler, & P. M. Miller (Eds.), *Progress in behavior modification* (Vol. 8). New York: Academic Press, 1979.

Eysenck, H. J. The effects of psychotherapy: An evaluation. *Journal of Consulting Psychology,* 1952, *16,* 319–324.

Eysenck, H. J. Learning theory and behavior therapy. In H. J. Eysenck (Ed.) *Behavior therapy and the neuroses.* London: Pergamon Press, 1960.

Fiedler, F. E. A comparison of therapeutic relationships in psychoanalytic nondirective, and Adlerian therapy. *Journal of Consulting Psychology,* 1950, *14,* 436–445.

Foa, E. G., & Steketee, G. S. Obsessive-compulsives: Conceptual issues and treatment interventions. In M. Hersen, R. M. Eisler, & P. M. Miller (Eds.), *Progress in behavior modification* (Vol. 8). New York: Academic Press, 1979.

Ford, D. H., & Urban, H. B. Psychotherapy. In P. R. Farnsworth, O. McNemar, & Q. McNemar (Eds.), *Annual review of psychology* (Vol. 18). Palo Alto, Calif.: Annual Reviews, 1967.

Garfield, S. L. A note on patients' reasons for terminating therapy. *Psychological Reports,* 1963, *13,* 38.

Garfield, S. L. Evaluating the psychotherapies. *Behaviour Therapy,* 1981, *12,* 295–307.

Garfield, S. L.& Affleck, D. C. An appraisal of duration of stay in outpatient psychotherapy. *Journal of Nervous and Mental Disease,* 1959, *129,* 492–498.

Garfield, S. L., & Kurtz, M. Evaluation of treatment and related procedures in 1216 cases referred to a mental hygiene clinic. *Psychiatric Quarterly,* 1952, *26,* 414–424.

Gendlin, E. T., & Rychlak, J. F. Psychotherapeutic processes. In P. H. Mussen & M. R. Rosenzweig (Eds.), *Annual review of psychology* (Vol. 21). Palo Alto, Calif.: Annual Reviews, 1970.

Goldstein, A. P., & Dean, S. J. (Eds.), *The investigation of psychotherapy: Commentaries and readings.* New York: Wiley, 1966.

Hamburg, D. A. (Ed.) Report of an ad hoc committee on central fact-gathering data. New York: American Psychoanalytic Association, 1967.

Herbert, E. W., Pinkston, E. M., Hayden, M. L., Sajwaj, T. E. Pinkston, S., Cordua, G., & Jackson, C. Adverse effects of differential parental attention. *Journal of Applied Behavior Analysis,* 1973, *6,* 15–30.

Hersen, M. Single-case experimental designs. In A. S. Bellack, M. Hersen, & A. E. Kazdin (Eds.), *International handbook of behavior modification and therapy.* New York: Plenum Press, 1982.

Hersen, M, & Barlow, D. H. *Single-case experimental designs: Strategies for studying behavior change.* New York: Pergamon Press, 1976.

Hollingshead, A. B., & Redlich, F. C. *Social class and mental illness: A community study.* New York: Wiley, 1958.

Holt, R. R., & Luborsky, L. *Personality patterns of psychiatrists* (Vol. 1). New York: Basic Books, 1958.

Imber, S. D., Nash, E. H., & Stone, A. R. Social class and duration of psychotherapy. *Journal of Clinical Psychology,* 1955, *11,,* 281–284.

Kazdin, A. E. Methodological and interpretive problems of single-case experimental designs. *Journal of Consulting and Clinical Psychology,* 1978, *46,* 629–642.

Klerman, G. L., DiMascio, A., Weissman, M. M., Prusoff, B., & Paykel, E. S. Treatment of depression by drugs and psychotherapy. *American Journal of psychiatry,* 1974, *131,* 186–191.

Knapp, P. H., Levin, S., McCarter, R. H., Wermer, H., & Zetzel, E. Suitability for psychoanalysis: A review of 100 supervised analytic cases. *Psychoanalytic Quarterly,* 1960, *29,* 459–477.

Krasner, L. The therapists as a social reinforcement machine. In H. Strupp & L. Luborsky (Eds.), *Research in psychotherapy* (Vol. 2). Washington, D.C.: American Psychological Association, 1962.

Lacey, J. I. Psychophysiological approaches to the evaluation of psychotherapeutic process and outcome. In E. A. Rubenstein & M. B. Parloff (Eds.), *Research in psychotherapy* (Vol. 2). Washington, D.C.: American Psychological Association, 1962.

Landis, C. A statistical evaluation of psychotherapeutic methods. In L. E. Hinsie (Ed.), *Concepts and problems in psychotherapy.* New York: Columbia University Press, 1937.

Lang, P. J., & Lazovik, A. D. Experimental desensitization of a phobia. *Journal of Abnormal and Social Psychology,* 1963, *66,* 519–525.

Lang, P. J., Lazovik, A. D., & Reynolds, D. J. Desensitization, suggestibility and pseudo-therapy. *Journal of Abnormal Psychology,* 1965, *70,* 395–402.

Leitenberg, H. The use of single case methodology in psychotherapy research. *Journal of Abnormal Psychology,* 1973, *82,* 87–101.

Luborsky, L., Chandler, M., Auerbach, A. H., Cohen, J., & Bachrach, H. M. Factors influencing the outcome of psychotherapy. *Psychological Bulletin,* 1971, 75, 145–185.

Luborsky, L., Singer, B., & Luborsky, L. Comparative studies of psychotherapies: Is it true that "everyone has won and all must have prizes?" *Archives of General Psychiatry,* 1975, *32,* 995–1008.

Marks, I. M. Review of behavioral psychotherapy, I: Obsessive-compulsive disorders. *American Journal of Psychiatry,* 1981, *138,* 584–592.

Marshall, E. Psychotherapy works, but for whom? *Science,* 1980, *207,* 506–508.

Matarazzo, J. D. Psychotherapeutic processes. In P. R. Farnsworth, O. McNemar, & Q. McNemar (Eds.), *Annual review of psychology (Vol. 16).* Palo Alto, Calif.: Annual Reviews, 1965.

Matarazzo, J. D., & Wiens, A. N. *The interview: Research on its anatomy and structure.* Chicago: Aldine-Atherton, 1972.

Matarazzo, R. Research on the teaching and learning of psychotherapeutic skills. In A. E. Bergin & S. L. Garfield (Eds.), *Handbook of psychotherapy and behavior change: An empirical analysis.* New York: Wiley, 1971.

Mavissakalian, M., & Mitchelson, L. *Short-term outcome of 32 agoraphobic patients undergoing behavioral and pharmacologic treatments: Study I.* Unpublished manuscript, 1982. (a)

Mavissakalian, M., & Michelson, L. *Agoraphobia: Behavioral and pharmacologic treatments: Patterns of change: Study II.* Unpublished manuscript, 1982. (b)

McNair, D. M., Callahan, D. M., & Lorr, M. Therapist "type" and patient response to psychotherapy. *Journal of Consulting Psychology,* 1962, *26,* 425–429.

McNair, D. M., Lorr, M., & Callahan, D. M. Patient and therapist influences on quitting psychotherapy. *Journal of Consulting Psychology,* 1963, *27,* 10–17.

Meltzoff, J., & Kornreich, M. *Research in psychotherapy.* New York: Atherton Press, 1970.

Noblin, C. D., Timmons, E. O., & Reynard, M. C. Psychoanalytic interpretations as verbal reinforcers: Importance of interpretation content. *Journal of Clinical Psychology,* 1963, *19,* 479–481.

Olson, R. P., Ganley, R., Devine, V. T., & Dorsey, G. C. Long-term effects of behavioral versus insight-oriented therapy with inpatient alcoholics. *Journal of Consulting and Clinical Psychology,* 1981, *49,* 866–877.

Orlinski, D. E., & Howard, K. I. The relation of process to outcome psychotherapy. In S. L. Garfield & A. E. Bergin (Eds.), *Handbook of psychotherapy and behavior change.* New York: Wiley, 1978.

Parloff, M. B. Can psychotherapy research guide the policymaker? A little knowledge may be a dangerous thing. *American Psychologist,* 1979, *34,* 296–306.

Parloff, M. B., Goldstein, N., & Iflund, B. Communication of values and therapeutic change. *Archives of General Psychiatry,* 1960, *2,* 300–304.

Parloff, M. B., Iflund, B., & Goldstein, N. *Communication of "therapy values" between therapist and schizophrenic patients.* Paper presented at the American Psychiatric Association Annual Meeting, Chicago, 1957.

Parloff, M. B., Waskow, I. E., & Wolfe, B. Research on therapist variables in relation to process and outcome. In S. L. Garfield & A. E. Bergin (Eds.), *Handbook of psychotherapy and behavior change* (2nd ed.). New York: Wiley, 1978.

Paul, G. L. *Insight versus desensitization in psychotherapy.* Stanford, Calif: Stanford University Press, 1966.

Paul, G. L. Insight vs. desensitization in psychotherapy two years after termination. *Journal of Consulting and Clinical Psychology,* 1967, *31,* 333–348. (a)

Paul, G. L. Strategy of outcome research in psychotherapy. *Journal of Consulting Psychology,* 1967, *31,* 104–118. (b)

Razin, A.M. The A-B variable: Still promising after twenty years? In A. S. Gurman & A. M. Razin (Eds.), *Effective psychotherapy: A handbook of research.* New York: Pergamon Press, 1977.

Rehm, L. P., & Kornblith, S. J. Behavior therapy for depression: A review of recent developments. In M. Hersen, R. M. Eisler, & P. M. Miller (Eds.), *Progress in behavior modification* (Vol. 7). New York: Academic Press, 1979.

Reisman, J. M. *The development of clinical psychology.* New York: Appleton Century Crofts, 1966.

Riess, B. F., & Brandt, L. W. What happens to applicants for psychotherapy? *Community Mental Health Journal,* 1965, *2,* 175–180.

Rioch, M. J., Elkes, C., Flint, A. A., Usdansky, B. S., Newman, R. G., & Silber, E. *Pilot project in training mental health counselors* (U.S. Public Health Service Publication No. 125, 1965). Washington, D.C.: U.S. Government Printing Office, 1965.

Rogers, C. R. The necessary and sufficient conditions of therapeutic personality change. *Journal of Consulting Psychology,* 1957, *21,* 95–103.

Rogers, C. R., & Dymond, R. *Psychotherapy and personality change.* Chicago: University of Chicago Press, 1954.

Rosenbaum, J., Friedlander, J., & Kaplan, S. Evaluation of results of psychotherapy. *Psychosomatic Medicine,* 1956, *18,* 113–132.

Rosenthal, D. Changes in some moral values following psychotherapy. *Journal of Consulting Psychology,* 1955, *19,* 431–436.

Rosenthal, D., & Frank, J. D. The fate of psychiatric clinic outpatients assigned to psychotherapy. *Journal of Nervous and Mental Disease,* 1958, *127,* 330–343.

Rotter, J. B. Psychotherapy. In P. R. Farnsworth & Q. McNemar (Eds.) *Annual review of psychology* (Vol. 11). Palo Alto, Calif.: Annual Reviews, 1960.

Rubinstein, E. A., & Lorr, M. A comparison of terminators and remainers in out-patient psychotherapy. *Journal of Clinical Psychology,* 1956, *12,* 345–349.

Rubenstein, E. A., & Parloff, M. B. (Eds.), *Research in psychotherapy* (Vol. 2). Washington, D. C.: American Psychological Association, 1962.

Rush, A. J., Beck, A. T., Kovacs, M., & Hollon, S. Comparative efficacy of cognitive therapy and pharmacotherapy in the treatment of depressed outpatients. *Cognitive Therapy and Research,* 1977, *1,* 17–37.

Saslow, G., & Matarazzo, J. D. A technique for studying changes in interview behavior. In E. A. Rubenstein & M. B. Parloff (Eds.), *Research in psychotherapy* (Vol. 3). Washington, D.C.: American Psychological Association, 1962.

Schaffer, L., & Meyers, J. K. Psychotherapy and social stratification: An empirical study of practice in a psychiatric outpatient clinic. *Psychiatry,* 1954, *17,* 83–93.

Schofield, W. *Psychotherapy, the purchase of friendship.* Englewood Cliffs, N.J.: Prentice-Hall, 1964.

Shepard, M., & Gruenberg, E. The age of neurosis. *Millbank-Memorial Quarterly Bulletin,* 1957, *35,* 225–265.

Sloane, R. B., Staples, F. R., Cristol, A. H., Yorkston, N. J., & Whipple, K. *Psychotherapy versus behavior therapy.* Cambridge, Mass.: Harvard University Press, 1975.

Smith, M. L. & Glass, G. V. Meta-analysis of psychotherapy outcome studies. *American Psychologist,* 1977, *32,* 752–760.

Smith, M. L., Glass G. V., & Miller, T. I. *The benefits of psychotherapy.* Baltimore: John Hopkins University Press, 1980.

Staples, F. R., Sloane, R. B., Whipple, K., Cristol, A. H., & Yorkston, N. J. Differences between behavior therapists and psychotherapists. *Archives of General Psychiatry,* 1975, *32,* 1517–1522.

Staples, F. R., Sloane, R. B., Whipple, K., Cristol, A. H., & Yorkston, N. J. Process and outcome in psychotherapy and behavior therapy. *Journal of Consulting and Clinical Psychology,* 1976, *44,* 340–350.

Strupp, H. H., Hadley, S. W., Gomes, B., & Armstrong, S. H. *Negative effects in psychotherapy: A review of clinical and theoretical issues together with recommendations for a program of research.* Report to the National Institute of Mental Health, 1976. (Contract No. 278-75-0036 [ER])

Strupp, H. H. & Luborsky, L. (Eds.), *Research in psychotherapy* (Vol. 2). Washington, D.C.: American Psychological Association, 1962.

Sullivan, P. L., Miller, C., & Smelzer, W. Factors in length of stay and progress in psychotherapy. *Journal of Consulting Psychology,* 1958, *22,* 1–9.

Tennov, D. *Psychotherapy: The hazardous cure.* New York: Abelard-Schuman, 1975.

Tharp, R. G., & Wetzel, R. J. *Behavior modification in the natural environment.* New York: Academic Press, 1969.

Truaz, C. B., & Mitchell, K. M. Research on certain therapist interpersonal skills in relation to process and outcome. In A. E. Bergin & S. L. Garfield (Eds.), *Handbook of psychotherapy and behavior change.* New York: Wiley, 1971.

Waskow, I. E. Fantasied dialogue with a researcher. In I. E. Waskow & M. B. Parloff (Eds.), *Psychotherapy change measures: Report of the clinical research branch outcome measures project.* Rockville, Md.: National Institute of Mental Health, 1975.

Waskow, I. E., & Parloff, M. B. (Eds.), *Psychotherapy change measures: Report of the clinical research branch outcome measures project.* Rockville, Md.: National Institute of Mental Health, 1975.

Weiss, J., & Schaie, K. W. Factors in patient failure to return to clinic. *Diseases of the Nervous System,* 1958, *19,* 429–430.

Weissman, M. W., Klerman, G. L., Prusoff, B. A., Sholomskas, D., & Padian, N. Depressed outpatients: Results one year after treatment with drugs and/or interpersonal psychotherapy. *Archives of General Psychiatry,* 1981, *38,* 51–55.

Whitehorn, J. C., & Betz, B. J. A study of psychotherapeutic relationships between physicians and schizophrenic patients. *American Journal of Psychiatry,* 1954, *111,* 321–331.

Whitehorn, J. C., & Betz, B. J. A study of psychotherapeutic relationships between physicians and schizophrenic patients when insulin is combined with psychotherapy and when psychotherapy is used alone. *American Journal of Psychiatry,* 1957, *113,* 901–910.

Whitehorn, J. C., & Betz, B. J. Further studies of the doctor as a clinical variable in the outcome of treatment with schizophrenic patients. *American Journal of Psychiatry,* 1960, *117,* 215–223.

Wilson, G. T., & Rachman, S. Meta analysis and the evaluation of psychotherapy outcome: Limitations and liabilities. *Journal of Consulting and Clinical Psychology,* 1983, *51,* 54–64.

Wolpe, J. *Psychotherapy by reciprocal inhibition.* Stanford, Calif.: Stanford University Press, 1958.

Zeiss, A. M., Lewinsohn, P. M., & Munoz, R. F. Nonspecific improvement effects in depression using interpersonal skills training, pleasant activity schedules, or cognitive training. *Journal of Consulting and Clinical Psychology,* 1979, *47,* 427–439.

II

Assessment

2

Diagnostic and Nosological Issues in Psychotherapy Research

PETER E. NATHAN

Everything that can be thought at all can be thought clearly. Everything that can be said can be said clearly.

Wittgenstein, *Tractatus Logicophilosophicus*

INTRODUCTION

This chapter considers the manifold ways in which syndromal diagnosis— epitomized by DSM-III, the third edition of the *Diagnostic and Statistical Manual of Mental Disorders*—impacts on psychotherapy research and psychotherapy researchers. Since syndromal diagnosis, or diagnosis from coherent signs and symptoms, has been the principal basis for classifying psychiatric and psychological disorders through recorded history, the chapter provides, when appropriate, historical perspective on the recent evolution of syndromal diagnosis through the successive editions of the DSM. But DSM-III also represents a marked departure, in important ways, from what has gone before. Hence, much of our focus will be prospective, in anticipation of what DSM-III and its new departures will mean for psychotherapy research in the future.

A brief overview of DSM-III's new form, content, and, especially, assumptions and goals leads into a more detailed consideration of the syndromes themselves. Since DSM-III differs markedly from its predecessors,

---

PETER E. NATHAN ● Graduate School of Applied and Professional Psychology, Rutgers University, Busch Campus, Piscataway, New Jersey, 08854.

appreciation of the principles that guided the drafters of the new system is essential for the psychotherapy researcher. Most important, of course, is that the researcher know the changes that have come to the diagnostic categories in which he or she is interested. Especially important and significant changes have been introduced into the major syndromes of schizophrenia, the neurotic disorders, the substance-use disorders, and the personality disorders.

In our discussion of issues commonly raised by psychotherapy researchers about syndromal diagnosis, the section that concludes the chapter, we consider whether the new departures of DSM-III by themselves will contribute to solution of the pressing problems commonly posed or whether other solutions are required—or are possible.

DEFINITIONS

This chapter concerns itself with syndromal diagnosis and the impact it has had on psychotherapy research. Diagnosis is both "the process of determining by examination the nature and circumstances of a diseased condition" (the strictly *medical* usage of the word), and "scientific determination: a description which classifies precisely" (the more broadly *scientific* definition of the term). A syndrome is "a group of symptoms that together are characteristic of a specific condition, disease, or the like." Both definitions are from the *Random House Dictionary of the English Language* (Stein, 1966). Accordingly, syndromal diagnosis involves determining by examination the nature and circumstances of a disease condition by identifying groups of symptoms that are characteristic of that disease condition.

It is unnecessary—and unwise—to take the phrase *disease condition* literally in this context. Many psychologists would be resistant to such a literal interpretation, justifying that resistance by pointing to viable behavioral etiologies for many syndromes. Accordingly, it seems preferable to view the phrase *disease condition* metaphorically, or as a euphemism for disorder or dysfunction, since it implies tissue damage, destruction, or injury, and many or most of the disorders included in DSM-III do not meet that definitional criterion. Similarly, many will prefer to view *groups of symptoms* as complaints or target behaviors instead, since neither conveys the implication of physical disease process that *symptoms* does.

Syndromal diagnosis is neither behavioral nor dynamic assessment—comparable activities that behavioral and psychoanalytic clinicians undertake at the start of a relationship with a client. Although behavioral and dynamic assessment, like syndromal diagnosis, derive from specific theoretical positions on categorization and understanding, those positions are at variance both with each other and with syndromal diagnosis. Behavioral and dynamic assessment, moreover, typically represent initial steps in a (behavioral or analytic) intervention process, while syndromal diagnosis does not. As well, the basic aim of syndromal diagnosis is precise description, while understanding and evaluation bulk larger in the two other approaches to assessment. Further, syndromal diagnosis is not medical diagnosis, nor is it diagnosis of disease process (Nathan, 1981a), although certain of the diagnostic entities included in the DSM are of organic etiology. Finally, syndromal diagnosis of behavioral, psychological, and psychiatric disorders may not necessarily yield best understanding of these disorders or even best description of them. It is simply the most widely accepted approach to description. As discussed below, there are other approaches to description than syndromal diagnosis. None, however, has reached the level of acceptance that syndromal diagnosis enjoys.

DSM-III: NEW GOALS, ASSUMPTIONS, AND METHODS

The first two editions of the DSM, which appeared in 1952 and 1968, were welcomed by mental health professionals because these manuals brought some order to the chaos of conflicting psychiatric taxonomies of first decades of the 20th century. The absence of a comprehensive taxonomy enjoying consensual support was particularly strongly felt during and after World War II, when it was found that many of the most common wartime psychiatric syndromes would not be described according to available diagnostic systems, which emphasized only the more severe, usually psychotic, syndromes (Matarazzo, 1982).

The eagerness with which the first edition of the DSM was welcomed was tempered, shortly after its publication, by recognition that the instrument was not well enough organized or structured to improve diagnostic reliability (which had been low before the DSM because clinicians had no diagnostic system on which they could agree). Unfortunately, the poor diagnostic detail of the 1952 edition of the DSM did little to enhance diagnostic agreement, despite the generally universal agreement on the new system in this country. Minor modifications were made when the second edition of the DSM was published in 1968; unfortunately, structural alterations in procedures for diagnosis and greater detail in diagnostic criteria were not provided; as a result, reliability remained so low that the instrument's validity and utility were unproven (Beck, 1962; Matarazzo, 1978; Nathan, 1967). It was clear that a radically new, syndrome-based instrument designed to enhance advances in diagnostic understanding (e.g., Feighner, Robins, Guze, Woodruff, Winokur, & Munoz, 1972; Spitzer, Endicott, & Robins, 1975) was required. The result, in 1980, was the third edition of the DSM.

In fact, the 1980 DSM is markedly—in some cases, dramatically—different from its predecessors. To begin with, the instrument is both larger and more comprehensive than they were. More diagnostic labels, far more material descriptive of each syndrome, far greater attention not only to diagnostic criteria but also to differential diagnosis, prognosis, treatment, and etiology are all a part of the new instrument.

Notable too are the principles that guided the drafters of the DSM-III: a commitment to identify and describe all syndromes that could come to the attention of clinicians or represent problems for patients; a descriptive, phenomenologic emphasis in the nomenclature and a demand for empirical support for the taxonomy (as against a nomenclature reflective of theory, clinical opinion, or widespread belief); and belief in as wide a process of consultation during creation of the instrument as possible.

Specific distinctions between DSM-I, DSM-II, and DSM-III include the following:

1. Multiaxial diagnoses, by which diagnostic judgments are more closely linked to their environmental referents (e.g., premorbid stress and adjustment levels) to enhance the value of the diagnostic process for treatment planning
2. Operational criteria, by which diagnostic decision making can now be accomplished more reliably
3. Extensive testing, evaluation, and modification of the instrument's format, structure, content, and guiding principles prior to publication, to involve as wide a segment of the professional community as possible in the process of its development
4. Separate diagnostic recognition, for the first time, of the substance use disorders, the gender identity disorders, the psychosexual disorders, and disorders of impulse control
5. Expansion in the range and variety of childhood and substance use disorders
6. Separation of what were the neurotic disorders in DSM-II into separate diagnostic categories according to phenomenology

Multiaxial Diagnosis: Enhanced Utility?

DSM-III's multiaxial diagnostic system, which requires diagnostic judgments on five separate diagnostic axes, was designed to meet the criticism leveled at DSM-III's predecessors that syndromal diagnosis was of little value for treatment planning or other clinical purposes beyond simple categorization.

The multiaxial system calls first for recognition of most psychiatric and psychological conditions—the bulk of disorders described in DSM-III—on Axis I; when more than one condition is observed, all are to be entered. Multiple entries are entirely appropriate on this axis.

Characterological problems (those listed as personality disorders in DSM-II) and transient developmental disorders of childhood (or specific developmental disorders) are to be listed on Axis II, to differentiate them from the more serious disorders of Axis I. Axis III calls for physical disorders and conditions relevant to Axis I and II diagnoses. The call for these physical disorders in Axis III is a new one. It might focus clinicians more effectively on the interplay between physical and psychological, behavioral, or psychiatric conditions.

Axes IV and V call for clinicians to assess, on 7-point scales, the severity of psychosocial stressors impinging on the conditions listed in Axis I and the patient's highest level of adaptive functioning during the past year. It is along these dimensions that the clinician is to provide information on which treatment might well be based and by which its effectiveness might more effectively be judged. The major problem with these dimensions is reliability, that old bugaboo of DSM-I and -II. Although early data suggest that Axis IV and V reliabilities might be adequate (Spitzer & Forman, 1979), the judgments required are difficult, since it is impossible to specify their dimensions with the degree of detail of, for example, the operational criteria.

To the extent that multiaxial diagnosis paints a fuller picture of the patient than a simple diagnostic label, it ought to help psychotherapy researchers to match patients in comparison groups more carefully and completely. Knowing that a patient falls in diagnostic category A on Axis I and that he or she suffers from a concurrent serious physical disability (*and* that the Axis I condition was influenced by a severe environmental stressor) provides much useful information for comparison-group matching to reduce intersubject differences impacting adversely on psychotherapy outcome data. As well, in single-subject design studies that tailor treatment to patient, the fuller information provided by this system will doubtless aid the process. Finally, reliable Axis 5 data will permit judgments about therapy outcome reflecting differences in adequacy of pre- and posttreatment functioning, a more sensitive measure of change than posttreatment adjustment alone.

Operational Criteria: Increased Reliability?

DSM-III's operational criteria specify both the signs and symptoms required for each diagnosis and the decision processes by which they are to be integrated for diagnostic purposes. They are designed to heighten diagnostic reliability by setting out the common set of procedures and observations on which clinicians are to agree for each diagnosis.

An important virtue of the operational criteria is their largely empirical basis. They were derived, in most cases with few changes, from research diagnostic criteria generated from large samples of psychiatric patients, then tested and clarified by researchers at the Washington University School of Medicine and the New York State Psychiatric Institute (Feighner *et al.*, 1972; Spitzer, Endicott, & Robins, 1975). Encouragingly, early data from initial tests of the criteria with both adults and children suggest that the reliability of most diagnoses derived from the operational criteria is significantly greater than the reliability of the same diagnoses when DSM-II procedures are used (Cantwell, Mattison, Russell, & Will, 1979; Cantwell, Russell, Mattison, & Will, 1979; Spitzer, Forman, & Nee, 1979).

The significance of this change in diagnostic procedure for psychotherapy research is great. For the first time, the psychotherapy researcher can be relatively secure in the knowledge that experienced clinicians who adhere to the DSM-III operational criteria will seek the same diagnostic signs and symptoms when they diagnose research patients and will apply the same decision rules when they process their observations for diagnostic purposes. While it will still be necessary to sample resultant diagnoses for adherence to standards and procedures, the operational criteria offer a ready-made criterion against which the work of research diagnosticians can be compared.

While DSM-III's apparently successful effort to generate enhanced reliability only indirectly affects validity and utility, the effort is nonetheless necessary before validity and utility can be tested. Unreliable diagnosis rules out diagnostic validity and utility; reliable diagnosis makes it possible to determine the extent and nature of validity and utility. The implications of enhanced validity and utility for psychotherapy research are discussed below.

Pilot Testing and Consultation: Greater Consensus?

DSM-I and DSM-II were developed "behind closed doors" by a small group of senior psychiatrists who drew on their extensive clinical experience to shape a diagnostic system that strongly reflected a single theoretical approach to clinical work, the psychoanalytic. Details of the development of DSM-I are provided in the forward to DSM-III as well as two other recent review chapters (Matarazzo, 1982; Nathan & Harris, 1982).

Contrasting with the closed process that led to DSM-I and -II was the successful effort to involve hundreds of psychologists, psychiatrists, social workers, and others as formal or informal consultants to the Task Force on Nomenclature and Statistics of the American Psychiatric Association, the body responsible for the development of DSM-III. Advisory committees, composed largely but not entirely of psychiatrists, launched the developmental process by preparing initial drafts of sections corresponding to the 14 major syndromes of DSM-III; these committees numbered from 4 to 18. When the drafts were forwarded to the 19-member task force, they were critiqued and evaluated, then sent to various liaison groups within and outside the American Psychiatric Association. Among the latter were groups representing the national associations of psychologists and social workers as well as multidisciplinary interest groups (for example, the American Group Therapy Association, the American Psychoanalytic Association, and the Association for Advancement of Behavior Therapy). These consultations, which were sometimes heated, resulted in significant changes in the final document. The liaison committee from the American Psychological Association, for example, argued successfully for removal—from a statement of Guiding Principles, to accompany the final document—of an affirmation that mental disorders are a subset of medical disorders, a position with which the nonmedical mental health disciplines could not have lived.

Beyond the impact the advisory and consultative groups had on the evolving DSM-III, an extensive series of field trials, which began more than 3 years before the final document was published, shaped the final instrument. Pilot field trials and then a large-scale, 2-year field trial (sponsored by NIMH and involving almost 500 clinicians at more than 200 public and private settings) resulted in additional changes in the wording, structure, organization and, in a few cases, conceptual bases for most of the syndromes originally described in drafts of the document.

The process by which DSM-III was developed called on the collective wisdom of a very large and heterogeneous group of mental health professionals. As a consequence, the instrument reflects, to a far greater extent than its predecessors, the views on and experiences with syndromal diagnosis of mental health clinicians from a variety of disciplines, backgrounds, and theoretical persuasions. As well—and as importantly—the final form of the instrument reflects the actual experience of clinicians working with it in the field.

The principal value of this extensive prepublication consultation and field testing for psychotherapy researchers lies in the greater assurance they will have that the document and the process for diagnosis it prescribes more accurately reflect clinical practice. This attribute—DSM-III's consonance with clinical practice—means that clinicians will be familiar with the terms and procedures of the instrument and able to use them with conviction, support and, most of all, enhanced reliability.

The Syndromes

Some of the changes in syndromes introduced by DSM-III have little or no impact on psychotherapy research, while others have great significance. The change in the basis for categorization of the organic brain disorders from temporal to syndromal, for example, is not terribly important to psychotherapy researchers because psychotherapy is not a treatment of choice for the majority of such patients. Further, the marked changes introduced into the childhood disorder schema in DSM-III are also likely to be of only passing interest to most psychotherapy researchers. To those psychotherapy researchers who do investigate child intervention modes and, as a result, require this information, we apologize for our decision to detail only changes in syndromes that affect adults. Readers interested in DSM-III's section on childhood disorders can find an extensive analysis of changes in this section elsewhere (Nathan & Harris, 1982).

What follows in this section is an effort to highlight changes in the remaining DSM-III syndrome clusters of interest to psychotherapy researchers planning to study the outcome of psychotherapy offered to patients whose behavior falls within these clusters. Of course, at least as important to psychotherapy researchers are the generic changes in the philosophy of classification underlying DSM-III described above—changes designed to enhance reliability, increase validity and utility, and augment the instrument's empirical bases.

Substance Use Disorders

DSM-III accords the behavioral accompaniments of drug and alcohol abuse and dependence separate recognition as substance use disorders. This taxonomic decision recognizes the growing importance of these syndromes from both the clinical and public health perspective. The decision also has a destigmatization consequence (a recurrent theme in DSM-III) because it eliminates the "guilt by association" that has characterized the official diagnostic treatment of the addictions since 1952. DSM-I, published that year, labeled the Addictions (which included alcoholism and drug addiction) as sociopathic personality disturbances, along with the sexual deviations and the dyssocial and antisocial reactions. The moral disapprobation this categorization system conveyed, by bringing together those behaviors of which society most disapproved, was obvious to all. DSM-II transmitted a similar message when it included alcoholism and drug dependence, the sexual deviations, and the personality disorders together. Although the first two syndrome groupings were no longer defined explicitly as sociopathic reactions, the implication that all were unacceptable remained.

In part, separation of this group of disorders in DSM-III recognizes empirical findings that alcoholism and drug dependence are not always associated with other psychopathological conditions, that one can be an alcoholic without carrying concurrent psychiatric diagnoses (Mendelson & Mello, 1979; Nathan & Hay, 1983). In part, the separation reflects the desire of the drafters of DSM-III to destigmatize, to the extent possible, when doing so accords with the descriptive thrust of the instrument. Finally, setting these disorders off from others acknowledges their increased incidence in contemporary American society and their consequent increased importance to the clinician (Nathan, 1980).

The most important change in the diagnostic treatment of these disorders is the two-stage differentiation of abusers from those who are dependent—of problem drinkers from alcoholics and of recreational drug users from drug addicts. According to the operational criteria, abusers use substantial quantities of drug or alcohol over extended periods of time, with the result that they are demonstrably impaired personally, socially, or vocationally. Dependent persons show the drug-related behaviors of the abuser but demonstrate physical dependence and tolerance as well. This distinction, commonly drawn previously by experienced clinicians, makes explicit what had been implicit before. The change also permits useful distinctions to be drawn among drugs which cause both abuse and dependence (including alcohol, the barbiturates, the opioids, amphetamine, and cannabis), those that can only be abused because they do not cause physical dependence (phencyclidine and the hallucinogens), and those that cannot be abused but do cause dependence (tobacco).

Relatively little psychotherapy research of a conventional sort has been done with alcoholics or drug addicts. Group studies of alcoholics comparing conventional "milieu" treatment and broad-spectrum behavioral treatment have been reported, however, (Nathan, 1981b; Nathan & Briddell, 1977; Pomerleau, Pertschuk, Adkins, & d'Aquili, 1978; Sobell & Sobell, 1973). These studies were controversial because, in some instances, they suggested that behavioral treatment with controlled drinking rather than abstinence as the prime treatment goal was more promising. Among the critics of this research have been those questioning its diagnostic adequacy, asking whether the subjects of these studies were actually chronic alcoholics or whether they were not, instead, problem drinkers. Completed before DSM-III was published in 1980, this research has raised important diagnostic questions which could have been answered, in part, if subjects had been categorized according to the new DSM-III criteria. The distinction between alcohol abusers and addicts is more generally important also because problem drinkers may be drawn from a different population than alcoholics; that is, the two may differ in etiology as well as in behavior (Goodwin, 1979).

Schizophrenia

The changes introduced by DSM-III affecting the concept of schizophrenia and its diagnosis are, in some ways, as notable as Kraepelin's brillant original synthesis of the disorder a century ago. In fact, many of the changes introduced into DSM-III hearken back to Kraepelin's efforts, in that way emphasizing the lasting distinction of his initial contribution.

The net results of the DSM-III treatment of schizophrenia are (1) to heighten the reliability of the diagnosis of the disorder and (2) to reduce the incidence of the diagnosis markedly, perhaps by as much as 30%. Because psychotherapy for schizophrenia has been widely studied, these nosological changes have great importance for readers of this book.

DSM-III's changes include (1) a more narrowly defined concept of schizophrenia, which now requires a period of active psychosis—characterized by delusions, hallucinations, and/or formal thought disorder and deterioration in functioning—that lasts at least 6 months; (2) deletion of the borderline schizophrenia label and its replacement, in the personality disorder grouping, with the labels borderline personality and schizotypal personality; (3) removal of the schizoaffective psychosis label from the schizophrenic disorder grouping to a residual category outside the schizophrenic spectrum; and (4) reduction in the number of schizophrenic disorder labels from 12 in DSM-II to 5 in DSM-III.

These changes in the concept of schizophrenia bring American diagnostic practices into closer alignment with those of Europe, where the diagnosis of schizophrenia has always been significantly lower (Spitzer, Williams, & Skodol, 1980). A side benefit of this change will be a reduction in the likelihood that a person experiencing a brief reactive psychosis or a psychotic reaction in the face of extreme environmental stress will be diagnosed schizophrenic and have to carry that diagnosis the remainder of his or her life. The potential for a stigmatizing mislabeling, then, is greatly reduced.

These changes present real problems for psychotherapy researchers, despite their obvious virtues otherwise, since they render the results of many prior studies of psychotherapy with schizophrenics not comparable to the studies which will use DSM-III diagnostic criteria Many patients diagnosed as schizophrenic in prior studies no longer merit that diagnosis. Prior studies investigating the efficacy of psychotherapy with acute schizophrenics, for example, almost certainly included individuals who would now be given the schizophreniform or schizoaffective labels, neither of which remains a part of the schizophrenic spectrum. Hence, conclusions about potential gains from psychotherapeutic intervention with schizophrenics based on the broad concept of schizophrenia in vogue for the past 50 years in this country are no longer necessarily valid.

The research worker will have to return to earlier studies to attempt his or her own rediagnosis. The results of studies whose schizophrenic subjects had carried that diagnosis for several years and had not maintained prominent symptoms of affective disorder are probably as valid now as before, although the adequacy of the initial diagnostic judgments must be verified. But studies of acute and/or reactive schizophrenic patients are suspect, since many patients in these DSM-II categories are not schizophrenic according to DSM-III criteria.

Prospectively, the changes introduced by DSM-III to the concept of schizophrenia are generally positive. Diagnostic reliability, a problem in many prior studies of schizophrenics, will almost certainly be enhanced, both because of the empirically derived operational criteria and because narrowing of the concept reduces the range and kind of choices available to the clinician and makes more clear necessary and sufficient observations for the diagnosis. Also the results of outcome studies of schizophrenics done in this country will now be valid in England, Europe, and elsewhere, and vice versa, in that way broadening the usefulness of resultant data. On balance, then, the return to original Kraepelinian conceptions of schizophrenia implicit in DSM-III's reconceptualization of the schizophrenic spectrum appears to bode well for the psychotherapy researcher—but only after he or she becomes familiar with the changes and their implications for his or her work.

Affective Disorders

DSM-III brings together the entire spectrum of affective syndromes—mania and depression of mild, moderate, and severe varieties experienced on either an acute or a chronic basis—within a single major syndrome grouping, the affective disorders. By contrast, DSM-I and DSM-II located affective disorders several places in their taxonomies.

The affective disorders in DSM-III are divided into three subgroups: the major affective disorders, other specific affective disorders, and atypical affective disorders. The first of these groupings includes the manic-depressive psychoses of DSM-II. In DSM-III, these conditions are differentiated into the bipolar affective disorders and major depression, in recognition of the fact that it is possible to suffer from severe depression without being manic-depressive. This new distinction reflects empirical findings separating bipolar from unipolar affective disorders on the basis of phenomenology, etiology, treatment, and prognosis (Winokur, Clayton, & Reich, 1969).

The other specific affective disorders include conditions which were previously categorized with the neuroses (dysthymic disorder) and personality disorders (cyclothymic disorder). The first describes a reactive depression of moderate intensity while the second refers to a prevailing approach to life that involves rapid and short-lived emotional ups and downs.

The advantages of a single locus for the categorization of disorders that all embrace disturbance in mood include the opportunity to make finer distinctions among conditions which share phenomenology. Mild, moderate, and severe depression can more consistently and reliably be differentiated, since the unitary framework within which the distinctions are made is consistent. This new diagnostic capability allows the psychotherapy researcher to draw distinctions among potential subjects on the basis of severity and intensity of mood disturbance which could not be made reliably before. By the same token, the chance to draw these distinctions makes it likely that data from earlier psychotherapy studies of depressives will no longer be comparable to those from new studies. This situation is like that affecting psychotherapy outcome studies of schizophrenia.

The new system also makes it easier to differentiate between individuals suffering from clinical depressions (e.g., those categorized by DSM-III) and those whose unhappiness is benign and not of clinical significance. Since psychotherapy studies often explore intervention modes with college students whose "depression" is by self-report and does not appear to be of clinical import, DSM-III's new diagnostic capability offers the psychotherapy researcher the opportunity to determine whether or not the procedures he or she has developed for treating mild or moderate depression do so for persons whose behavior is properly described as affectively disordered.

The system presents at least one troublesome conceptual problem. The dysthymic disorder category, to include persons who would have been given the depressive neurosis diagnosis in DSM-II, is also to describe persons who would have been diagnosed depressive personality by the 1968 instrument. Hence, there is room for considerable confusion in psychotherapy outcome studies based on DSM-III categories over whether patients given the dysthymic disorder label are reactive depressives or are suffering from chronic, mild to moderate depression of the characterological variety. Whether the distinction between "neurotic" and "characterological" depressive behavior makes any difference in response to treatment or prognosis is uncertain; what is sure is that this diagnostic overlap offers real potential for confusion.

Anxiety Disorders

A brief note in boldface in the summary of the new nomenclature located at the beginning of DSM-III observes that the neurotic disorders of DSM-II are now "included in Affective, Anxiety, Somatoform, Dissociative, and Psychosexual Disorders." The note adds, though, that the DSM-II labels may still be used by diagnosticians, since they remain in a part of the ICD-9 diagnostic system, in widespread use worldwide. This acknowledgment of the strong hold the concept of neurosis continues to have on American psychiatry does not obscure the purpose of the reordering, renaming, and reorganizing of these disorders in DSM-III: to carry through the fundamental intent of the drafters of the document to base the entire nosology on description and phenomenology rather than opinion and theory. (The psychoanalytic theory of neurosis was the basis for unification of these disorders in DSM-I and -II.)

DSM-III's anxiety disorders, which include three DSM-II neuroses, have been changed in two important ways: by according increased diagnostic attention to the phobic disorders (one was listed in DSM-II; four are included in DSM-III) and by highlighting the posttraumatic stress disorders (for which no comparable diagnosis existed in DSM-II), The greater diagnostic emphasis on the phobic disorders in DSM-III reflects the success behavior therapists have had in understanding and treating these disorders, while the post traumatic stress disorders are highlighted because of their clinical importance as sequelae of the Vietnam experience for many veterans of that war.

The significance of these changes for psychotherapy researchers is greatest for the behavior therapy researcher who wishes to compare standard and innovative behavioral treatment packages for agoraphobia, social phobia, and simple phobia. A psychotherapy research area with a considerable and important history for behavior therapy dating from Paul's (1966) classic comparison of insight and desensitization for public speaking phobia, treatment studies of phobic behavior must now draw diagnostic distinctions among subjects like those drawn by DSM-III If subjects of the earlier studies that shaped this research area could be differentiated according to the new system as well, those studies' findings would have more impact on prospective research on treatment of phobic behavior. It now seems clear, for example, that effective behavioral treatment for agoraphobia and social phobia differs (Rachman & Wilson, 1980).

Somatoform Disorders and Dissociative Disorders

Most of the labels in these two diagnostic groupings stem, respectively, from DSM-II's hysterical neurosis, conversion type, and hysterical neurosis, dissociative type. To this end, the first three somatoform disorders—somatization disorder, conversion disorder, and psychogenic pain disorder—encompass separate elements of the earlier conversion hysteria: loss of or change in physical functioning due to psychological factors, the experience of pain out of proportion to physical findings, repeated, diverse somatic complaints with no physical explanation. The fourth somatoform disorder is hypochondriasis, another somatically focused behavior that bears an uneasy relationship to available physical evidence. Psychogenic amnesia, psychogenic fugue, and multiple personality, three of the dissociative disorders, all derive from DSM-II's dissociative neurosis. All represent abrupt but (usually) temporary changes in level of consciousness or psychomotor functioning.

Although objects of great attention by Freud and the other early psychoanalytic theorists, these syndromes no longer elicit much attention from psychotherapy researchers. Those psychotherapy researchers contemplating study of these conditions would do well to appreciate the behavioral bases on which two related syndromes have become six and to ensure that their recourse to historical precedent to plan their studies takes these changes into account.

Psychosexual Disorders

The psychosexual disorders have been the objects of striking changes and reconceptualizations in DSM-III; unlike most of the other syndromes receiving this kind of attention in DSM-III, however, the psychosexual disorders were recast less for empirical reasons than for sociopolitical ones.

A major change reflecting contemporary views on sexual behavior more than new discoveries from the laboratory is the decision to include the psychosexual disorders as a separate major diagnostic category rather than, as before, a species of sociopathy (DSM-I) or a variant of personality disorder (DSM-II). Now, it seems, society is willing to accept the view that these disorders are not always the result of psychiatric disturbance. Another sociopolitical change has led to the renaming of the sexual deviations, as they were called in DSM-II; in DSM-III, they are the paraphilias, which is a label that does not carry the moral opprobrium implicit in the former label. Another change reflecting contemporary societal mores, a most important one, is the decision to include homosexuality as an identified sexual disorder, albeit not as a paraphilia but in a separate category of its own. Only some of the nation's homosexuals are affected by the decision, though, because only those whose homosexuality causes them distress merit the diagnosis ego-dystonic homosexuality, while those who are satisfied with their sexual identification will remain undiagnosed.

An important change with clinical justification is placement of the psychosexual dysfunctions (e.g., inhibited sexual excitement, inhibited sexual desire, inhibited female orgasm, and inhibited male orgasm) as a subgroup of psychosexual disorders. A decision made partly as a consequence of the increased attention paid these disorders by popular scientific writers like Masters and Johnson and Helen Singer Kaplan, their presence corrects an important omission from DSM-II, where these disorders were mentioned only in passing as psychophysiologic disorders.

The significance of these changes to psychotherapy researchers depends, of course, on the syndromes to which they direct their attention. Behavior therapists who evaluate treatment for the sexual deviations will now study the paraphilias, but little else will be changed—unless they were developing treatments for homosexuality. However, following Davison's (1978) strong injunction, which found wide acceptance among behavior therapists, fewer behavior therapists now treat homosexuals who want to become heterosexual. Hence, the peculiar status of homosexuality as an "elective disorder" should not affect psychotherapy researchers very much.

Those persons who have been investigating treatment approaches to what are now called the sexual dysfunctions now have more specific diagnostic criteria and wider professional recognition for their efforts. Although some of the conditions in which they are interested have been renamed, either for consistency with the rest of the nomenclature or in an effort to remove offending connotations, here again nothing fundamental has changed.

The impact of removal of the sexual disorders from their place in DSM-II alongside the other disorders of which society most disapproves has no direct impact on psychotherapy researchers. If this action by the drafters of DSM-III reflects concomitant changes in public views on these disorders, though, it may presage easier access to research subjects, their greater willingness to cooperate with researchers, and greater success in developing and implementing prevention and intervention programs based on empirical research findings.

Psychological Factors Affecting Physical Condition

The psychophysiologic reactions of DSM-II—the psychosomatic disorders—have been eliminated in DSM-III. To make the same diagnosis in DSM-III now requires joint use of Axes I and III. On Axis I will be the diagnostic statement, psychological factors affecting physical condition; on Axis III will be the psychosomatic disorder in question. This change recognizes the role of the new Axis III, its function in identifying all physical conditions of relevance to Axis I and Axis II diagnoses. It also acknowledges the difficulty many clinicians have had linking, with assurance, a physical disorder and its psychological causation. The new system could help ease the problem by making clear both that a physical disorder exists and that it is affected by psychological factors whose precise role in causation is unknown.

Researchers whose interests lie in behavioral medicine—who are engaged in research on treatment of stress-related disorders like gastric ulcers, eczema, and asthma, for example—will need to become familiar with this new system. Also, and importantly, they must be alert to concurrent Axis I disorders that could affect either the expression of the physical disorder, its etiology, or its treatment. A schizophrenic patient who also suffers from psoriasis, for example, will clearly represent a very different treatment problem than someone who carries no concurrent psychiatric diagnosis. Researchers comparing treatment outcome for groups of patients with a given psychosomatic disorder, following divergent interventions, must be alert to such major confounding factors as concurrent psychotic illness; whether they will attend to less serious coexisting diagnoses—characterological ones, for example—is uncertain.

Personality Disorders

Now in a distinct and separate diagnostic category for the first time (no longer the "parent" diagnostic grouping for alcoholism, drug dependence, and the sexual deviations), the personality disorders have also been moderately affected in other ways by the revolutionary changes of the third edition of the DSM.

Most important of these changes involving the personality disorders are those stemming from the effort to "cleanse" the concept of schizophrenia—to limit the diagnosis to persons who have shown signs and symptoms of schizophrenia for 6 or more months. As a consequence, the personality disorder grouping now includes two diagnostic possibilities that fell within the schizophrenic spectrum in DSM-II. The DSM-II labels schizophrenia, simple type, and schizophrenia, latent type now accord with the DSM-III labels schizotypal personality disorder and borderline personality disorder. While this change will require some adjustment on the part of the researcher, it makes extremely good sense. No longer will patients who are not psychotic have to suffer the stigma of schizophrenia because a clinician believes it merited on the basis of what might happen at some time in the future. And no longer will researchers find that they have included in their samples of schizophrenics a few who have never shown overt psychosis but have qualified because of the availability of the latent or simple label.

DIAGNOSTIC ISSUES THAT AFFECT PSYCHOTHERAPY RESEARCH

There is a surprising paucity of discussion of the diagnostic process—including its problems as well as its potential—in either psychotherapy research reports or review articles written by psychotherapy researchers. Beyond observing, *de rigueur*, the inadequacies of diagnosis and recommending rigorous controls to guard against unreliable diagnosis, most of those who have commented on DSM-II diagnosis have chosen to devote most space to personality tests and measures that are *not* designed to yield syndromal diagnoses rather than to the interview procedures that are. Part of this apparent reluctance to "do something" about the reliability problems of syndromal diagnosis reflects the widespread conviction that clinical diagnosis has been the psychiatrist's responsibility while research and personality evaluation are the domain of the psychologist. In fact, persons of very diverse backgrounds and training are now involved in the diagnostic enterprise; researchers can no longer assume, if they ever did, that diagnosis will be done competently by those specifically trained to do it.

Psychotherapy researchers may also have felt that, once reliability has been ensured, the problems of diagnosis have little real relevance to outcome research and that, as a result, other issues are much more important to them. Yet, as we hope our review suggests, this position is no longer tenable. The increased specificity of DSM-III diagnoses guarantees that the impact of the changes in the major syndromes in DSM-III will be major for psychotherapy researchers. Said another way, it is time that those responsible for psychotherapy research begin to take responsibility not only for the adequacy of the diagnoses with which they must work but also for the impact the diagnoses have on their findings.

A few behavioral scientists have written about diagnostic issues that affect psychotherapy research. What we discuss to the end of this section is drawn from some of what these persons have written. Since most of what has appeared has been written by several persons and/or published in several places, our acknowledgments reflect this fact. Our discussion here has been stimulated by discussions elsewhere of these issues by the following authors: Bergin and Garfield (1971); Garfield (1978); Garfield and Bergin (1978); Kazdin and Wilson (1978); Nathan and Lansky (1978); and Rachman and Wilson (1980).

Perhaps the most important issue—it is raised by virtually every observer of the psychotherapy research scene—is how to ensure that sufficient data on subjects beyond their valid diagnosis be gathered (1) to permit matching of experimental and control or comparison groups not simply on the basis of equivalent diagnoses, but on such variables with impact on psychotherapy outcome as SES, level of education, intelligence, severity and chronicity of the disorder, concurrent treatment, existence and identity of environmental supports, and so on and (2) to enable those who read the subsequent research report to know to what extent their own subjects or patients who share these diagnoses are comparable along these additional dimensions, since all affect response to treatment.

Information on these variables is of obvious importance, both for the integrity of a research design that requires a comparison of two or more groups of patients who share a common diagnosis but may differ along other relevant dimensions and for the consumer of psychotherapy research who wishes to use the published data he or she has read but is uncertain whether it applies to his or her patients of the same diagnosis. Despite their importance, most reports leave out crucial information on these variables. Although DSM-III will correct these omissions partly by demanding somewhat fuller vital information from the multiaxial system, it remains up to the researcher to ensure that this information is available both to himself or herself for matching or assignment purposes and to the research consumer.

A related issue is that of base rates. *The* base rate problem—the necessity to ensure that the sensitivity of the diagnostic procedure exceeds the naturally occurring frequency of patients with the given diagnosis in the population from which the patients are drawn—is not the question here. Rather, the issue is the base rates of concurrent predictors of response to psychotherapy in the specific environment and subject pool tapped and within the diagnostic grouping sampled. Said another way: Who are the agoraphobic patients drawn from the local community mental health center and how do they differ from patients with other diagnoses at the mental health center in demography *and* in concurrent predictors? Also how do they differ from agoraphobic patients in cities and towns outside the local area along the same dimensions? Given that DSM-III now helps ensure that, diagnostically, a local agoraphobic will share important clinical features with agoraphobics elsewhere, the question remains whether base rates of SES, level of education, intellectual capacity, family support, vocational opportunities, and other variables of relevance to psychotherapy in outcome differ among subject locales. The answer, of course, is that they do—which means that one cannot assume that patients with the same diagnosis from community mental health centers, state hospitals, or outpatient clinics, even in the same city or town, will be comparable otherwise. Some of these same issues will doubtless be addressed by the authors of this volume's chapter on behavioral assessment (Chapter 3); behavioral clinicians typically take a more informed view of the importance of these extradiagnostic personal characteristics and circumstances.

Another recurrent theme among psychotherapy researchers is the lack of precision that usually accompanies description of the diagnostic process. The usual descriptive statement may identify the diagnosticians ("two board-eligible psychiatrists and an ABEPP psychologist," "the clinical director, a psychiatrist, and three very experienced psychiatric nurse clinicians") and indicate that all had agreed on the diagnosis in question. Also, the observation that DSM-II criteria for diagnosis were followed may be included (recently, some reports have specified RDC criteria; from 1980 on, DSM-III criteria will presumably be indicated). Where the examination took place may also be specified. Usually *not* provided, however, are data on how long the diagnostic examination lasted, how intensive it was, what additional information the diagnostician had available to him or her, and whether sources of information outside the diagnostic setting or besides the patient, including family and friends, were elicited.

A final issue of concern is the question of when the diagnosis was originally conferred and whether it has been subsequently confirmed. Most researchers, not simply psychotherapy researchers, appear content simply to assume that a diagnostic process has been taken seriously by those responsible for the diagnosis. The question of how much time has elapsed since the original diagnosis is not generally raised. Yet the question is an important one, of course, since behavior tapped for diagnostic purposes changes over time, with the consequent necessity to alter the diagnosis. This problem is less pressing when it comes to patients who have remained in an institutional setting for a long time; generally, these patients carry the diagnosis of schizophrenia, which does not often change. The problem is more important with outpatient subjects whose diagnosed disorders do not involve psychosis. Questions of remission or regression, of changes in chronicity or severity, are central to meaningful assignment to experimental and control or comparison groups.

CONCLUSIONS

It should be clear to the reader that the author of this chapter is, in general, sympathetic to the aims and goals of DSM-III and pleased at their implementation. The guiding emphases on a descriptive, phenomenologic nosology that is both empirically based and the product of broad professional consensus appear to have served the instrument well. It promises to be much more reliable, easier to use, and more widely accepted by a broader range of mental health professionals than its predecessors.

Important questions about DSM-III and syndromal diagnosis remain. The most immediate of these is whether DSM-III will prove to be valid and useful now that its enhanced reliability permits such an assessment. Will it be useful to those who are charged with planning treatment? Will it help us understand the etiologic factors in a given patient's condition? Above all, for those reading this commentary, will it enable psychotherapy researchers to assign patients more reliably to experimental and control or comparison groups, and will it permit them to draw valid conclusions about what psychotherapy for what kind of patient under what circumstances?

Beyond this pressing question—to which answers ought to be available within a few years, now that the system is in place and appears to be reliable— is a more fundamental one. That question is whether syndromal diagnosis, diagnosis from the signs and symptoms of behavioral disorder, represents a useful approach to psychological, psychiatric, and behavioral disorder/dysfunction. Useful for the diagnosis and treatment of the physical disorders, where knowledge of signs and symptoms of a physical condition generally leads to effective treatment, syndromal diagnosis is of less obvious utility for the conditions listed in DSM-III. The absence, to this time, of confirming data attesting to the value of syndromal diagnosis for the disorders leads one naturally to ask whether some other taxonomic scheme (e.g., premorbid level of functioning, response to treatment, adaptability to change or response to a stressor, antecedents to or consequences of the targeted behavior) might prove more useful than syndromes as organizing factors. The fact that a viable alternative to syndromal diagnosis has not gained acceptance from any substantial segment of the mental health community suggests that one may not exist; but it is important again to raise the question and to suggest that psychotherapy researchers, along with their colleagues, consider the matter anew.

In the final analysis, none of these issues—and few of the issues addressed elsewhere in this chapter—are as important to the psychotherapy researcher or the psychotherapist as the issue of effective interventions. If the psychotherapeutic intervention tested on groups of patients is not robust, it matters little whether experimental, control, or comparison groups were well or poorly diagnosed. If psychotherapy is of little value to schizophrenics, a narrower concept of schizophrenia is irrelevant to the psychotherapy researcher. If the personality disorders are so satisfying to those who manifest them that nothing can make them change their behavior, improved specificity of diagnosis of these conditions will excite no one. Hence, with enthusiasm for the diagnostic advances DSM-III represents but sobered by the challenge facing the psychotherapy researcher to develop more robust interventions, I bring this chapter to a close.

REFERENCES

American Psychiatric Association. *Diagnostic and statistical manual of mental disorders* (1st ed.). Washington, D.C.: Author, 1952.

American Psychiatric Association. *Diagnostic and statistical manual of mental disorders* (2nd ed.). Washington, D.C.: Author, 1968.

American Psychiatric Association. *Diagnostic and statistical manual of mental disorders* (3rd ed.). Washington, D.C.: Author, 1980.

Beck, A. T. Reliability of psychiatric diagnoses: A critique of systematic studies. *American Journal of Psychiatry,* 1962, *119,* 210–216.

Bergin, A. E., & Garfield, S. L. (Eds.), *Handbook of psychotherapy and behavior change.* New York: Wiley, 1971.

Cantwell, D. P., Mattison, R., Russell, A. T., & Will, L. A comparison of DSM-II and DSM-III in the diagnosis of childhood psychiatric disorders. IV. Difficulties in use, global comparison, and conclusions. *Archives of General Psychiatry,* 1979, *36,* 1227–1228.

Cantwell, D. P., Russell, A. T., Mattison, R., & Will, L. A comparison of DSM-II and DSM-III in the diagnosis of childhood psychiatric disorders. I. Agreement with expected diagnosis. *Archives of General Psychiatry,* 1979, *36,* 1208–1213.

Davison, G. C. Not can but ought: The treatment of homosexuality. *Journal of Consulting and Clinical Psychology,* 1978, *46,* 170–172.

Feighner, J. P., Robins, E., Guze, S. B., Woodruff, R. A., Winokur, G., & Munoz, R. Diagnostic criteria for use in psychiatric research. *Archives of General Psychiatry,* 1972, *26,* 57–63.

Garfield, S. L. Research problems in clinical diagnosis. *Journal of Consulting and Clinical Psychology,* 1978, *46,* 596–607.

Garfield, S. L., & Bergin, A. E. (Eds.), *Handbook of psychotherapy and behavior change* (2nd ed.). New York: Wiley, 1978.

Goodwin, D. W. Genetic determinants of alcoholism. In J. H. Mendelson & N. K. Mello (Eds.), *The diagnosis and treatment of alcoholism.* New York: McGraw-Hill, 1979.

Kazdin, A. E., & Wilson, G. T. *Evaluation of behavior therapy: Issues, evidence, and research strategies.* Cambridge, Mass.: Ballinger, 1978.

Matarazzo, J. D. The interview: Its reliability and validity in psychiatric diagnosis. In B. B. Wolman (Ed.). *Clinical diagnosis of mental disorders: A handbook.* New York: Plenum Press, 1978.

Matarazzo, J. D. *The reliability of psychiatric and psychological diagnosis: A revolution in the making?* Manuscript in preparation, Oregon Health Sciences University, 1982.

Mendelson, J. H., & Mello, N. K. (Eds.), *The diagnosis and treatment of alcoholism.* New York: McGraw-Hill, 1979.

Nathan, P. E. *Cues, decisions and diagnoses.* New York: Academic Press, 1967.

Nathan, P. E. Alcoholism and DSM-III. *Advances in Alcoholism,* 1980, *1,* 15.

Nathan, P. E. Symptomatic diagnosis and behavioral assessment: A synthesis? In D. H. Barlow (Ed.), *Behavioral assessment of adult disorders.* New York: The Guilford Press, 1981. (a)

Nathan, P. E. Nonproblem drinking outcomes: The data on controlled drinking treatments. *Advances in Alcoholism,* 1981, *2,* 12. (b)

Nathan, P. E., & Briddell, D. W. Behavioral assessment and treatment of alcoholism. In B. Kissin & H. Begleiter (Eds.), *Biology of alcoholism* (Vol. 5). New York: Plenum Press, 1977.

Nathan, P. E., & Harris, S. L. *The diagnostic and statistical manual of mental disorders:* History, comparative analysis, current status, and appraisal, In C. E. Walker (Ed.), *Handbook of clinical psychology: Theory, research, and practice.* Homewood, Ill.: Dow Jones-Irwin, 1983.

Nathan, P. E., & Hay, W. M. Alcoholism: Psychopathology, etiology, and treatment. In H. E. Adams & P. B. Sutker (Ed.) *Comprehensive handbook of psychopathology.* New York: Plenum Press, 1983.

Nathan, P. E., & Lansky, D. Common methodological problems in research on the addictions. *Journal of Consulting and Clinical Psychology,* 1978, *46,* 713–726.

Paul, G. L. *Insight versus desensitization in psychotherapy.* Stanford, Calif.: Stanford University Press, 1966.

Pomerleau, O., Pertschuk, M., Adkins, D., & d'Aquili, E. Treatment for middle income problem drinkers. In P. E. Nathan & G. A. Marlatt (Eds.), *Alcoholism: New directions in behavioral research and treatment.* New York: Plenum Press, 1978.

Rachman, S. J., & Wilson, G. T. *The effects of psychological therapy.* London: Pergamon Press, 1980.

Sobell, M. B., & Sobell, L. C. Individualized behavior therapy for alcoholics. *Behavior Therapy,* 1973, *4,* 49–72.

Spitzer, R. L., Endicott, J., & Robins, E. Clinical criteria for psychiatric diagnosis and DSM-III. *American Journal of Psychiatry,* 1975, *132,* 1187–1192.

Spitzer, R. L., & Forman, J. B. W. DSM-III field trials: II. Initial experience with the multiaxial system. *American Journal of Psychiatry,* 1979, *136,* 818–820.

Spitzer, R. L., Forman, J. B. W., & Nee, J. DSM-III field trials: I. Initial interrater diagnostic reliability. *American Journal of Psychiatry,* 1979, *136,* 815–817.

Spitzer, R. L., Williams, J. B. W., & Skodol, A. E. DSM-III: The major achievements and an overview. *American Journal of Psychiatry,* 1980, *137,* 151–164.
Stein, J. (Ed.) *Random House dictionary of the English Language.* New York: Random House, 1966.
Winokur, G., Clayton, P., & Reich, T. *Manic depressive illness.* St. Louis: Mosby, 1969.

3

Issues for Behavioral Assessment in Psychotherapy Research

W. ROBERT NAY and RICHARD E. NAY

INTRODUCTION

Emerging from the application of principles surrounding behavior therapy or behavior modification, the term *behavioral assessment* is suggested to be an alternative to "traditional" approaches for understanding human behavior. While the demarcation between traditional and innovative approaches to behavioral assessment seemed to be quite clear at the outset (Goldfried & Kent, 1972) a variety of recent conceptual and applied reviews of the area indicate that such rigid differences are in fact illusory. Indeed, it is clear that the term *behavioral assessment* encompasses broad categories of conceptualization and operations and that considerable disagreement exists within the literature as to what it entails. For example, as Mash (1979) argues:

> Behavioral assessment has frequently been described in contrast to traditional personality theories and therapies where intrapsychic determination, inferred and stable personality traits, individual differences, cross-situational stability of behavior, indirect behavior samples, and assessment for diagnosis rather than treatment are given greater import. In fact, some would say that we now have a better description of what behavioral assessment is not, than of what it is, and there is some truth to this suggestion. (p. 24)

He goes on to emphasize that the strength of behavioral assessment as an approach is in similarities of concepts, methods and practices that character-

---

W. ROBERT NAY ● Behavioral Medicine Associates, 6870 Elm Street, Suite 100, McLean, Virginia 22101.   RICHARD E. NAY ● Psychology Department, University of South Carolina, Columbia, South Carolina 29169.

istically occur, and not in any rigid "boundaries" that isolate behavioral assessment from other approaches to understanding human behavior. Other authors addressing this question have agreed and emphasize that behavioral assessment tends to focus on certain procedures for segmenting human behavior from among person by situation events. Thus, O'Leary (1979) argues that the behavioral assessor focuses on observable behavior while not excluding cognitions and affect and emphasizes current behavior in the here and now while not ignoring the important role that historical events may play in understanding some behavioral expression in the present. Even the hallmark description of behavioral assessment as focusing upon samples and not "signs" of behavior, a polemic against so called "trait" approaches to personality, has recently been questioned (Haynes, 1979). Rather simplified accounts of "trait theory" have been offered as suggesting that an individual's behavior must be invariant across situation and time, while behavioral assessment focuses on the particular situation within which the individual functions and is not interested in such cross-situational stability. Obviously, these issues are not as simple as that. For example, Hogan, DeSoto, and Solano (1977) note that no trait theorist would imply that the expression of an individual's behavior is not influenced by the situation within which he or she functions. These authors emphasize that traits refer to stylistic consistencies and that it is not imperative for a trait theorist to assume rigid, situational invariance. On the other hand, behaviorally oriented theorists are finding overwhelming support for a careful explication of the relationship of a particular behavior to a particular situation under specified conditions. As Haynes (1979) argues, behavioral assessors should focus their efforts on identifying which categories of subject $\times$ behavior $\times$ situation *do* demonstrate stability across temporal replications. In fact, the weight of evidence for situational specificity across response channels, across modes of cues or sets provided to subjects, across settings, and across levels of obtrusiveness of assessment provide problems for behavioral assessment as well as its more traditional counterpart. Kazdin (1979) argues that situational specificity suggests that behavior may vary considerably as a function of changes in the assessment procedures and the conditions with which they are associated. Much as the "conceptual breath of behavioral approaches to understanding human behavior has been expanded greatly in the recent past to include a much more comprehensive view of a person and his environment" (Nay, 1979), the boundaries of behavioral assessment have also expanded considerably.

A variety of method options that should be of value to the conduct of psychotherapy outcome research are provided by behavioral assessment. The individual's self-report, the reports of significant others, the observation of behavior in environments ranging from controlled to natural, as well as assessment of psychophysiological responses to important life events are included. If behavior is indeed situationally specific, this suggests that assessment of such multiple response channels and situations (Kazdin, 1979) must be considered as measurement questions are asked prior to conducting research. However, while the multifaceted technology of behavioral assessment is available for research, its utility depends entirely on the purposes to which it is applied, in this case the inferences that the researcher desires to make. As noted by Kanfer and Nay (1981):

> Before an assessment technique is employed, the clinician or researcher should specify what decisions will be made on the basis of the assessment information or to what domains of populations, settings, times, situations the information will be generalizable. (p. 28)

The authors noted that methods do not exhibit fixed or specific reliability or validity coefficients or constant properties of generalizability. These characteristics depend on the purpose of the assessor and the desired inferences. The authors then emphasized the importance of a construct validity formulation as a foundation for planning and carrying out research within the clinical enterprise. Construct validity has recently been shown to be applicable to behavioral assessment (e.g., Cone, 1977; Nay, 1979) and asks important questions with regard to the theoretical assumptions that underlie any research question. In this context, it would seem to us that three important questions should be asked by the investigator who is planning to employ behavioral assessment within psychotherapy research: (1) *What* will be measured? (2) *Where, when,* and under what *conditions* will this phenomenon be assessed? and (3) *How,* or using what methodology, will the phenomenon be encoded or in some other way segmented from the stream of behavior? We submit that it is impossible to address any of these questions in systematic fashion without addressing them in order, and this order must begin with the theoretical assumptions of the investigator.

This chapter will first discuss some important issues that the first two questions raise for the researcher and offer a review of methods that may resolve problems of application. The third question has to do with methods. A discussion of issues attendant to four classes of behavioral assessment will be presented. A review of issues surrounding psychophysiological assessment is beyond the scope of this chapter.

WHAT TO ASSESS

First, we turn to the issue of what is to be assessed? This clearly has to do with the questions the research asks. In the outcome area, these questions are frequently formulated around the question: Does the treatment work? Unfortunately, when such an ambiguous question is asked, this often leads to a "mixed bag" of assessment procedures that may be selected for unspecified reasons. Merely putting together a collection of self-report and observational instruments to assess whether or not a program of systematic desensitization directed at dental phobics "works" will typically result in outcome data that are perplexing and impossible to interpret. If the investigator has asked the research question poorly, given a lack of conceptual preparation, a mixture of significant and nonsignificant results across response dimensions may lead the author to overly interpret or misinterpret findings that, implicitly, are impossible to understand. In asking the question "what," we offer some guidelines that should aid in selecting a behavioral assessment device that will produce information that is potentially capable of answering the questions of the research.

Why is this research being conducted? As Gottman and Markman (1979) note, research may be designed to generate questions; to understand or describe some phenomenon that is of interest; or, given this understanding, research may be designed to test systematically a hypothesis that frequently has to do with cause-and-effect relationships. Obviously, research directed at each of these broad questions may have unique implications for the researcher who now must select settings and instruments for assessment. Research aimed at testing a hypothesis will require measures carefully tailored to assess well-specified, targeted behaviors predicted to change as a function of some treatment. Clearly, the investigator should be able to justify why a given method $\times$ setting was selected given the specific outcome predicted by the hypothesis, and the measure must be capable of validly and reliably detecting the response dimension of interest. In contrast, research aimed at generating hypotheses or describing some phenomenon should be focused at measuring major classes (e.g., social skills) that, based on preliminary findings and previous research, permit the studied behavior to be detected by at least one dimensional measure (e.g., peer nomination, sociogram, observation of social approach behavior).

Although research aimed at description may require a more broad-band approach to method selection, each method should be justifiable for some conceptual reason; this should be made clear to the audience for that research. Even the most preliminary descriptive research aimed at understanding a construct such as "social skills" must implicitly adopt a working definition that focuses assessment toward some portion of the stream of behavior. This conceptual posture should then provide a coherent superstructure on which preliminary questions can be based.

THE CONTEXT OF THE ASSESSMENT

The question of where, when, and the conditions of assessment define the context for use of a given methodology. Obviously these questions cannot be answered without reference to a clear conceptual framework for what is being assessed. Behavior does not occur outside of a situation, therefore we must explore the environmental domains within which we expect the behavior to occur and to which we may wish to generalize. Also, we cannot consider what we wish to measure without specifying some time parameter around which we will pull the target from the stream of behavior of which it is a part. It is typically impractical or impossible to sample occurrences of behavior on a continuous 24-hour basis. Deciding to assess a given target at particular times obviously is not an arbitrary matter; the schedule for making assessments may have important implications for what is detected and ultimately what is encoded. For example, deciding to assess the incidence of smoking behavior for some group by systematically observing those individuals in their job settings may not provide information that is generalizable to the home and other life environments. Obviously, the nature of the task performed at the job site, rules and restrictions with respect to smoking, availability of social role models, and the like are among the myriad of work factors that may not be present as the individual traverses the home and community environment. These criticisms would not apply if the investigator were merely interested in the effects of some variable (e.g., job related stress, social facilitation, a corporate antismoking campaign) on the incidence of smoking within the job site. Thus, both the setting within which a behavior is assessed as well as the temporal parameters for assessing that event make up an *occasion* for assessment, and a particular occasion is not appropriate or inappropriate for a given target outside of the particular research questions asked. While it is certainly beyond the scope of this chapter to define characteristics and issues surrounding the various settings available for assessment, several important issues that have to do with the where and when of evaluation for research are worth reviewing, along with some particular settings that may be of use to the experimenter.

Ecological Validity

Like so many issues in psychology that seem to swing with the pendulum of the times, the issue of external validity has frequently been focused around the setting of research assessment. At times, we have seen the well-controlled albeit sterile laboratory environment touted as the *sine qua non* for conducting research. Many psychotherapy researchers have attempted to emulate the experimental psychologist in controlling for and holding constant certain variables while manipulating the independent variable of interest, all conducted within the confines of a laboratory environment. In recent years, psychologists of many stripes have begun to challenge the "ecological validity" and representativeness of such controlled and seemingly contrived environments (Allport, 1968; Gilmour & Duck, 1980). The issue of so-called ecological validity

(Brunswik, 1955) must be considered by the psychotherapy researcher who often has a clear choice of evaluating a clinical target under highly controlled, manipulable conditions that display what Carlsmith, Ellsworth, and Aronson (1976) would call "experimental realism," as opposed to performing assessments in the natural, unmanipulated environment within which the client/subject ordinarily functions (e.g., "mundane realism"). Berkowitz and Donnerstein (1982) have recently refocused this argument around the purposes of the experiment. These authors argue that research done in the more artificial yet controlled environment of the laboratory does not necessarily lack generalizability or external validity. The authors state:

> External validity, as we have argued, does not necessarily require ecological validity. The experimental results may tell us something about the conduct of a broad range of people in natural situations even though the subjects in laboratory settings are not physically representative of this population or the real world situations in which they are embedded. (p. 255)

They argue that while ecological validity is certainly important to determine population estimates or to establish behavioral norms in an environment of interest, the interpretation or meaning that the subject of some experimental manipulation acquires is the important determinant of the degree to which behavior in the laboratory may parallel behavior exhibited in other settings. Thus, subjects may assign much the same meaning to a videotape of child behavior that calls for parental responses to taped situations as they would to the actual situations in the real world. Berkowitz and Donnerstein suggest that participants be debriefed to determine how they have interpreted a given experimental manipulation, arguing that no "blanket" statement can be made about whether laboratory-induced behavior will or will not arise in other settings that are physically very different.

Regarding this issue, we would argue that clinical experimenters often confound the questions that should arise from any psychotherapy enterprise. One extremely important question has to do with *what the subject has learned as a result of the intervention.* This category of question might be called establishing the nature and limits of the client's repertoire under *any* conditions. For example, it is very important to know whether or not clients have accurately learned to self-monitor some behavior of interest or know how to employ a given self-instructional technique for reparenting their child. Similarly, knowing that a client can accurately reproduce information presented in relaxation training or achieved via biofeedback must precede questions of application in alternative settings. A second and much different question is: *Does the client engage in a particular behavior in some defined setting (or in all settings)?* This question has to do with a number of factors, including the client's motivation to employ newly learned behaviors in a particular environment, the nature of the environment and any constraints that may inhibit reproduction of that behavior, and the extent to which a given behavior is appropriate to and instrumental within that environment. We would argue that the foregoing two questions are frequently confounded in psychotherapy research, with a particular experimental manipulation evaluated as a "failure" because client behavior in the natural environment does not change in some predicted direction. If such a failure of generality is discovered within a research enterprise, it is most important that the researcher also be able to address the first question—that is: Did the clients learn to reproduce accurately the behavior that was trained? If the answer to this first question is positive while the outcome of the second is negative, it may be that the clinical intervention was successful but a separate and different intervention must be applied to deal with (1) motivating the subject to employ the behavior and/or (2) altering or diminishing environmental constraints that prohibit the subject from employing the behavior and/or (3) identifying the set of skills that are appropriate for the setting within which the client functions. Thus, the evaluation of outcome of a given treatment is complex and requires a careful and thoughtful stating of the questions being addressed by a given assessment and the appropriate setting within which that question can be answered. Clearly, when a controlled, fixed, clinical/laboratory setting is appropriate for a given question, the costs of assessment are reduced and a variety of practical and technological issues surrounding natural settings are circumvented. This point can be brought into focus by considering the different issues involved in assessing the outcome of training 60 parents within the same clinical environment versus going out to 60 different homes with a team of observers to assess them *in vivo*. Those researchers, including the authors, who have collected observational data in institutions, homes, schools, and other natural settings will attest to the logistical and practical difficulties involved. These costs may be well worth assuming if the research questions and purpose of the undertaking require high levels of ecological validity; however, a careful consideration of the issues already addressed should be considered when planning psychotherapy outcome research to ensure that the appropriate and least costly setting for answering a given question is entertained.

Even in natural, so-called ecologically valid settings, the experimenter must sometimes impose certain controls in order to facilitate maximal client interaction and observations (e.g., Forehend, Wells, & Greist, 1980; Nay, 1979; Sanders & Glynn, 1981). For example, Forehand *et al.* (1980) curtail television use, telephone calls, and leaving the two-room setting (except for an emergency) once observation of parents in the home has begun. Christenson, Johnson, Phillips, and Glasgow (1980) maintain that such restrictions on setting behavior may lead to an unrepresentative sample of family interaction. This trade-off of somewhat controlled natural settings for better-quality behavioral observation is an important consideration that must be dealt with according to the particular circumstances under scrutiny. We are now dealing with the issue of reactivity and its implication for valid and reliable assessment. In essence, the impact of the outside observer on setting behavior should be kept to a minimum. A number of factors, from not properly acclimating the observer to the setting (Ivancic, Reid, Iwata, Faw, & Page, 1981) to overly obstrusive methods of recording behavior (Ciminero, Nelson, & Lipinski, 1977), have been noted within the literature. It is surprising that so little well-controlled research has examined reactivity (Kent & Foster, 1977). Of this research, virtually no adequate study of reactivity over long-term observations has been completed (Reid & Patterson, 1976). Overall, it seems that some investigators have found reactivity due to observer presence (e.g., Mercatoris, Hahn, & Craighead, 1975; Hay, Nelson, & Hay, 1980), while others have found no such evidence (e.g., Ivancic *et al.* 1981; Johnson & Bolstad, 1975). The issue simply has not been settled and is raised to underscore the fact that any assessments may provoke reactivity, regardless of whether they are conducted in the laboratory or natural environment. In a later section, some ways of reducing the possibility of reactivity for observational methods will be mentioned.

Analogue Assessments

A variety of environments that can be employed at lower cost within the clinical or experimental domain have been developed in recent years. In effect, such behavioral analogues permit the researcher actually to evoke important targeted behaviors as opposed to waiting for them to occur under natural conditions, thus reducing the time requirements for assessments/therapy. These "analogue" measures (e.g., McFall, 1976; Nay, 1977, 1979) are particularly valuable when the target of behavior of interest occurs at low rate in the natural environment making naturalistic observations impractical as well as when the assessor is required to collect extensive information in a brief amount of time. Certainly, the reduced cost and increased control that can be exerted by the experimenter over these environments make them worthy of consideration. A brief review of analogue options available to the researcher along with important issues that should be considered will now be offered.

In an attempt to consolidate and catalogue the various approaches to such "clinical contrived" analogues currently in use, Nay (1977; 1984) developed the following five operationally defined analogue categories: (1) paper and pencil, (2) audiotape, (3) videotape, (4) enactments, and (5) role plays. A short discussion of each method follows.

Paper-and-Pencil Analogues

The paper-and-pencil analogue requires the respondent to write or describe verbally how he or she would respond to some "natural" situation that is presented in written form. In effect, the client is emitting a behavior which is dissimilar to but assumed to covary with the particular target behavior in the natural environment (Haynes, 1979). Often, the client is asked to describe what he or she would say or do in response to stimulus situations, either in a narrative or multiple-choice format (e.g., Nay, 1975). The stimuli can be presented to the clients in a variety of ways. For example, Clavelle and Turner (1980) gave subjects a series of 31 cards, each describing a particular situation. The client was given a choice of which situation to respond to first, thus providing more perceived freedom in the assessment process. The possibilities for stimulus situations are almost limitless for this analogue and can be tailor-made to suit the needs of a particular client. Scoring should be standardized, especially if responses are open-ended, to ensure reliability among multiple raters. A major advantage of the paper-and pencil analogue lies in ease of administration to large numbers of clients, with little assessment time involved. Although the relationship between written analogue and target behaviors in the natural environment has not been adequately assessed, Haynes (1979) reviews numerous studies suggesting that such analogues are sensitive to manipulations and can discriminate between populations. Certainly they permit the experimenter to test the limits of the client's knowledge.

Audiotape Analogues

The audiotape analogue differs in that the situations are presented on audiotape. Typically, a narrator introduces a scene, followed by statements of any number of actors representing viewpoints relevant to a particular problem. For example, a teacher and student may be discussing the student's habitual tardiness. The respondent verbally reacts to the actor's behavior as if the situation had actually occurred in the natural environment. These responses are amenable to either direct observation or tape recording for later coding. Alternatively, the response could be written in narrative format or in the form of ratings (e.g., Sattin, 1980).

The audiotape analogue is a very inexpensive and efficient way to sample a client's verbal responses to virtually any class of stimulus situation, although social situations like dating tend to be the most common targets (e.g., Arkowitz, Lichtenstein, McGovern, & Hines, 1975; Westefeld, Galassi, & Galassi, 1980). A problem lies in the inherent inability of the experimenter to observe nonverbal behaviors like facial expression and body movements.

Videotape Analogues

The videotape analogue enhances its audiotape counterpart by presenting both audio and video information to the client, who then makes verbal or non-verbal responses to the stimuli. Typically, after gathering important situational data from the client's natural environment about the presenting problem, an attempt is made to have the videotaped situation parallel this natural environment to as great a degree as is practically possible. The authenticity of the particular actors, settings, and props seems important in portraying a realistic scenario to which the client will respond (Nay, 1979). Typically, a narrator on the videotape introduces each scene, the "actors" portray their roles, and then the client makes a response directed at the actors themselves. To ensure adequate reliability and validity, the client should be allowed to practice the task with some guided example items.

Videotape analogues have been used in assessing a wide range of target behaviors (e.g., Conger & Keane, 1981) as well as for training behavioral observers (Farkas & Tharp, 1980). If care is taken in duplicating the natural environment, videotape analogues can offer a much more credible stimulus situation than can paper-and-pencil or audiotape instruments, thereby enhancing the videotape's relative validity. The reliability and validity data concerning this analogue procedure are at best scant, but overall they seem favorable (Haynes, 1979).

Enactment Analogues

The enactment analogue requires the client to actively interact, in the clinical setting, with the stimulus persons or objects that normally exist in the client's natural environment. The experimenter merely monitors the client's responses. For example, Adubato, Adams, and Budd (1981) observed parents interacting with their own problem children in an institutional/clinical setting. Also, married couples are common targets for such enactments (e.g., Carter & Thomas, 1973). Several investigators have used modified clinical settings that closely resemble the natural environment in order to increase the salience and immediacy of the situation for the client. For example, Strickland, Bigelow, Lawrence, and Liebson (1976) carefully simulated a bar setting to monitor the drinking episodes of alcoholics.

Overall, the most common example of an enactment is the Behavioral Approach Test (BAT). Typically, clients are brought into a laboratory/clinical setting and instructed to approach a feared object as the approach behavior, self-report and/or physiological responses (e.g., Borkovec, Weerts, & Bernstein, 1977) are assessed. For most enactments a formal scoring system is developed to evaluate the behavioral dimension of interest. One problem is that some natural environments are very expensive to simulate or even impossible to reconstruct in the clinical environment.

Researchers are now beginning to scrutinize the enactment itself, and the parameters (e.g., demand characteristics) that may affect client performance. Overall, the enactment offers a potentially efficient and valid method of obtaining behavioral samples in a clinical setting. In particular, the enactment is proving very useful in the areas of sexual dysfunction, marital interaction, and especially social skills.

Role-Play Analogues

The role-play analogue requires the client to visualize or covertly rehearse stimuli under standardized conditions or overtly role-play criterion situations with staff members (confederates) and/or other clients portraying relevant persons. The role play is typically used where full-scale enactments are, for financial or methodological reasons, not possible. In addition, the role play offers the researcher the opportunity to observe the client in unnatural settings to see how the client's behavior generalizes to less familiar environments. Obviously, the role play is very flexible in its application, although it has been used most extensively in assessing social skill behaviors such as assertiveness (Rahaim, Lefebvre, & Jenkins, 1980; Van Hasselt, Hersen, & Bellack, 1981; Westefeld, Galassi, & Galassi, 1980).

Almost any situation can be presented in a role-play format, and no special or elaborate setting is required; hence, role play is procedurally much simpler and less expensive than an enactment. However, as Nay (1984) points out, the degree of success of the role-paly simulation depends on the similarity between role-play stimuli and natural events, the client's inherent ability to role play and relate to the scene, and the instructions provided.

Indeed, as Van Hasselt *et al.* (1981) point out, no matter how much face validity role-play tests seem to have (e.g., convenient, economical) their value is at best speculative if they are "psychometrically unsound." Recently, investigators have begun to look at the validity of the role play and the results are not entirely optimistic. Bellack, Hersen, and Turner (1979) found that role-play tests were not (externally) valid for chronic psychotic patients and college students. The authors hasten to add, however, that such findings do not invalidate the role play as an assessment device but rather should serve as an impetus for further research in this area. Also, Van Hasselt *et al.* (1981) suggest that many role-play instruments involve "overly restrictive" formats, possibly not lasting long enough to evoke "critical interactive responses." The reliability data on role-play tests, scarce as they are, are also somewhat questionable (e.g., Van Hasselt *et al.*, 1981). In sum, the latter authors state that "before definitive statements regarding the validity of role-play tests can be made, several other psychometric properties (e.g., test length and item variance) of these instruments must be examined " (p. 213).

In summary, the various analogue settings offer considerable flexibility to the clinical researcher where time, cost, practicality, or ethical constraints limit access to natural environments. The potential trade-off of these benefits for ecological validity is a question that remains to be answered. At present, we suggest that analogues are certainly useful in assessing limits of knowledge or skill repertoire following some intervention.

MEASURES

The final set of questions has to do with selection of a *methodology* for detecting the incidence of some target-in-setting. We will not review again the importance of selecting a method that best meets the conceptual requirements of the research question. Once a behavior and setting have been selected (Questions 1 and 2), certain method options may be ruled out due to practical constraints, ethical requirements, or cost-effectiveness. The issue of reactivity must also be considered, in that the investigator has some control over the intrusiveness of the assessment given the methodology selected. We will now briefly review four categories of method in behavioral assessment, ordered crudely in terms of the possibilities for reactivity effects. Obviously some combination of methods might be entertained. We will focus on important issues the clinical investigator should keep in mind in selecting a given method rather than detailed presentations of the methods themselves. Further information can be found elsewhere (e.g., Ciminero, Calhoun, & Adams, 1976; Hersen & Bellack, 1976; Nay, 1979) and elsewhere in this text.

Product-Related Measures

At the lowest end of the reactivity scale are the so-called *unobstusive or product-related measures* (Webb, Campbell, Schwartz, & Sechrest, 1966). Such measures do not require that an assessor be present as a particular target behavior occurs. In essence, a "permanent or temporary effect generated by the subject's behavior" (Haynes, 1979) is encoded in some fashion. Webb *et al.* (1966) arbitrarily categorized these unobstrusive measures as erosion measures, trace measures, and archive measurement. In addition, Nay (1979) discusses "task" measures.

By *erosion measures,* we are referring to physical changes in or damage to a particular setting as being indicative of a behavior that occurred previously. For example, looking at the relative dirtiness of an area of floor in some standardized way might indicate some trend in physical density or grouping behavior in an adult hospital setting. Marks on a wall and wear and tear to furniture are other examples of such behavioral results.

When using such erosion measures, one must be very careful that some extraneous variable does not contaminate the relation between the product measure and the targeted behavior. For example, using "floor dirtiness" as a measure over time could be contaminated by the hiring of a new janitor who cleans the floor more frequently. To counteract such artifacts, many investigators advocate employing a myriad of erosion-measure samplings, each one representing a different physical location/target behavior modality in a particular setting.

Also, measures of erosion, decay, and damage must be evaluated with regard to normal wear and tear, which might otherwise be expected regardless of the clients' behavior. Hence, the particular setting should be normed before any assessment in order to have a more standard baseline against which later erosion measures can be compared.

The next class of product-related measures, *trace measures,* are defined by Webb *et al.* (1966) as those objects deposited by clients in a setting or the ordinary debris left by client interaction with or consumption of material. An example would be the systematic measurement of bar-soap size as indicative of (excessive) hand-washing behavior or movement of books on a bookshelf as an indicator of reading behavior. The possibilities are limited only by the creativity of the investigator, and such measures are very economical. Trace measurement must employ a standardized methodology for assessing number, extent, or other relevant characteristics of particular trace materials in a setting.

Archive measurement would include such records as tombstones, information in the Congressional Record, voting records, and records of accidents, all of which can indicate more "global" trends in behavior over a period of time. Examples of more direct relevance to clinicians would be the utilization of medical charts and pharmacy records as indicants of a client's arthritic pain (Varni, 1981) or job performance records maintained at the job site (Winett & Neale, 1981). Hospital records, such as time between intake and discharge, are also very helpful. However, some authors (e.g., Martin & Lindsey, 1976) warn that many archival measures (e.g., hospital discharge) can be related to multiple variables.

As a fourth product-related measure, Nay (1979) discusses so-called *task measures.* Examples would include the number of completed papers or other assignments, accuracy of printing, or erasures in math problems. The reader is referred to Hawkins, Axelrod, and Hall (1976) for a more complete discussion of the many task products that could be unobtrusively assessed.

When employing unobtrusive measures, it is important that the baseline assessments of the erosion, trace, or archival product be conducted prior to the implementation of the treatment. In addition, it is necessary to collect the dependent measures on an ongoing basis throughout treatment, at posttreatment, and as a follow-up measure (Glasgow, Swaney, & Schafer, 1981; Nay, 1979; Varni, 1981). Also, unobtrusive measures should, as a general rule, be used in conjunction with other classes of assessment to ensure that an adequate, reliable and comprehensive view of the client's behavior is obtained.

While there has recently been more interest in employing product-related measures, there seems to be a wide discrepancy in the reporting of interrater agreement methods and calculation procedures (Haynes, 1979). Systematic reliability assessment is important, since the use of unobtrusive, product-related methods does not necessarily exclude the possibility of observer error (O'Leary & Kent, 1973).

Although the degree of reactivity associated with unobtrusive measures is probably lower than any of the other common clinical assessment procedures, it is nonetheless a factor to be reckoned with. Haynes (1979) points out that reactivity may be related to the obtrusiveness of the assessment procedure and the specific contingencies associated with the outcome. For example, the examination of a student's academic record can be carried out with minimal disruption to the direct behavior of the student, hence producing almost insignificant reactive effects. In contrast, the examination/measurement of a client's fingernail length is probably more reactive because it cannot be carried out without the knowledge of the client and involves at least a minimal disruption of the client's natural behavior. It seems safe to assume that greater reactive effects would occur for a particular target behavior when the subject knows that that behavior is being scrutinized, no matter what type of assessment procedure is employed.

In the realm of behavioral assessment, there have been many recent technological advances which increase the efficiency, accuracy, storage capabilities, and comprehensiveness of behavioral observations. Such methods generally fall under the rubric of the *mechanical approaches.*

The mechanical methods can vary from very simple, such as a wrist frequency counter (Lindsley, 1968), to very advanced, as evidenced by the common use of telemetry equipment (Miklich, Chai, Purcell, Weiss, & Brady, 1974). In all cases, the common theme is the monitoring of specific target behaviors using a mechanical apparatus.

For the recording of simple, discrete frequency data, portable electronic counters (Wolack, Roccaforte, & Breuning, 1975) operated by independent observers have often been used. A variant of this is to make the counter a part of the target behavior object, such as a cigarette case with a built-in counter to record the frequency of smoking behavior (Azrin & Powell, 1978). Stopwatches have often been used to record elapsed time (Wolack *et al.*, 1975) A variant on this would be to attach the elapsed-time meter directly to the client

(Azrin, Rubin, O'Brien, Ayllon, & Roll, 1968). Multichannel recorders can be employed if the target involves a multiplicity of behaviors (e.g., Lovaas, Freitag, Gold, & Kassorla, 1965).

If the target behaviors require sound-level measurements, such as deviant vocal pitch levels, a variety of electronic band-pass filter systems have been employed to detect certain audio and ultrasonic frequencies (Hoats, 1975; Shriberg, 1971). Also, audiometric meters have been used to monitor noise levels in a classroom (Nay, 1979). Such devices, however, are often impractical because they physically limit client interactions and are fairly expensive.

If the behavioral target is physiological in nature, telemetry has been a common method of gathering data (Greenfield & Sternbach, 1972). The instrumentation, usually a transducer and a radio transmitter, makes it possible to monitor specific behavioral and psysiological events (heart rate, respiratory behavior, etc.) without requiring the observer to be in such close proximity to the client. In the interest of unobtrusiveness, such devices are commonly no larger than a pack of cigarettes and weigh only about 8 ounces (Miklich *et al.*, 1974).

Nay (1979) mentions that the telephone can be employed economically as an assessment device in collecting self-reported information from clients. Indeed, many researchers use the telephone to collect "take home" assessment data (e.g., questionaires) during an ongoing therapy process (Bernal, Klinnert, & Schultz, 1980; Christensen *et al.*, 1980). In addition, the telephone can be used for posttreatment/postassessment purposes within almost any format (Forehand *et al.*, 1980). A telephone record can be scored for certain targeted events (e.g., signs of "depression," suicidal intent). Telephone assessment can easily reach 100% interrater reliability levels when more than one interviewer is used (Bernal *et al.*, 1980).

By far the most widely used mechanical assessment options are the audio/ video recording devices. Such devices evoke reactive effects only to the extent that they are actually perceived by the clients in the setting. Although they do not usually facilitate the direct coding of behavioral data, audio/video recordings provide a completely stable, permanent record that can be saved for encoding at a later time. As Haynes (1979) points out, difficulties in an observation system can be ironed out, interobserver reliability assessments can be performed, and "new hypotheses" investigated without loss of the actual events.

When verbal targets are of interest, clients can be audiotaped very inexpensively and with rather minimal obtrusiveness. For example, Forehand, Wells, and Griest (1980) used a cassette tape recorder to monitor mother–child interactions in a home setting. In many such cases, however, merely the turning on of such a recording device, if done in the client's presence, can evoke reactive effects, especially if the client is the one responsible for activating the machine. To control for such reactivity, Christensen *et al.* (1980) hid the recording device and used a random timing device to activate it. This randomness also makes for a more representative sample of client behavior. A variant on this is to outfit the client with a transmitter capable of transmitting verbal behavior to a receiver/recorder outside the setting proper.

Videotape records (camera and microphone concealed) are in order when both verbal and nonverbal client behavior are of interest. Like audio recordings, videotapes may be viewed and coded at the researcher's convenience, and videotapes can also easily be used for assessment/therapy training purposes (Howe & Silvern, 1981). In addition, videotapes can be used as "focused feedback" for a client if the client is asked to scrutinize certain aspects of a previously taped intervention in which he or she was involved (Hersen & Eisler, 1973). As with the audio machines, video recorders can be used in conjunction with a timer, either time-specified or random, in order to reduce reactivity effects and ensure representative behavior samples (Anders & Sastek, 1976).

When audiovisual recordings of client behavior are made, the researcher should always obtain prior permission from the client (or legal guardian). The specific nature of such recordings should be made clear, preferably in the form of a written release. Clients should be permitted not only to erase any part of their tapes they deem overly private but also to turn off the recording device at their choosing (e.g., Christensen *et al.*, 1980). Although the researcher may have to compromise valuable data, he or she may nonetheless thwart undesirable ethical and legal problems.

There seems to be evidence of good reliability among the mechanical approaches (e.g., O'Leary & Kent, 1973), probably due to the standard and systematic nature of such assessment. Although the validity of mechanical devices as assessment tools has not been adequately investigated, most studies seem to support these procedures (e.g., Bernal, Gibson, William, & Pesses, 1971). One major disadvantage of mechanical measures, beyond the ones alluded to already, should be discussed. It is often difficult to discriminate certain words, sounds, or nonverbal behaviors even on the best audio/video recording equipment, especially when unplanned, extraneous setting variables are taken into account (e.g., loud noises or a move by the client to another part of the setting). Of course, the use of multiple recording devices, used simultaneously, can partially offset this problem. As a general rule: it is the characteristics of the particular targeted behavior that predict the usefulness and success of mechanical methods.

Although still fairly new and undeveloped areas of behavioral assessment, the mechanical approaches would seem to have much to offer the clinical researcher. They can be employed to reduce reactivity of observations, are economical, and permit increased flexibility for data analysis and may facilitate a more comprehensive view of behavior.

Self-Report Measures

Now we will turn to the methods calling for a paper and pencil or other endorsement made by the client to reflect over past behavior or to carry out ongoing self-observations. The entire realm of self-observational assessments can be placed into one of two categories: *written self-reports and self-observation.*

Paper-and-pencil self-reports continue to hold a prominent place among the assessments that clinicians frequently use. The client typically self-generates the data for assessment by responding to some type of structured written stimuli. These written protocols are then scored in some manner and subsequently interpreted to reveal certain aspects of the client's behavior. Such written assessments are highly standardized, easy and economical to administer, and frequently offer the researcher a starting point in outlining an assessment strategy.

Issues attendant to each of five basic categories of the written report will now be discussed.

Free-response measures permit the client to make relatively unrestricted written responses (e.g., open-ended questions and incomplete sentences). They are an excellent source of initial information pertaining to the presenting problem, and as such are often a prelude to more formal assessment procedures. Such measures are especially valuable in that they offer an account of the problem from the client's point of view. Regardless of the accuracy of such an account, a written self-statement may offer a wealth of information about the client's overall ability to communicate and organize his or her thoughts. Most free-response instruments are designed to meet the specific needs of a particular client and setting, and few (except for the sentence-completion formats) are published for general use. A major problem is that standarized scoring criteria must be developed. One solution is to administer the instrument to a large, highly defined population in order to determine a "baseline" of normative responses. Another approach is to submit the test items to a panel of judges, who would categorize them according to some particular category of relevance from which scoring "manuals" could be developed. Finally, a more "face-valid" approach could be employed, in which items would be placed into categories that somehow "look right." Because of such loose protocols, the free-response measures leave much to be desired for the clinical researcher.

Checklists typically ask the respondant simply to indicate presence, absence, or agreement (via a check or yes–no) in response to a fixed set of stimulus items. The Depression Adjective Checklist and the Gough Adjective Checklist (social behavior) are commonly used by many investigators (Harmon, Nelson, & Hayes, 1980; Wiens, Harper, & Matarazzo, 1980). Because of the nature of the data obtained, such checklists are amenable to almost any kind of statistical analysis. In addition, they are easily (self-) administered, can be computer-or hand-scored, and psychometric properties are readily established and often published. Finally, these measures are easy to construct and can be adapted to virtually any assessment question (Nay, 1979).

In the *rating approaches*, the client is typically asked to rate the stimulus item along some relevant dimension (e.g., degree of agreement or intensity), which permits fine discriminations. Such discriminations may be valuable in assessing slight intensity or frequency changes in a target behavior that otherwise might be overlooked. Rating approaches range from highly standardized and validated tests such as the MMPI (e.g., Clopton, Weiner, & Davis, 1980) to scales constructed and used within only one study for a very specific purpose, such as the Therapist Action Scale (Hoyt, Marmar, Horowitz, & Alvarez, 1981).

Many variants of the rating approach exist, such as the Likert Scale (e.g., the Comparative [arthritic] Pain Assessment Inventory; Varni, 1981), in which a dimension different than agreement is used. When using the Likert Scale approach, it is important that each potential response be defined operationally, using precise behavioral referents. Referents should be written to avoid "theoretical jargon and/or the need for complex clinical inferences" (Hoyt *et al.*, 1981, p. 111). It is necessary at least to anchor the endpoints and midpoint of the scale, so that the respondent is never confused as to the meaning of any of the scale numbers. Such clear definitions are necessary if one is to obtain meaningful data.

Another alternative to the Likert Scale is the Graphic Rating Scale (Wiggins, 1973), in which the respondent is presented with a geometric (vs. numerical) representation of the dimensions to be rated. Several precise, operationally defined anchors are placed beneath a straight line and the respondent is asked to make a rating by placing a mark on the line, which is broken up with hash marks. A variant of this approach, the continuous scale, provides the respondent with an unbroken line and asks him or her to make a mark on the line at any point, depending on his or her subjective rating of the dimensions of interest. The continuous scale is especially advantageous because it provides the respondent with increased rating flexibility and it detects very subtle discrepancies from the middle and end points, which the Likert Scale, by definition, cannot do. It is believed that these slight discrepancies reflect a directionality of opinion or attitude on a particular test item if perceptual–motor anomalies (e.g., sloppiness) can be ruled out.

If the researcher wishes to evaluate the affective meaning that persons ascribe to various objects in their environment, the Semantic Differential (SD) (Osgood, Suci, & Tannenbaum, 1957) is well suited. The Semantic Differential is well known and need not be described here. The interested reader should consult Osgood *et al.*, (1957) or a sourcebook of readings such as Snider and Osgood (1969).

In certain assessment situations a highly idiosyncratic rating system can be constructed to assess some relevant dimension that none of the other written methods can adequately address. An excellent example of such a scale is the subjective degree of discomfort scale (SUDS) developed by Wolpe (1973). The client subjectively rates his or her fear of some object (e.g., test taking) on a scale from 0 to 100 SUDS units. Points (SUDS units) along the dimension are anchored in terms of specific, behavioral states of the client (e.g., tightness of certain muscles, heart rate).

The psychometric properties of any written self-report instrument should be assessed within three major categories: internal consistency, stability, and validity. Internal consistency is concerned with the degree to which items within the instrument are selected to measure some core dimension or theme, as opposed to surveying a wide variety of discrete and perhaps mutually exclusive categories of life events. Only written instruments in which all of the statements are related to some core problem area (e.g., difficulties in social situations) would be expected to show internal consistency. In instruments assessing a wide variety of problems areas (e.g., problem checklists), internal consistency, by definition, would not apply.

Stability deals with the degree to which a given instrument, applied over several occasions with the same population, yields the same score. If such stability is not demonstrated, the cause is probably an instrument defect (method variance), although other factors are possible also (e.g., poor motivation, etc.). Stability is assessed using a correlation coefficient, with $r = .85$ being a generally acceptable stability criterion level according to the literature (Pfeiffer & Heslin, 1975).

Validity is concerned with the issue of whether or not an assessment instrument measures what it purports to measure. At the outset it should be noted that validity is inherently limited by reliability. That is, a measure must show some degree of consistency (reliability) before it can be expected to measure anything. Here we will only touch on each type of validity in scant detail, as many relevant test construction references are available to the interested reader.

The presence of content or "face" validity implies that the items chosen for the self-report instrument seem to be related to the behavioral dimension of interest. For example, test item domains such as "destructive behavior" and "severe temper tantrums" are categories that would make sense in a rating scale which attempts to assess the behaviors of problem children. Typically, some sort of empirical approach (Goldfried & Sprafkin, 1976) is used in selecting the items for the instrument. Judges or "experts" are often used in selecting relevant items, as are the subjects themselves (e.g., their preferences out of an item pool). Usually some *a priori* criteria of rater (judges or subjects) endorsement is used to determine which items should be included in the final assessment device (e.g., at least 50% of the raters had to endorse the particular item). The item selectors rely not only on their own theoretical backgrounds for formulating test items, often using items from previous rating systems and process–outcome studies (Hoyt *et al.*, 1981). Situational analyses (Goldfried & Kent, 1972) are also useful in item selection.

Concurrent validity deals with how subjects' responses on the proposed instrument compare with their responses elicited by other assessment methods that presumably tap the same behavior dimension. The practical utility of this type of validity is limited by the choice of "criterion" measures used for comparison. Poor comparison criterions lead to ambiguous validity assessments. Nay (1979) points out that

> in validating an instrument, the BCA must be sure that variability in responding attributable to unreliable use of that instrument (e.g., poor training or raters, lack of standard instructions) is held to a minimum so that the correlations obtained are relatively unconfounded by method variance and the results are interpretable. (p. 112)

Indeed, for many paper-and-pencil assessment techniques, the specific method of presentation of the items has been shown to influence test scores (hence validity) much more than previously thought (Phillips, Phillips, & Shearn, 1980). Schriesheim and Denisi (1980) suggest not grouping items together on a test that measures similar constructs. Grouping can allow subjects to be overly reliant on their "implicit theories" in answering the questions, thus impairing both convergent and discriminant validity. Convergent and discriminant validity are intimately related to the concept of concurrent validity (see Campbell & Fiske, 1959).

Construct validity is associated with the extent to which a test measures a particular psychological construct. It is evaluated by such methods as cluster analysis (e.g., Hoyt *et al.*, 1981) and group comparisons (e.g., Lobitz & Johnson, 1975) in order to determine how the test items relate to the theoretical constructs that the test purports to measure. Put another way, construct validation deals with the extent to which the test constructor's theoretical presuppositions and personal hypotheses (in choosing appropriate test items) about the assessment targets are accurate.

One problem that has consistently plagued the formulators of written assessment instruments is the possibility that self-reports may be unduly biased by certain response sets (client expectancies or situational demands of the tests themselves). The most common of these response sets, social desirability, is the idea that certain individuals tend to respond in a socially desirable manner regardless of their true feelings (Crowne & Marlowe, 1964). It is a common finding of standardized personality assessment devices that subjects tend to rate favorable, general personality interpretations as significantly more accurate descriptions of their personalities than unfavorable ones (Weinberger & Bradley, 1980). In addition, some investigators claim that this "Barnum effect" is differentially more pronounced for projective tests than objective tests (e.g., Snyder, Larsen, & Bloom, 1976). Although some instruments have been developed to measure a client's tendency toward social desirability (e.g., two-alternative forced-choice formats; Pfeiffer & Heslin, 1975), the latter still remains a rather disconcerting "thorn in the side" of test constructors.

Another bothersome response set is acquiesence, or the tendency of some respondents to agree that a statement or behavior description is always self-descriptive regardless of its content (Wiggins, 1973). One common method of reducing acquiescence is to balance "affective direction" of test items randomly, so that unconditional "agreement" on the client's part will be reflected in conflicting (opposite) responses. Finally, level of client motivation (manipulated by instructions to the client) seems to be an important factor affecting a client's tendency to "fake" an empirically derived instrument. However, the question of whether faking significantly influences the validity of a test remains unanswered (Thornton & Gierash, 1980).

Self-Monitoring

For many and varied reasons, behavioral targets cannot always be observed by independent or "outside" assessors. In fact, covert targets (e.g., thoughts, feelings) are preceivable only by the client. Thus, it may be necessary to have the client self-monitor his or her own behavior. Everything from weight and caloric intake (e.g., Rosen, 1981) to hair-pulling urges (McLaughlin & Nay, 1975) has been reliably self-monitored. Overall, self-monitoring has been employed not only as an assessment device but also as a promotor of certain therapeutically desirable behaviors (Rostow, 1980).

Many different methods of recording self-observational data are available, including mechanical methods, product-related methods, and written (aside from standard paper-and-pencil) endorsements.

It is common to supply the client with a mechanical or electronic device for indicating the occurrence of a target behavior. As examples, (wrist) counters have been used (e.g., Rostow, 1980) for frequency counts, and stopwatches and stove timers are common methods of obtaining duration measures (e.g., Ciminero et al., 1977). In essence, any means by which the client actively manipulates some material object to count previously defined behaviors is considered a mechanical procedure. In addition, mechanical devices are often used to actually measure qualitative dimensions of target behaviors such as blood pressure (Engel, Gaardner, & Glasgow, 1981). There are several notable advantages to these mechanical self-assessments. They are usually easy to perform and require a minimal time commitment on the part of the client. They can be carried out in most settings, thus giving a truly representative and global sampling of the target. One very nice by-product is that the client receives immediate feedback for his or her behavior, which is very important if reactive behavioral change is desired.

The mechanical methods are not without their disadvantages. As stated previously, such methods necessarily limit the data to simple frequency counts or duration measures. Very important situational information (e.g., controlling conditions, cognitive and physical referents) is thus sacrificed. Also, the required mechanical equipment may not be available or cost-effective or the client may refuse to or be unable to use such equipment for cosmetic or other reasons. Finally, the client must fully understand how to use the given mechanical device (Engel *et al.*, 1981) or the obtained data will obviously be questionable.

The product-related self-assessments involve monitoring products or debris (or other material effect) that may be evaluated as an indicant of behavioral frequency (Webb *et al.*, 1966). For example, McLaughlin and Nay (1975) had their client record the number of hairs collected as a measure of hair pulling. Similarly, cigarette butts or amount of candy left in a jar at day's end could both serve as accurate indicates of task-related efforts.

One of the advantages of the behavioral product is precise and indisputable evaluation of client behavior. Also, such methods require little effort on the part of the client, as a once-per-day "tally" is sufficient. Nay (1979) points out that these methods are probably the least reactive of any of the observational methods because assessment takes place at a time other than when the behavior occurs.

Several disadvantages are inherent in the product-related methods. As before, information regarding "controlling conditions" and other contextual referents is lost. Also, the client does not obtain immediate feedback about the occurrence of the target, which precludes any possible beneficial, reactive effects. Obviously, covert behaviors (e.g., thoughts and images) do not usually produce an overt product, so such behavior is not amenable to product-related assessment. However, the researcher should be cognizant of possible overt events (e.g., nail biting) being linked to covert events (e.g., anxiety; see Nelson & McReynods, 1971). The product must be carefully evaluated to ensure that it is indeed a consequence of the target behavior and not some other extraneous variable. Finally, product measures may not be entirely sensitive to subtle changes in frequency of the target behavior, such changes often being very reinforcing to the client (Kazdin, 1974).

The written methods would include "any occasion when the client is asked to make a written response about his behavior" (Nay, 1979, p. 126). The two major types of written self-assessments are "nonspecific" and "specific" records, although sometimes clients are asked to combine both.

Nonspecific records are merely a "diary" of all behaviors occurring within a specified time frame, with no regard for any specific class or category of behaviors. This approach seems especially useful when the client cannot accurately pinpoint his or her problem or life situation. Feelings of depression have been common targets for nonspecific records (e.g., Harmon *et al.*, 1980; Lewinsohn, Sullivan, & Grosscup, 1980). Many self-report "logs" (Winett & Neale, 1981) are also of a nonspecific nature. Nonspecific recording may help the behavioral change agent (BCA) to "target" certain behaviors as a focus for later (specific) self-recording by the client. Nonspecific records are also useful in helping the client to verbalize otherwise abstruse thoughts or feelings. This approach has little value in research.

Specific recording procedures exclude all events other than one or more target behaviors. In addition, the client may also be asked to record information as to time, location, physical and cognitive events, and other conditions that describe the behavior; however, such information is recorded only on the incidence of the target. Usually, specific records involve coding sheets or rating forms which the client completes as the target behavior occurs. Many obesity-control programs employ such coding sheets so that the client can actively monitor such parameters as weight, calorie intake, and foods eaten (e.g., Carrol, Yates, & Gray, 1980; Rosen, 1981.) Glasgow *et al.*, (1981) had their clients carry around a notebook in which each instance of nail biting behavior was recorded. The "specific" method has even been used in monitoring anorexia nervosa (McGlynn, 1980). It is often very helpful to the client to graph or chart the frequency of a target (over days, weeks, etc.), so that the client can observe progress and receive behavioral feedback.

There are several important advantages to written record keeping. The written approach allows the client to collect information other than mere frequency counts, which can be very useful in defining controlling conditions. Also, many formats of data gathering from diaries to checklists, are possible, and all give immediate feedback to the client. Importantly, written, especially nonspecific recording, allows the client to describe more accurately the target in the precise context in which it occurs. Even some "specific" approaches encourage the client to record some of the more subjective (cognitive) aspects of behavior (Carroll *et al.*, 1980). Finally, written recordings are probably more reinforcing to a client simply because he or she has more of an investment in a written account. Overall, written accounts make more sense to a client than the other more impersonal, mechanistic procedures.

Several disadvantages in written recording are evident. In some settings (e.g., sports or certain social settings), it is difficult or impossible for a client to make a written response. Written responses often disrupt ongoing activities in an overly obtrusive fashion, thus provoking considerable reactivity in clients. Finally, and most important, the researcher should be certain that the client does not view the written task as aversive, as the client may not accurately perform the task. As long as some care is taken in assigning a particular written method to a client, accurate and reliable information regarding the client's behavior can be obtained. Generally, speaking, the format of written record keeping makes this approach particularly suitable for research.

The setting in which a client is asked to self-observe may be determined by a number of factors, such as the nature of the problem itself (e.g., setting-specific or setting-independent). For example, in his treatment of obese clients, Rosen (1981) construed a restaurant setting as much different (in terms of situational variables) than a home setting. The researcher must carefully evaluate all facets of a client's problem before focusing self-observation toward one or two specific settings. Even if treatment is only directed at one high-rate setting (e.g., the home), it may still be useful to have the client collect data for the targeted behavior in other settings to determine if some generalization is occurring.

An important question regarding behavioral sampling is how many "samples" of behavior (events or times) portray a representative view of the client's behavior. Although no hard and fast rules exist, we offer the following guidelines: (1) With low-rate behaviors (e.g., episodes of physical aggression) the number of baseline days should be increased to ensure an accurate assessment. Conversely high-rate behaviors require only an abbreviated time span for acceptable accuracy. (2) Sampling should be performed as frequently as possible, both within days and across days. Increasing the number of observations within days leads to more accurate overall hourly frequency estimates. Extending observation across days helps to reveal daily and weekly trends more clearly. (3) Finally, the variability of the target across days should be considered. Many researchers (e.g., Engle *et al.*, 1981) suggest that data should be collected until variation in frequency (across days) is minimized, usually at some preselected criterion level (e.g., no more than 5% variation in range of scores over some period). One very favorable result of frequent initial sampling of behavior is that "peak" periods of target behavior (or target "cycles") can be ascertained. Such data can be helpful later on in that valuable assessment time can be focused primarily on times when the target is known to occur (Rostow, 1980).

Reactivity and self-observation seem intimately related. Indeed, as Nay (1979) observes, "the usefulness of self-monitoring as an assessment device is concomitantly limited as reactivity effects are enhanced and . . . the manner in which reactivity is viewed depends on the goals of self-observation" (p. 133). The issue of reactivity has pervaded the entire range of the self-monitoring literature, including smoking behavior (Martin & Fredericksen, 1980), clinical depression (Harmon *et al.*, 1980), obesity (Carroll *et al.*, 1980), self-control in retarded persons (Litrownik & Freitas, 1980), and self-control behaviors (Rosenbaum, 1980). While some studies have shown seemingly dramatic, long-lasting therapeutic effects following self-monitoring (e.g., Mahoney, 1977), few studies have employed follow-up measurement to assess the durability of self-monitoring effects. It seems that the (reactive) effects of self-monitoring tend to dissipate with time (Kanfer, 1977).

The literature does, however, reach considerable consensus when it comes to defining the conditions where reactivity is maximized. The nature of the specific target seems to have important consequences for reactivity. It is obvious to say that there is a difference between recording an "urge" to do something and actually recording a behavior once it has occurred. Letting the behavior occur and merely recording it only serves to further reinforce the client in that behavior. In contrast, recording an "urge" (prior behavior) permits the client to predict and possibly control the onset of that behavior. This speculation, while plausible, has received only limited support in the self-monitoring literature (e.g., McFall, 1970). Whether to alter the covert antecedents (the "urges") or the overt behavior itself largely depends on the goal of intervention. Interestingly, different clinical populations tend to benefit defferentially from self-monitoring and its reactive effects. Self-monitoring by itself has been shown to contribute to the treatment of depression (Harmon *et al.,* 1980), while its therapeutic benefits for a moderately retarded population appear much less robust (Litrownik & Freitas, 1980). It is, however, amazing that any such benefits occur when one considers that retarded persons have been shown to display very inadequate self-monitoring skills (Zegiob, Klukas, & Junginger, 1978).

Another characteristic of the target that enters into the "self-monitoring equation" is the valence of the behavior for the client. If a behavior is negatively valued (negatively valenced), it may not increase in frequency regardless of how often the client self-monitors the event. The converse also holds true for postively valenced behaviors (Kanfer, 1970). Empirical support for Kanfer's valence model remains to be obtained, as the evidence thus far has been equivocal (e.g., Litrownik & Freitas, 1980). Closely associated with valence is the issue of motivation. In essence, a client will be more motivated to change a behavior of positive or negative valence than one of neutral valence (Kazdin, 1974). Much recent research has supported this speculation (e.g., Harmon *et al.,* 1980; Nelson, Lipinski, & Black, 1976).

Another factor that may affect reactive behavior change is the provision of a standard by which the self-monitoring client can evaluate performance. Indeed, such standard setting is a prerequisite for behavioral change in many of the self-regulation models. Kazdin (1974) showed that monitoring oneself or being monitored by another are equally reactive. In a related vein, Hayes and Cavior (1977) found that monitoring of two or three behaviors provoked significantly less reactivity than self-recording only a single target. Kazdin

(1974) also found that performance feedback (a digital display of self-recorded behavioral frequency) significantly increased the frequency of the desired target. For example, such feedback has been shown to promote marked improvement in a smoker's accuracy in estimating the carbon monoxide concentration in exhalations (Marin & Frederiksen, 1980) and to greatly facilitate weight loss in obese clients (Carroll *et al.,* 1980).

In summary, reactivity, a somewhat inconsistently demonstrated product of self-monitoring, seems to be enhanced by a number of factors including the valence of the behavior, the setting of specified predetermined standards for the client's performance, and performance feedback. Knowledge of these factors may help the researcher to employ self-monitoring in a manner consonant with research goals.

Finally is it very important that self-derived information be reliably gathered. By reliable, we simply mean that a client's self-recorded data should consistently coincide with recordings made by an independent (one hopes unobtrusive) observer. Because of the practical limitations of using an independent observer at times (e.g., low-frequency behaviors, covert behaviors, etc.), it is comforting to know that clients can themselves learn to self-record at acceptable levels of reliability. In fact, reliabilities (percentage of agreement between clients and independent observers) are typically above 90% (e.g., Santogrossi, O'Leary, Romanczyk, & Kaufman, 1973). Interestingly, many investigators have found either minimal or no relationship between reliability and reactivity (Hayes & Cavior, 1977; Herbert & Baer, 1972).

It is important that reliability in client responses be assessed prior to any formal data collection. Several recommendations are offered by Nay (1979) to help in establishing acceptable levels of reliability: (1) It is often very helpful to the client if the researcher verbally reviews and/or models the observation task with clients, given that the task demands are commensurate with the clients' capabilities and level of motivation. In a related vein, it is of utmost importance that a convincing rationale for the necessity of self-monitoring be provided, so that clients are motivated to be accurate and reliable (Lewinsohn, Sullivan, & Grosscup, 1980). (2) The client should be given ample opportunity to practice self-monitoring the targeted behavior. In fact it is generally advisable for the researcher to permit practice of the self-monitoring task until a previously specified criterion level of performance is reached (Engle *et al.,* 1981). Imagery or role playing can be used to evoke the target in the client if it does not occur spontaneously. (3) If the target is overt, simultaneous recordings can be made with the client as a reliability check. Engle *et al.* (1981) determined that their clients maintained acceptable levels of reliability by making simultaneous target recordings (with the client) on a weekly basis. If the client records the target improperly, the researcher can then provide immediate corrective feedback. (4) If continued self-monitoring (after therapy) is a desired goal, it is important that continued incentives for monitoring be provided in the client's natural environment. Self-control training can be very useful in this regard (Martin & Frederiksen, 1980; Nay, 1976; Rosenbaum, 1980; Rosen, 1981).

Although many investigators concede that it is impossible to assess reliability in self-monitoring of covert targets (e.g., Simkins, 1971), some suggest that overt behaviors linked to covert targets be assessed (Nelson & McReynolds, 1971), such as assuming that a covert smoking "urge" reliably precedes the overt act of smoking.

Overall, the prospects of continued use of self-monitoring as a measure in therapy research are favorable. It is safe to say that self-monitoring offers an economical alternative in assessment (Schrauger & Osberg, 1981), not to mention its humanistic appeal of putting the responsibility on the client (Rosen, 1981).

Direct Observation by Others

Although unobtrusive, mechanical, and self-assessments all fall under the general rubric of observation, there are other observational assessments that are unique in that an independent observer transduces the target of interest under well-defined rules. Issues surrounding these observational methods will now be considered.

Nonspecific Approaches. Observational methods of recording and encoding behavior generally fall into one of two major categories: nonspecific and specific approaches. The nonspecific or "specimen" record-keeping approach is defined by Wright (1960) as "an unselected sequential, narrative description of behavior and the conditions within which it occurs." In effect, the observer records all behavior within certain time parameters and is in no way selective as to observational content. The idea here is to obtain a complete, "global" assessment picture, which includes all specific behaviors observed, the effects of those behaviors, and cause–effect relations between persons and objects as behavior commences.

Nonspecific records should include the time of observation, the specific behaviors observed, and relevant situational and interpersonal relationship information. In regard to the latter, Michelson and Dilorenzo (1981) specify the great need for such (interpersonal) social-interaction assessment, especially with problem children. Employing several observers, each observing during preselected time periods, can reduce the possibility of bias or incomplete written accounts due to fatigue or boredom. Also, simultaneous audio or video recordings can be made to ensure good reliability, thus ensuring a comprehensive record (Ivancic *et al.,* 1981; Nay, 1979).

The significant advantages of the nonspecific method include the following: (1) the observer can use any of several thousand English words to describe the behavioral subtleties of the client; (2) nonspecific written accounts accurately describe behavior in the situational context in which it occurs (e.g., environmental contingencies); and (3) nonspecific assessments often serve as a good starting point for further observational assessment by defining categories that should be the targets of behavior change.

Issues to be addressed in using the nonspecific approach would include at least the following: (1) The written account is necessarily limited by the observer's ability to articulate succinctly in the spoken language. He or she simply may not be able to "find the words" to describe certain of the behaviors observed, in which case information is lost. (2) Nonspecific accounts are not very practical with a large number of interactors. Biases may color the observer's record as he or she is probably forced to choose the apparently most salient behaviors out of the entire range of behaviors emitted by all the persons in the setting. (3) It is very difficult to reach any suitable level of interobserver reliability when each observer has his or her own way of describing a particular behavior.

Specific Approaches. A solution to some of the problems encountered in the nonspecific approach is the so-called specific or "on–off" method of behavioral observation. In the latter method, predetermined coding definitions/systems are employed which enable the observer to record the frequency of certain prechosen, scorable behaviors.

Typically, a fixed observation interval is used, and the observer merely notes whether or not the behavior occurred ("on") or did not occur ("off"). Because the observation interval is fixed, one only has to add up all the intervals to define "total observation time," so that specific behaviors can be related directly to time (e.g., Forehand, Wells, & Griest, 1980). It is beneficial either to use an easily remembered coding scheme or to have the coding symbols already written on the scoring sheet. If the latter is used, the observer only has to make a mark (e.g., a check or slash) on the sheet next to the coding symbol of the behavior observed. Such coding schemes can vary from delineating discretely defined behaviors (Howe & Silvern, 1981) to the use of major, broad-based categories such as "antecedent vocalizations" (Ivancic *et al.*, 1981).

When information beyond mere frequency of occurrence is desired, the "field format" variant (Weick, 1968) of this on–off approach requires that observers systematically attend to several aspects of a setting. Using such an approach, one could gather information about the conditions under which a behavior occurred, such as physical proximity of interactors, temporal information, and clients or objects that the behavior seems to be related to. Farkas and Tharp (1980), for example, had their "naive" student observers report on more subjective features of the setting, such as the relationships between observed behaviors and personality constructs. Virtually any category of information can be assessed using this approach.

Michelson and Dilorenzo (1981) used an interesting variant on the use of multiple observers. Four observers each observed one target child for a brief time interval, after which they would "rotate" to a new child until all children had been rated once. Interrater reliabilities were consistently above the 90% agreement level, and this is not uncommon for well-defined observation categories. Many studies employ a "random check," using at least two observers simultaneously, to evelute interobserver agreement (Adubato et al., 1981; Forehand et al., 1980; Hay, Nelson, & Hay, 1980).

Overall, the specific approach does well in reducing method variance, as well-defined, standardized coding schemes are employed. The individual codes can often be directly added together to form a succinct and comprehensive "summary score" (Christenson et al., 1980). Howe and Silvern (1981) recommend defining each of the behavior codes "concretely," using everyday language, not scientific jargon. Such well-defined behavioral categories also serve to ensure good interrater reliabilities, something not usually found in the more subjective, nonspecific approaches. Also, clear behavioral definitions make it very practical to share observational protocols with other settings. One major problem is that what can be detected and encoded by the observer is limited by the coding system. Clinically relevant (new) behaviors that are not part of the coding scheme will be excluded, with the result of losing some very important assessment data. This objection points out the importance of thorough preassessment observation and screening in order to adequately determine which behaviors should be assessed for outcome evaluation.

General Issues

Although observations in the natural environment seem desirable, they simply are not always practical or possible. For example, Patterson and Harris (1968) point out that many target behaviors do not occur with a high enough frequency in the natural enviornment to be of any real clinical value. Often in such cases, the natural setting can be modified so that the desired targets occur at much more frequent rates, making them more readily amenable to study (Weick, 1968).

If the situation permits, the client can be observed in his or her ordinary or "natural" surroundings. The natural environment could be virtually any setting in which the client's target behaviors are known to naturally occur.

Given the possibility of reactivity effects, several investigators have offered some general guidelines for observers in natural settings (Nay, 1979; O'Leary, Romanczyk, Kass, & Dietz, 1979). First the observer should attempt to become as neutral a stimulus as possible and avoid interactions with the persons to be observed, particularly children. For examples, Sanders and Glynn (1981) had their observers avoid all eye contact with target clients and totally ignore the children in the setting. Next, "the observer should physically position himself away from ordinary paths of movement, yet have unobstructed visual access to the setting" (Nay, 1979, p. 184). Importantly, the observer should hold no apparent stimulus value, such as atypical dress or demeanor, which could serve to increase the observer's obtrusiveness in the setting. Particularly in institutional settings, the observer should be absolutely certain to follow all the rules and regulations, formal or informal, that are characteristic of the setting. In such cases, it is optimal if the assessment/therapy program can actually be incorporated into the typical routines of the institution (Ivancic *et al.*, 1981). These authors also point out that the observer should be present in the setting for a few days prior to actual observation to aid in reducing reactivity. Finally, the observer should enter the setting in such a way as to minimize disruption of ongoing behavior (e.g., between classes or prior to a meal). While these suggestions may not eliminate reactivity effects, they do make for a much more unobtrusive and standardized observational protocol.

Several other issues attendant to observation of natural settings should be considered. As mentioned before, many target behaviors occur at too low a rate, given time constraints, to be observed in the natural environment. In order to evoke such targets, the setting may be modified in some way (Forehand *et al.*, 1980; Sanders & Glynn, 1981) or one of the analogues already mentioned might be employed. Another problem is that extraneous variables, unusual events, and other environmental variables (e.g., a sick family member or a recent family crisis) may alter the natural environment so that data obtained therein are not typical of behavior in that setting. Such problems can easily be overcome through a more representative, extensive behavior sample. Finally, such environmental variables as poor lighting or acoustics or an overly large physical setting can have very deleterious effects on an otherwise well-structured observational program (Forehand *et al.*, 1980).

One possible methodological variant of purely naturalistic observation is observation by "significant other" (Romanoff, Lewittes, & Simmons, 1976). Typically, someone who is already a member of the setting is trained to carry out the observations. Peers, family, or staff members are most commonly employed in this role. Parent–child observational dyads (parent as observer) seem especially common and well documented in this area (Christensen *et al.*, 1980; Forehand *et al.*, 1980; Sanders & Glynn, 1981). Unfortunately, as Romanoff *et al.* (1976) point out, little empirical research on such observation by "significant other" has been carried out.

Some issues for "significant other" observation are worth mentioning. The independent "other" may provoke as much reactivity as an independent (outside) observer if the client is aware that observations are being carried out. Also, significant others may not be objective in their observations and may impose a degree of bias on ratings. Christensen *et al.* (1980) made frequent, random phone calls to parents to encourage them to keep their data current and standardized. Finally, family or close friends may not be willing to endure observational training sessions.

Another variation on this theme is that of *participant observation*, an area much more adequately researched (Balaban, 1973; Hay, Nelson, & Hay, 1980). Essentially, participant observation employs a trained observer who is introduced into and subsequently becomes a member of a clinical or research setting. An adequate amount of time for the observer's adaptation to the setting must be allotted. Although use of participant observers is commonly construed as a method of reducing reactivity, several recent investigations suggest that even participant observers may induce noticeable reactive effects (both "observee" and "observer" reactivity) in the setting (Forehand, 1973; Hay, Nelson, & Hay, 1977; Hay *et al.*, 1980). From a legal and ethical viewpoint, it is important that the researching participant–observer obtain permission from each member of the setting before observation is carried out.

A very important yet often overlooked issue in observation assessment is that behavior often varies across settings. Such variation is better understood when viewed in terms of an ecological approach to assessment, which suggests that the "behavior of any one element (e.g., person) in a particular environment is influenced by and in turn influences a variety of other elements in that system in an interdependent fashion" (Nay, 1979, p. 190). Viewing any one such element in isolation can often lead to a very misleading and limited view of the target person's behavior. Lichstein and Wahler (1976) suggest a number of drawbacks to a "nonecological" viewpoint. First, focusing exclusively on any one setting will likely produce a sample of behavior that is representative only of that setting and no others. Next, the characteristics of any one setting are apt to change over time, thus possibly provoking changes in target behaviors. Finally, the use of only one setting severely limits the external validity of the data obtained, and this seems in direct contrast to the researcher's frequent goal of generalizability (of therapy) across diverse settings. Goggin, Flemenbaum, and Anderson (1980), using an ecological assessment in a hospital setting, suggest that "control groups such as patients . . . on other inpatient wards, should be used to determine the degree to which hospital settings induce similar personality patterns on staff and patients" (p. 200).

SUMMARY

In summary, three major questions that should underlie the selection of a behavioral assessment protocol each suggest certain issues that have been addressed in this chapter. We have stressed the importance of careful planning to ensure that the instruments selected permit the investigator to answer the research questions posed. Questions that are well articulated with careful specification of the theoretical biases of the investigator will lead to artful selection of the instrument that can best detect and encode the behavior of interest. We have offered a variety of behavioral assessment methods that provide a high degree of flexibility and choice to the researcher, and we have discussed both theoretical and practical issues of use. For the clinical researcher, behavioral assessment offers a degree of measurability and precision that cannot be realized with instruments that assess global constructs/personality traits. We believe that multimethod assessment across major response channels, a hallmark of behavioral assessment, offers the hope of achieving a truly comprehensive view of person-in-situation.

REFERENCES

Adubato, S. A., Adams, M. K., & Budd, K. S. Teaching a parent to train a spouse in child management techniques. *Journal of Applied Behavior Analysis,* 1981, *14,* 193–205.

Allport, G. W. The historical background of modern psychology. In G. Lindzey & E. Aronson (Eds.), *The handbook of social psychology* (Vol. 1). Reading, Mass.: Addison-Wesley, 1968.

Anders, T. F., & Sastek, A. M. The use of time lapse video recording of sleep-wake behavior in human infants. *Psychophysiology,* 1976, *13,* 155–158.

Arkowitz, H., Lichenstein, E., McGovern, K., & Hines, P. The behavioral assessment of social competence in males. *Behavior Therapy,* 1975, *6,* 3–13.

Azrin, N. H., & Powell, J. Behavior engineering: The reduction of smoking behavior by a conditioning apparatus and procedure. *Journal of Applied Behavior Analysis,* 1968, *7,* 577–581.

Azrin, N. H., Rubin, H., O'Brien, F., Ayllon, T., & Roll, D. Behavior engineering: Postural control by a portable operant apparatus. *Journal of Applied Behavior Analysis,* 1968, *1,* 99–108.

Balaban, R. M. The contribution of participant observation to the study of process in program evaluation. *International Journal of Mental Health,* 1973, *2,* 59–70.

Bellack, A. S., Hersen, M., & Turner, S. M. The relationship of role playing and knowledge of appropriate behavior to assertion in the natural environment. *Journal of Consulting and Clinical Psychology,* 1979, *45,* 679–685.

Berkowitz, L., & Donnerstein, E. External validity is more than skin deep. *American Psychologist,* 1982, *37,* 245–257.

Bernal, M. E., Gibson, D. M., William, D. E., & Pesses, D. I. A device for recording automatic audio tape recording. *Journal of Applied Behavior Analysis,* 1971, *4,* 151–156.

Bernal, M. E., Klinnert, M. D., & Schultz, L. A. Outcome evaluation of behavioral parent training and client-centered parent counseling for children with conduct problems. *Journal of Applied Behavior Analysis,* 1980, *13,* 677–691.

Borkovec, T. D., Weerts, T. C., & Bernstein, D. A. Assessment of anxiety. In A. R. Ciminero, K. S. Calhoun, & H. E. Adams (Eds.), *Handbook of behavioral assessment.* New York: Wiley, 1977.

Brunswik, E. Representating design and probabilistic theory in functional psychology. *Psychological Review,* 1955, *62,* 193–217.

Campbell, D. T., & Fiske, D. W. Convergent and discriminant validation by the multitrait-multimethod matrix. *Psychological Bulletin,* 1959, *56,* 81–105.

Carlsmith, J. M., Ellsworth, P. C., & Aronson, E. *Methods of research in social psychology.* Reading, Mass.: Addison-Wesley, 1976.

Carroll, L. J., Yates, B. T., & Gray, J. J. Predicting obesity reduction in behavioral and non-behavioral therapy from client characteristics: The self-evaluation measure. *Behavior Therapy,* 1980, *11,* 189–197.

Carter, R. D., & Thomas, E. J. A case application of a signaling system (SAM) to the assessment and modification of selected problems of marital communications. *Behavior Therapy,* 1973, *4,* 629–645.

Christensen, A., Johnson, S. M., Phillips, S., & Glasgow, R. E. Lost effectiveness in behavioral family therapy. *Behavior Therapy,* 1980, *11,* 208–226.

Ciminero, A. R., Calhoun, K. S., & Adams, H. E. (Eds.), *Handbook of behavioral assessment.* New York: Wiley, 1977.

Ciminero, A. R., Nelson, R. O., & Lipinski, D. P. Self monitoring procedures. In A. R. Ciminero, K. S. Calhoun, & H. E. Adams (Eds.), *Handbook of behavioral assessment.* New York: Wiley, 1977.

Clavelle, P. R., & Turner, A. D. Clinical decision-making among professionals and paraprofessionals. *Journal of Clinical Psychology,* 1980, *36,* 833–838.

Clopton, J. R., Weiner, R. H., & Davis, H. G. Use of the MMPI in identification of alcoholic psychiatric patients. *Journal of Consulting and Clinical Psychology,* 1980, *48,* 416–417.

Cone, J. D. The relevance of reliability and validity for behavioral assessment. *Behavior Therapy,* 1977, *8,* 411–426.

Conger, J. L., & Keane, S. P. Social skills intervention in the treatment of isolated or withdrawn children. *Psychological Bulletin,* 1981, *90,* 478–493.

Crowne, D. P., & Marlowe, D. *The approval motive: Studies in evaluative dependence.* New York: Wiley, 1964.

Engel, B. T., Gaardner, K. R., & Glasgow, M. D. Behavioral treatment of high blood pressure I. Analyses of intra- and inter-daily variations of blood pressure during a one-month, baseline period. *Psychosomatic Medicine,* 1981, *43,* 255–270.

Farkas, G. M., & Tharp, R. G. Observation procedure, observer gender, and behavior valence as determinants of sampling error in a behavior assessment analogue. *Journal of Applied Behavior Analysis,* 1980, *13,* 529–536.

Forehand, R. Teacher recording of deviant behavior: A stimulus for behavior change. *Journal of Behavior Therapy and Experimental Psychiatry,* 1973, *4,* 39–40.

Forehand, R., Wells, K. C., & Griest, D. L. An examination of the social validity of a parent training program. *Behavior Therapy,* 1980, *11,* 488–502.

Gilmour, R., & Duck, S. *The development of social psychology.* London: Academic Press, 1980.

Glasgow, R. E., Swaney, K., & Schafer, L. Self-help manuals for the control of nervous habits: A comparative investigation. *Behavior Therapy,* 1981, *12,* 177–184.

Goggin, J. E., Flemenbaum, A., & Anderson, D. An ecological analysis of A-B and its relationship to field dependence. *Journal of Clinical Psychology,* 1980, *36,* 195–200.

Goldfried, M. R., & Kent, R. N. Traditional versus behavioral personality assessment: A comparision of methodological and theoretical assumption. *Psychological Bulletin,* 1972, *77,* 409–420.

Goldfried, M. R., & Sprafkin, J. N. Behavioral personality assessment. In J. T. Spence, R. C. Carson, & J. W. Thibaut (Eds.), *Behavioral approaches to therapy.* Morristown, N.J.: General Learning Press, 1976.

Gottman, J. M., & Markman, H. J. Experimental designs in psychotherapy research. In S. Garfield & A. E. Bergin (Eds.), *Handbook of psychotherapy and behavior change*. New York: Wiley, 1979.

Greenfield, N. S., & Sternbach, R. A. *Handbook of psychophysiology*. New York: Holt, 1972.

Harmon, T. M., Nelson, R. O., & Hayes, S. C. Self-monitoring of mood versus activity by depressed clients. *Journal of Consulting and Clinical Psychology*, 1980, *48*, 30–38.

Hawkins, R. P., Axelrod, S., & Hall, R. V. Teachers as behavior analysts: Precisely monitoring student performance. In J. A. Brigham, R. P. Hawkins, J. Scott, & J. F. McLaughlin (Eds.), *Behavior analysis in education: Self-control and reading*. Dubuque, Iowa: Kendall-Hunt, 1976.

Hay, L. R., Nelson, R. O., & Hay, W. M. Some methodological problems in the use of teacher as observers. *Journal of Applied Behavior Analysis*, 1977, *10*, 345–348.

Hay, L. R., Nelson, R. O., & Hay, W. M. Methodological problems in the use of participant observers. *Journal of Applied Behavior Analysis*, 1980, *13*, 501–504.

Hayes, S. L., & Cavior, N. Multiple tracking and the reactivity of self-monitoring: I. Negative behaviors. *Behavior Therapy*, 1977, *8*, 819–831.

Haynes, S. N. Behavioral variance, individual differences and trait theory in a behavioral construct system: A reappraisal. *Behavior Therapy*, 1979, *1*, 41–49.

Herbert, E. W., & Baer, D. M. Training parents as behavior modifiers. *Journal of Applied Behavior Analysis*, 1972, *5*, 139–149.

Hersen, M., & Bellack, A. (Eds.), *Behavioral assessment: A practical handbook*. New York: Pergamon, 1976.

Hersen, M., & Eisler, R. M. Behavioral approaches to study and treatment of psychogenic ties. *Genetic Psychology Monographs*, 1973, *87*, 289–312.

Hoats, D. L. A pair of electronic devices for detecting restricted-bond audio and ultasonic frequencies. *Behavior Research Methods and Instrumentation*, 1975, *7*, 542–544.

Hogan, R., DeSoto, C. B., & Solano, C. Traits, tests and personality research. *American Psychologist*, 1977, *32*, 255–264.

Howe, P. A., & Silvern, L. E. Behavioral observation of children during play therapy: Preliminary development of a research instrument. *Journal of Personality Assessment*, 1981, *45*, 168–181.

Hoyt, M. F., Marmar, C. R., Horowitz, M. J., & Alvarez, W. F. The therapist action scale and the patient action scale: Instruments for the assessment of activities during dynamic psychotherapy. *Psychotherapy: Theory, Research, and Practice*, 1981, *18*, 109–116.

Invancic, M. T., Reid, D. H., & Iwata, B. A., Faw, G. D., & Page, T. J. Evaluating a supervision program for developing and maintaining therapeutic staff-resident interactions during institutional care routines. *Journal of Applied Behavior Analysis*, 1981, *14*, 95–106.

Johnson, S. M., & Bolstad, O. D. Reactivity to home observations: A comparison of audio recorded behavior with observers present or absent. *Journal of Applied Behavior Analysis*, 1975, *8*, 181–185.

Kanfer, F. H. Self monitoring: Methodological limitations and clinical application. *Journal of Consulting and Clinical Psychology*, 1970, *35*, 148–152.

Kanfer, F. H. The many faces of self-control, or behavior modification changes its focus. In R. B. Stuart (Ed.), *Behavioral self-management: Strategies, techniques, and outcomes*. New York: Brunner/Mazel, 1977.

Kanfer, F. H., & Nay, W. R. Behavioral assessment: Toward an integration of epistemological and methodological issues. In C. M. Franks & G. T. Wilson (Eds.), *Behavior therapy and its foundations* (Vol. 1). New York: Guilford Press, 1981.

Kazdin, A. E. Self-monitoring and behavior change. In M. J. Mahoney & C. E. Thoresen (Eds.), *Self control: Power to the person*. Monterey, Calif.: Brooks-Cole, 1974.

Kazdin, A. Situational specificity: The two-edged sword of behavioral assessment. *Behavioral Assessment*, 1979, *1*, 57–75.

Kent, R. N., & Foster, S. L. Direct observational procedures; methodological issues in natural settings. In A. R. Ciminero, K. S. Calhoun, & H. E. Adams (Eds.), *Handbook of behavioral assessment*. New York: Wiley, 1977.

Lewinsohn, P. M., Sullivan, J. M., & Grosscup, S. J. Changing reinforcing events: An approach to the treatment of depression. *Psychotherapy: Theory, Research, and Practice*, 1980, *17*, 322–333.

Lichstein, K. L., & Wahler, R. G. The ecological assessment of an autistic child. *Journal of Abnormal Child Psychology*, 1976, *4*, 31–54.

Lindsley, O. R. A reliable wrist counter for recording behavior rates. *Journal of Applied Behavior Analysis*, 1968, *1*, 77–78.

Litrownik, A. J., & Freitas, J. L. Self-monitoring in moderately retarded adolescents: Reactivity and accuracy as a function of valence. *Behavior Therapy*, 1980, *11*, 245–255.

Lobitz, G. K., & Johnson, S. M. Normal versus deviant children: A multimethod comparison. *Journal of Abnormal Child Psychology*, 1975, *3*, 353–374.

Lovaas, O. I., Freitag, G., Gold, V. J., & Kassorla, I. C. Recording apparatus for observation of behaviors of children in free play settings. *Journal of Experimental Child Psychology*, 1965, *2*, 108–120.

Mahoney, M. J. Reflections on the cognitive-learning trend in psychotherapy. *American Psychologist*, 1977, *32*, 5–13.

Martin, J. E., & Frederiksen, L. W. Self-tracking of carbon monoxide levels by smokers. *Behavior Therapy*, 1980, *11*, 577–587.

Martin, P. J., & Lindsey, C. J. Irregular discharge as an unobtrusive measure of . . . something: Some additional thoughts. *Psychological Reports*, 1976, *38*, 627–630.

Mash, E. J. What is behavioral assessment. *Behavioral Assessment*, 1979, *1*, 23–29.

McFall, R. Effects of self-monitoring on normal smoking behavior. *Journal of Consulting and Clinical Psychology*, 1970, *35*, 135–142.

McFall, R. Analogues. In M. Hersen & A. Bellack (Eds.) *Behavioral assessment: A practical handbook*. New York: Pergamon, 1976.

McGlynn, F. D. Successful treatment of anorexia nervosa with self-monitoring and long distance praise. *Journal of Behavior Therapy and Experimental Psychiatry*, 1980, *11*, 283–286.

McLaughlin, J. G., & Nay, W. R. Treatment of trichotillomania with positive coverants and response cost. *Behavior Therapy*, 1975, *6*, 87–91.

Mercatoris, M., Hahn, L. G., & Craighead, W. E. Mentally retarded residents as paraprofessionals in modifying mealtime behavior. *Journal of Abnormal Psychology*, 1975, *84*, 299–302.

Michelson, L., & Dilorenzo, T. M. Behavioral assessment of peer interaction and social functioning in institutional and structured settings. *Journal of Clinical Psychology*, 1981, *37*, 499–503.

Miklich, D. R., Chai, H., Purcell, K., Weiss, J. H., & Brady, K. Naturalistic observation of emotions preceding low pulmonary flow rates. *Journal of Allergy and Clinical Immunology*, 1974, *53*, 102.

Nay, W. R. *Behavioral intervention: Contemporary strategies*. New York: Gardner Press, 1976.

Nay, W. R. Analogue measures. In A. R. Ciminero, K. S. Calhoun, & H. E. Adams (Eds.), *Handbook of behavioral assessment*. New York: Wiley, 1977.

Nay, W. R. *Multimethod clinical assessment*. New York: Gardner, 1979.

Nay, W. R. Analogue assessment: Practice and conceptual issues. In A. R. Ciminero, K. S. Calhoun, & H. E. Adams (Eds.), *Handbook of behavioral assessment* (Vol. 1). New York: Wiley, 1984.

Nelson, C. M., & McReynolds, W. T. Self-recording and control of behavior: A reply to Simkins. *Behavior Therapy,* 1971, *2,* 594–597.

Nelson, R. O., Lipinski, D. P., & Black, J. L. The reactivity of adult retardates self-monitoring: A comparison among behaviors of different valences, and a comparison with token reinforcement. *Psychological Record,* 1976, *26,* 189–201.

O'Leary, K. D. Behavioral assessment. *Behavioral Assessment,* 1979, *1,* 31–36.

O'Leary, K. D., & Kent, R. Behavior and modification for social action: Research tactics and problems. In L. A. Hamerlynck, L. C. Handy, & E. J. Mash (Eds.), *Behavior change: Methodology, concepts, and practice.* Champaign, Ill.: Research Press, 1973.

O'Leary, K. D., Romanczyk, R. G., Kass, R. E., Dietz, A., & Santogrossi, D. *Procedures for classroom observation of teachers and children.* Unpublished manuscript, State University of New York at Stony Brook, 1971.

Osgood, C. E., Suci, G. J., & Tannenbaum, P. H. *The measurement of meaning.* Urbana: University of Illinois Press, 1957.

Patterson, G. R., & Harris, A. *Some methodological considerations for observation procedures.* Paper presented at the meeting of the American Psychological Association, San Francisco, September 1968.

Pfeiffer, J., & Heslin, R. *Instrumentation in human relations training.* Iowa City: University Associates, 1975.

Phillips, W. M., Phillips, A. M., & Shearn, C. R. Objective assessment of schizophrenic thinking. *Journal of Clinical Psychology,* 1980, *36,* 79–89.

Rahaim, S., Lefebvre, C., & Jenkins, J. O. The effects of social skills training on behavioral and cognitive components of anger management, *Journal of Behavior Therapy and Experimental Psychiatry,* 1980, *11,* 3–8.

Reid, J. B., & Patterson, G. R. Follow-up analyses of a behavioral treatment program for boys with conduct problems: A reply to Kent. *Journal of Consulting and Clinical Psychology,* 1976, *44,* 297–302.

Romanoff, B. D., Lewittes, D. J., & Simmons, W. L. Accuracy of "significant others'" judgments of behavior change. *Journal of Counseling Psychology,* 1976, *23,* 409–413.

Rosen, L. W. Self-control program in the treatment of obesity. *Journal of Behavior Therapy and Experimental Psychiatry,* 1981, *12,* 163–166.

Rosenbaum, M. A schedule for assessing self-control behaviors: Preliminary findings. *Behavior Therapy,* 1980, *11,* 109–121.

Rostow, C. D. The effect of self- vs. external-monitoring and locus of control upon the pacing and general adjustment of psychiatric inpatients. *Behaviour Research and Therapy,* 1980, *18,* 541–548.

Sanders, M. R., & Glynn, T. Training parents in behavioral self-management: An analysis of generalization and maintenance. *Journal of Applied Behavior Analysis,* 1981, *14,* 223–237.

Santogrossi, D., O'Leary, K., Romanczyk, R., & Kaufman, K. Self-evaluation by adolescents in a psychiatric hospital school token program. *Journal of Applied Behavior Analysis,* 1973, *6,* 277–287.

Sattin, D. B. Possible sources of error in the evaluation of psychopathology. *Journal of Clinical Psychology,* 1980, *36,* 99–104.

Schrauger, J. S., & Osberg, T. M. The relative accuracy of self-predictions and judgments by other in psychological assessment. *Psychological Bulletin,* 1981, *90,* 322–325.

Schriesheim, C. A., & Denisi, A. S. Item presentation as an influence on questionnaire validity: A field experiment. *Educational and Psychological Measurement.* 1980, *40,* 175–182.

Shriberg, L. D. A system for monitoring and conditioning model fundamental frequencies of speech. *Journal of Applied Behavior Analysis,* 1971, *4,* 337–339.

Simkins, L. The reliability of self-recorded behaviors. *Behavior Therapy,* 1971, *2,* 83–87.

Snider, J., & Osgood, C. (Eds). *Semantic differential technique: A sourcebook*. Chicago: Aldine, 1969.

Snyder, C. R., Larsen, D., & Bloom, L. J. Acceptance of personality interpretations prior to and after receiving diagnostic feedback supposedly based on psychological, graphological, and astrological assessment procedures. *Journal of Clinical Psychology*, 1976, *32*, 258–265.

Strickland, D., Bigelow, G., Lawrence, C., & Liebson. Moderate drinking as an alternative to alcoholic abuse: A non-aversive procedure. *Behaviour Research and Therapy*, 1976, *14*, 279–288.

Thornton, G. C., & Gierasch, P. F. Fakability of an empirically derived selection instrument. *Journal of Personality Assessment*, 1980, *44*, 48–50.

Van Hasselt, V. B., Hersen, M., & Bellack, A. S. The validity of role play tests for assessing social skills in children. *Behavior Therapy*, 1981, *12*, 202–216.

Varni, J. W. Self-regulation techniques in the management of chronic arthritic pain in hemophilia. *Behavior Therapy*, 1981, *12*, 185–194.

Webb, E. J., Campbell, D. T., Schwartz, R. D., & Sechrest, L. *Unobtrusive measures: A survey of nonreactive research in the social sciences*. Chicago: Rand McNally, 1966.

Weick, K. E. Systematic observational methods. In G. Lindsey & E. Aronson (Eds.), *The handbook of social psychology* (2nd ed., Vol. 2). Reading, Mass.: Addison-Wesley, 1968.

Weinberger, L. J., & Bradley, L. A. Effects of "favorability" and type of assessment device upon acceptance of general personality interpretations. *Journal of Personality Assessment*, 1980, *44*, 44–47.

Westefeld, J. S., Galassi, J. P. & Galassi, M. D. Effect of role-playing instructions on assertive behavior: A methodological study. *Behavior Therapy*, 1980, *11*, 271–277.

Wiens, A. N., Harper, R. G., & Matarazzo, J. D. Personality correlates of nonverbal interview behavior. *Journal of Clinical Psychology*, 1980, *36*, 205–215.

Wiggins, J. S. *Personality and prediction: Principles of personality assessment*. Reading, Mass.: Addison-Wesley, 1973.

Winnett, R. A., & Neale, M. S. Flexible work schedules and family time allocations: assessment of a system change on individual behavior using self-report logs. *Journal of Applied Behavior Analysis*, 1981, *14*, 39–46.

Wolack, A. H., Roccaforte, P., & Breaning, S. E. Converting an electronic calculator into a counter. *Behavior Research Methods and Instrumentation*, 1975, *7*, 365–367.

Wolpe, J. *The practice of behavior therapy*. New York: Pergamon, 1973.

Wright, H. Observational child study. In P. H. Mussen (Ed.), *Handbook of research methods in child development*. New York: Wiley, 1960.

Zegiob, L., Klukas, N., & Junginger, J. Reactivity of self-monitoring procedures with retarded adolescents. *American Journal of Mental Deficiency*, 1978, *83*, 156–163.

4

Observer Ratings

J. R. WITTENBORN

INTRODUCTION

According to Webster (1966), an observer is "one who observes, especially one engaged in or trained to habits of close and exact observations." Observation is defined as "the act of recognizing and noting some fact or occurrence, especially in nature, often involving the measurement of some magnitude." Rating is defined as "classification according to grade, rank or class." The diversity of interpretation permitted by these definitions is reflected in published reports of observer rating procedures.

It is inherently difficult to distinguish between an observation, a judgment, and an inference; in many rating procedures, all three of these functions are confounded. Some observer rating procedures provide explicit definition of the behavior, condition, or product that is to be rated and equate observable distinctions with specific steps on a rating continuum. Other procedures require the rater to infer abstract attributes or qualities and to rate distinctions on a judgmental continuum undefined by observable referents. Between these extremes, all manner and combinations of rating procedures may be found.

The involvement of judgmental considerations, particularly the use of inference, results in ratings which may vary with the rater. This possibility raises the issue of objectivity and the related question of verifiability (Wittenborn, 1972). To the extent that the rating is a direct response to public elements of the physical environment (e.g., the directly observable behavior of the patient in his immediate circumstances), the rating may be said to be objective.

J. R. WITTENBORN ● Interdisciplinary Research Center, Rutgers University, New Brunswick, New Jersey 08903. The preparation of this manuscript was supported in part by a grant from the Cape Branch Foundation, Dayton, New Jersey.

To the extent that the rating is based upon private considerations not available to direct observation by another observer, the rating may be regarded as subjective.

The objectivity or subjectivity of an observer rating, like its verifiability, is not wholly a function of the rating procedure *per se*. For the rater whose observations are unguided by appropriate concepts and taxonomical awareness, the appropriate response to some rating tasks may be a difficult inference involving several tenuous subjective judgments. For raters who are prepared with certain concepts, definitions, and taxonomic distinctions, however, an appropriate response to the same task may be regarded as simply noting a distinction among direct observations. Thus, whether a rating task should be regarded as an objective observation or a subjective inference may depend on the training and experience of the rater. Ordinarily, the more closely the rating distinctions are tied to the observable physical components of the event to be rated, the more objective the rating.

Verifiability may be expressed in terms of *intrarater* consistency; inconsistencies between an observer's rating of a videotape on one occasion and his rating on another detract from verifiability. Inconsistencies between two ratings by the same rater can be due to ambiguity in the language of the rating procedure; it is conventional to apply the concept of reliability (i.e., random instrumental error) when referring to such sources of discrepancies. The differences between two sets of reliable ratings by the same rater may reflect changes in the patient, changes in the rater, or changes in the situation.

The verifiability of observer ratings may also be expressed in terms of *interrater* differences on the same occasion. In addition to possible differences in experience and training, there may be interrater differences in observational acuity, in the rater's perception of the task, and in the motivation with which it is approached. Different raters may apply their observations to the required rating distinctions in different ways. In addition, ambiguous terminology, incomplete instructions, or vague concepts generate unreliability. These sources of instrumental error, like the instrumental error that contributes to intrarater inconsistencies, have a random, variable, unpredictable nature and should not be confused with the systematic error that may result from rater bias.

It is useful to distinguish between the verifiability of a rating and the reliability of the rating procedure *per se*. Usually the rater is the observer, but there are occasions when the rater is an interviewer who bases his ratings on observations reported by an informant. The use of informants always raises questions concerning the degree to which the informant is influenced by judgmental interpretations, and an intrinsically reliable rating device may fail to produce reliable results because of limitations of the observer.

The verifiability of ratings may be increased by simplifying the rating task, by minimizing possible uncertainties concerning the pertinent observations, and by restrictively standardizing the manner in which the rating procedure is used. When the rating procedure becomes too formalized and restrictive, however, the procedure begins to assume some of the characteristics of a mental test. With such restrictions, the behavior may no longer represent the class or qualities of behavior in which the examiner is interested.

Traditionally, observer ratings have been regarded, implicitly at least, as an assessment of behavior that is emitted in spontaneous response to situations that occur in the ordinary course of events. Such situations should be distinguished from contrived situations. A response to a contrived situation may or may not be indicative of responses to situations in which the examiner is interested. In the construction or selection of an optimal rating procedure, one must consider the kinds of questions one wishes the assessment to answer as well as the characteristics of the probable user and the situations in which the procedure will be used.

In most rating procedures, the observed, judged, or inferred alternative features of a behavioral quality or of a situation are arranged according to some sequence, and the rater must decide which of these features is most descriptively applicable. When these alternative features are arranged according to magnitude (such as size, frequency, or duration), the feature placed in Position 3 implies more of the quality than the feature placed in Position 2, and Position 2 implies more than Position 1. If Position 3 also implies more of the quality than Position 1, the continuum represented by this arrangement of descriptive features is transitive. With such an arrangement, the numbers are not merely designations; they have arithmetic qualities also (e.g., they can be averaged). When the steps express, if not define, a hypothetical continuum (e.g., anger) that has no tangible physical correlate, the transitive requirement may be unintentionally violated. Occasionally raters say that some of the steps do not belong on the same continuum or that the arrangement of steps does not scale. This apparent lack of a transitive internal consistency in an hypothetical continuum usually means that for some applications or according to some concepts, the hypothetical continuum is violated by the arrangement of some of the observational or judgmental steps.

There are obviously many advantages in gathering information in a form that lends itself to the conventions of arithmetic manipulation, and the makers of rating procedures have been resourceful in describing both observed and hypothetical attributes or events in quantified terms. All behavioral measures are of a relatively primitive nature, however, and are subject to severe limitations. In most cases a zero score or rating does not mean an absolute absence of the behavior in question. It simply means that zero is assigned the lowest score or rating provided by the procedure, and it is misleading to describe performances in terms of ratios of scores. For example, although it is possible to say that one score is twice as large as another, there is no reason for assuming that a score of 6 implies twice as much of a given attribute as a score of 3. Another limiting feature of behavior ratings is the probability that equal numerical increments on the rating scale do not represent equal increments in the behavior or attribute represented by the scale. Although it is perfectly proper to speak of the percent of a sample where the score was diminished or increased, it is meaningless to speak of percentage loss or percentage gain.

The number of units or alternatives used in developing a rating continuum may vary from 2 (as in a yes–no checklist) to any maximum which the developer feels some raters might be able to discriminate. Only a few rating scales involve more than 9 distinctions; currently, the use of 5 units appears to be most usual. It may be assumed, however, that the number of distinctions that can be discriminated by a rater in a given situation will vary with the behavioral content of the scale, the observational acuity of the rater, and the heterogeneity of the sample with respect to the quality being rated. For maximal discriminative efficiency, the continuum of steps should include the lowest level likely to be observed and the highest level expected. In this way the discriminative range will not be truncated by a "ceiling" that is too low or a "floor" that is too high to reflect the entire range of observations or judgments discriminable in the situation.

Many factors other than irrelevant content can detract from the potential validity of a rating. Validity does not accrue from the rating procedure *per se*. In addition to an appropriate behavioral content and sufficient self-consistency to be reliable, the degree to which a scale will distinguish in the desired manner depends on the characteristics of the sample under assessment and the training and orientation of the person who makes the assessment. The *de facto* validity of a reliable rating procedure is dependent upon the purpose for which and the manner in which it is used. The "halo" bias may reflect observational obtuseness, strong generalizing opinions about the person being rated, or preconceptions concerning the kinds of behavioral qualities that may be expected to "go together." There are also errors of leniency or parsimony, where some raters tend to avoid unfavorable rating while other raters tend to avoid favorable ratings. Some raters are excessively cautious and tend to avoid extreme ratings, either favorable or unfavorable. Explicit observational referents help protect the ratings from certain familiar errors of rater bias.

In general, a rating procedure is said to be valid if it meets one or more of three criteria applicable to most behavioral assessment devices. First, do recognized experts accept the procedure as reflecting the quality for which assessment is desired? Second, do the various alternative procedures for assessing a given quality tend to distinguish among subjects in a mutually consistent manner? Third, does the rating procedure distinguish among subjects in the same way as some external criterion that has *prima facie* validity?

HISTORICAL DEVELOPMENTS

In most respects, distinctions among observations based on ratings are more primitive than distinctions based on counts, and rating procedures may have antedated counting. There are early reports of rating procedures in the study of astronomy, in the assessment of animals, and in the descriptions of sensible temperature (e.g., the temperature of bath water), where the markings on "thermometers" ranged from "unbearably cold" to "intolerably hot" (Osbourne, 1877). Presumably the first systematic use of rating scales in psychology was described by Galton (1883), but an early 19th-century bronze or brass plate by which a child could gauge his maturity was found in New Harmony, Indiana (Ellson & Ellson, 1953). Galton's 9-step self-descriptive scale of clarity of imagery of the breakfast table suggests that he had little to learn from his successors about this type of scale. An early and noteworthy observer scale was developed some years later by Pearson (1906). Each step of this 7-step scale of mental ability described observable behaviors which were assumed to be indicative of the respective level of mental ability.

Thorndike (1910) developed a scale in which the rater matched a handwriting sample against a standard, carefully graded series of handwriting specimens. This procedure was met with a diversity of criticisms, including the claim that the style of handwriting in the standard scale did not represent the kind of handwriting currently taught. Subsequently, Ayers (1912) developed a handwriting scale that escaped most of the criticisms which had been directed at Thorndike's innovation, and the Ayers Scale continued in common use for several decades.

In 1915, Boyce surveyed methods for measuring teacher efficiency. It is apparent from this report that graphic rating scales were in common use at that time, and this principle continues to be in general use today. Graphic rating scales provide the rater with a line along which any number of gradations may have been provided to represent different degrees of the quality being rated. Usually, but not invariably, some verbal labels for different levels of the qualities are placed along the line, and sometimes the labels have been behaviorally explicit. Originally, graphic scales involved many of the features of good rating scales (Filer & O'Rourke, 1923) and had the additional advantage of allowing the rater to respond to whatever gradations or compromises of judgment he regarded as appropriately descriptive. The graphic principle has been combined with other features of rating scales in the construction of some important procedures (e.g., Fels Parent Behavior Rating Scales; Champney, 1941).

One of the continuing problems in the use of rating scales in a standard manner results from the difficulty of providing observational referents that are both pertinent to the purposes of the assessment and likely to have similar significance to various users. This problem of standard assessment was particularly acute during World War I, when it was necessary to assemble hastily a civilian army, including a suitable officer cadre. In consequence, a man-to-man rating procedure was used by Scott in 1917 to evaluate a potential officer candidate in terms of the degree to which he resembled other men who were known to the raters, and who were chosen to represent different levels of a desirable attribute (Scott & Clothier, 1923). Obviously, the merit of this procedure depended upon the correct selection of men to serve as standards. Although this man-to-man principle of scaling was generally used during the war, it found little postwar application in business and industry and is primarily of historic interest. Long ago, Hollingworth (1911) concluded that there is no general judicial capacity and that a person who is a good judge in one situation may prove unsatisfactory in another or that a rater who is accurate in the assessment of some qualities may not be satisfactorily accurate in assessing another trait.

In some situations, the rating task was met by use of a ranking procedure. For example, Hull and Montgomery (1919) wished to examine the validity of the then commonly held hypothesis that there was a useful degree of relationship between personality and certain characteristics of handwriting. A consensual ranking of personality qualities was developed by asking students to rank each of their fellows with respect to each of the qualities independently. The average of these rankings for a given trait was then correlated with the objectively determined qualities of handwriting (e.g., slope or thickness of line) that were considered to be related to one or more personality traits. Consensual ranking procedures of this kind can provide a proper and useful rating procedure when the sample is small enough and sufficiently known by all raters to make ranking feasible, but rank in one sample may have no relationship to rank in some other sample.

During the 1920s, several devices were developed to make interpersonal distinctions, and their use has persisted into recent decades. For example, a list of words implying approval or disapproval was prepared for teachers to use in a checklist fashion for assessing members of a class (Hepner, 1926). The Guess Who test was developed by Hartshorne and May (1929), and rating procedures for "self–ideal" comparisons were described by Knight and Franzen (1922).

Willoughby (1932) used a checklist approach to assess emotional maturity. On the basis of consensus among clinicians, 60 items were given a weight of from 1 to 9, and a combination of these weightings provided rating of an individual's emotional maturity. This procedure was subsequently applied in several investigations, including the Rogers and Dymond (1954) study.

Early users of rating scales were much aware of the hazard of unreliable scores and often included estimates of reliability with their reports of the development or use of scales. Webb (1915) reported that the average reliability of a group of rating procedures was .55, a value reported some years later by Voelker (1921) in a different survey. When Hartshorne and May (1929) used the checklist described by Hepner (1926), they found that the reliability varied from .64 to .88, depending on the personality quality being assessed. The Guess Who test also was found to be surprisingly reliable; Hartshorne and May (1929) reported reliability coefficients of .88. According to Sweet (1929), reliabilities for the use of the self–ideal procedure can vary from .79 to .94, depending on the attribute in question. In contrast with these encouraging reports, Symonds (1924) stated that in the rating of human traits one cannot expect, even under the best conditions, reliabilities of over .60 or .70. He used the formula for coarseness of grouping (Kelley, 1923) in order to estimate the degree to which the true reliability would be underestimated when the number of class intervals on which the rating was based was limited. On the basis of these statistical effects alone, Symonds recommended that since true reliability in the order of .60 may be expected, 7 different distinctions in a rating procedure are sufficient (1931). Most current users of rating scales would consider loss in reliability due to limitations in the number of steps or intervals to comprise only a small portion of the reliability problem. Nevertheless, reports such as Symonds' show that reliability and its determinants were an object of concern among the early users of rating scale procedures.

Miner (1917) used a graphic rating principle to examine the reliability with which student attributes of personality and character can be rated by judges who have had opportunity to observe the students in school and related situations. Although the rating mark could be placed on any position on the line, the line presented was divided into fifths and labeled to indicate quintiles. The reliability of the graphic ratings was not dependent on whether the line was divided in 5 or 10 units, but the pooled ratings of two judges were substantially more reliable than the ratings of a single judge (.57 for one judge as contrasted with .72 for two judges).

Moore (1933) developed a set of 40 symptom rating scales, each involving 6 steps of increasing pathological significance. The steps were documented by behavioral referents, either observational or judgmental in nature, and the ratings were made on the basis of interviews with the patient and with the patient's ward nurse. Intercorrelations were computed for a sample of 402 patients, and the mode of analysis he employed identified five syndromes or symptom clusters. Unfortunately, Moore's analyses involved methods that were soon to be superseded; possibly because of this, his study appears to have generated relatively little interest. He used a limited method of factor analysis developed by Spearman (1904) and later extended somewhat by Holzinger (1935). This method assumed that the intercorrelations of variables, and hence any meaning they might share, could be explained in terms of two factors: (1) a general factor which is shared by all the variables submitted to a given analysis and (2) a group factor to account for the fact that some variables are more highly correlated with each other than could be explained in terms of the general factor alone.

The methodological and conceptual advances of the decade between 1925 and 1935 provided a basis for continuing enrichments in psychology, particularly in the assessment of human behavior. Among those contributions related to the development and use of observer ratings are the method of paired comparisons (Thurstone, 1927) and a well-conceptualized and practical procedure for a rigorous multiple-factor analysis of intercorrelated variables (Thurstone, 1935).

The revolution in psychological thinking resulting from Thurstone's multiple-factor analysis (1935) has yet to run its course. As psychology has ventured from the laboratory and concerned itself with the wide world of human experience and reaction, it is obvious that most situations and procedures have many facets and must be considered from the standpoint of a discouragingly large and diverse body of potential observations. Whether these observations are in the form of elemental bits or summarizing bites, incisive thinking requires that they be reduced in number by some objective method of classification and that groups of observations which tend to distinguish between people in the same way be identified as related to a common group of variables and conceptualized, if only descriptively, as providing a distinguishable function.

By 1940, some psychologists were assisting psychiatrists by providing objective assessments of their patients' abilities, interests, and personalities; together, they had used this information in formulating diagnoses considered to be an important part of the care of psychiatric patients. Counselors were becoming generally acceptant of the possibility that student difficulties could not be understood or managed on the basis of such simple objective information as test results alone and were assuming a broadened interest and responsibility that could be called psychotherapeutic. With the influx of young veteran psychiatric patients and the encouragement of responsible participation of psychologists by the Veterans Administration, psychology was confronted with and accepted the reality of psychopathology as a part of its area of responsibility and competence.

This emerging position of much of psychology was soon to be expressed in many ways, including the content of rating scales. An early effort in this direction appeared in the form of the Fels Parent Behavior Rating Scales

(Champney, 1941). This scale, unlike most scales in the field of mental hygiene, does not offer a description of the current status of the patient but describes certain parental characteristics and situations that may have shaped the current behavioral status of the child. Although such an emphasis probably grew out of the nature–nurture controversy of the 1930s, it also anticipated the dynamic emphasis that was to become increasingly important in the perception of the troubled individual and his management. The Fels Scales were graphic scales, each referring to a well-described behavioral quality and providing verbal guides and documentation for 6 levels or degrees of the behavioral quality under observation. For the purpose of scoring, a 9-mm line was divided into 90 equal segments. With the possible exception that some of the scales may encourage bipolar thinking and provide descriptions that could be bipolar, the scales appear to be well constructed.

One of the first scales to emphasize currently observable behavior of psychiatric pertinence was a report by Malamud, Hoagland, and Kaufman (1946). These scales described current behavior which could be observed and rated by nurses as well as psychiatrists. The 20 items comprised by this system of rating were of a bipolar nature and could be rated as excessive to a pathological degree or deficient (e.g., from excitable to lethargic). The degree of a patient's disturbance was the total number of points that the patient deviated from normal, regardless of direction. The observational bases for the rating were of three kinds: observations during the interview, communication with the patient during the interview, and observations by ward personnel.

Wittman and Sheldon (1948) offered a classification of psychotic behavior reaction. Their *a priori* classes were affective exaggeration, paranoid projection, and schizoid withdrawal. A set of 16 symptom manifestations represented each of the 3 *a priori* classes of psychotic reaction, and these 48 items of symptomatic aberration were conceived to be more or less bipolar in manifestation. The ratings could be summed to indicate the degree to which the patient's psychosis was expressed in terms of each of the three *a priori* classes.

In 1947, Wittenborn undertook an intensive study of the nature, manifestations, and distribution of psychiatric symptoms. He assumed that disjunctive diagnostic distinctions misrepresented many patients and that a proper description of a patient on any one syndrome would require a context which included his standing on other major syndromes (Wittenborn, 1950). The goal of this effort was to prepare a procedure whereby a competent observer could (1) rate a standard set of currently discernible symptoms and related behaviors of a mental hospital patient, (2) score this set of ratings in a standard manner, and (3) prepare a profile that would indicate the degree to which the patient's most severe manifestations during a given period resembled each of the symptom patterns identifiable in samples of mental hospital patients. Thus, the rating procedure required that all patients be observed and rated with respect to a comprehensive set of symptoms.

Members of the Yale Department of Psychiatry as well as other New England psychiatrists participated in a series of individual interviews that involved a discussion of the kinds of observable symptoms providing a basis for each of the common diagnostic classifications. For each of the symptoms considered, there was a discussion of whether the symptoms might exist in degree and, if so, what behavioral cues could be expected if the symptom were present in minimal degree, maximal degree, and at various intermediate degrees. In this way, the content of symptom rating scales, their behavioral referents, and their format evolved. Eventually a lengthy set of scales was ready for pretesting. After several cycles of trial in clinical situations, review, discussion, revision, and further trial, a set of scales acceptable to a diversity of psychiatrist raters in several clinical settings was attained.

As a next step, a group of psychiatrists in the Veterans Administration Hospital rated their patients who had a functional disorder (Wittenborn, 1951). Ratings of currently most severe manifestations of a suitable body of patients were intercorrelated, submitted to a multiple factor analysis and orthogonal rotation as described by Thurstone (1935). The factors extracted were distinctive and interpretable as traits, each resembling a traditional diagnostic stereotype. To confirm these findings, 1,000 consecutive new admissions at Connecticut State Hospital at Middletown were rated by members of the diagnostic staff, and the ratings of patients with functional disorders were intercorrelated and factor analyzed in the same manner as the ratings from the Veterans Administration Hospital. Despite the fact that the second sample came from a different community, was observed and rated by different psychiatrists, and included women as well as men, the factor structure of the second sample confirmed the findings of the initial sample (Wittenborn & Holzberg, 1951; Wittenborn, Bell, & Lesser, 1951; Wittenborn, Mandler, & Waterhouse, 1951).

Further analyses provided no indications that major differences in rater training and background were reflected in the factor structure. The claim for objectivity remained unchallenged (Wittenborn, Herz, Kurtz, Mandell, & Tatz, 1952). The question of whether patients who were confidently assigned to a given diagnostic class required more than one syndrome to describe their symptomatic manifestations was examined for several different diagnostic groups. It was found that patients comprising a single diagnostic category were symptomatically diverse and that more than one syndrome was required for their description. The initial assumption that a descriptive diagnosis did not adequately describe symptomatic manifestations of patients received empirical support (Wittenborn & Bailey, 1952; Wittenborn & Weiss, 1952; Wittenborn, Holzberg, & Simon, 1953).

These scales were used as criteria in various studies of therapeutic effects (Wittenborn & Mettler, 1951; Wittenborn, Plante, Burgess, & Livermore, 1961; Wittenborn, Plante, Burgess, & Maurer, 1962; Wittenborn, Weber, & Brown, 1973). Later, the scales were revised and extended, and further confirmatory analyses of the factor structure of the enlarged set of scales were conducted (Wittenborn, 1962). An abridged form has been prepared for outpatient use (Wittenborn, 1976).

CURRENT RATING SCALES

By 1950, there was an accelerating interest in exploring the efficacy of various approaches to patient care. The general public was becoming acceptant of the realities of mental health problems, optimistic about the potential for significant improvement, and impatient for therapeutic progress. By 1960, effective antipsychotic medications and the popular acceptance of anxiolytic drugs had become a gratefully accepted reality, and confidence in the potential of antidepressant medications was growing. In substantial measure, these changes were both reflected and encouraged by increasing support in the form of federal grants. The FDA responded to the emergence of psychotropic drugs by formalizing its requirements for efficacy and safety, and most pharmaceutical companies were ready to support the research of clinicians who were interested in testing the efficacy of drugs. The excitement of a new field of therapeutic advancement plus the availability of support from both the government and industry stimulated research efforts by many persons who might otherwise never have undertaken scientific investigation.

The FDA required that claims of therapeutic efficacy be supported by evidence of patient changes that were verifiable and of demonstrable therapeutic relevance. Ratings of observable behavior were an obvious answer to the criterion-of-efficacy question, and rating scales that could be used conveniently and confidently by clinicians treating the patients were understandably preferred. The importance of the clinician's preference often resulted in the use of rating scales that asked the clinician to express his judgment concerning the patient's status. Rating scales that asked the clinician to report observable aspects of behavior were often avoided or used without the rater's enthusiasm.

Although all observations may involve judgmental components, it cannot be assumed that judgments concerning patients reflect all pertinent potential observations or are based on observations alone. Additional determinants that could influence the clinician's judgment include an awareness of the patient's history, his current circumstances, or his behavior in respects other than those required by the rating procedure. Possibly, as the clinician generates his judgmental ratings, these additional components cause him to weight the observable behaviors in a manner which varies from patient to patient. Because of these and other components in the formulations of a judgment, the meaning and validity of judgments can be a matter of uncertainty.

Many rating scales have been developed since 1950; their diversity indicates that investigators have seen their problems in different ways and have chosen different ways of meeting them. Table 1 provides a representative but far from exhaustive listing of these rating scales. Self-rating procedures have been omitted, but rating scale procedures that require judgments as well as those that emphasize observations have been included. Some of the scales, particularly those in the behavior expectation format, ask the rater to use observations to infer traits that could have a predictive potential. For the most part, however, the scales represented in Table 1 assess behavioral states, such as those that might be modified by an effective therapy and are not representative of scales designed to assess traits.

The rating procedures included in Table 1 are descriptive of behavior that may emerge spontaneously, particularly those aberrant behavioral qualities which are of interest, if not concern, to others. Elicited behaviors (i.e., those that might be provoked in a standard, contrived situation) are not represented. There are, however, many situations contrived for the purpose of eliciting pathological manifestations or other aberrant characteristics, including limitations that are pertinent to or expressive of the impairments of troubled individuals. Standard interview procedures are among the devices that do not have the formality of mental tests but do provoke behavior in a standard manner or in contrived situations. Such procedures vary appreciably in their design. Some are concerned with the content of the elicited response; others are concerned with such qualities of response as latency, duration, amplitude, susceptibility to distraction, and so on.

In Table 1, 22 entries describe rating procedures appropriate for the assessment of manifestations of troubled adults. Of these procedures, 15 refer to behaviors that may be called symptomatic. In contrast, 7 are diversified in nature and content. The Health–Sickness Rating Scale (Luborsky, 1962) comprises steps that are documented by illustrative cases arranged in such a manner that a continuum of increasing sickness is implicit. Three of the rating procedures (Honigsfeld & Klett, 1965; Lewinsohn, Mischel, Chaplin, & Barton, 1980; Rosen, Tureff, Daruna, Johnson, Lyons, & Davis, 1980) assess behavior in special situations (i.e., the psychiatric ward, group discussions, and the day room and gymnasium, respectively). In the Rosen *et al.* (1980) procedure, the assessment is based on behavior observed in a series of time samples. Two of the procedures (Platt, Hirsch, & Knights, 1981; Schooler, Hogarty, & Weissman, 1979) are distinctive because they involve ratings based on the reports of informants. The Liston, Yager, and Strauss (1981)

Table 1. Some Rating Procedures for Observations and Judgments

I. General use-adult

Name	Author, date	Setting for rating	Rater	Number of items	Number of steps	Anchoring basis	Content	Scoring
Wittenborn Psychiatric Rating Scales (WPRS)	Wittenborn, 1950	Mental hospital	Psychiatrist	52	3–4	Descriptions of observed behavior arranged by severity	Symptoms rated for maximal severity	Sum of ratings for each of 9 factor groupings
L–M Fergus-Falls Behavior Rating Scale	Lucero & Meyer, 1951	Mental hospital	Psychiatric aides	11	5	Observed behavior	Symptoms	
Multidimensional Scale for Rating Psychiatric Patients (MSRPP)	Lorr, Schaefer, Rubinstein, & Jenkins, 1953	Clinics	Professional staff	31	6	Judgmental on graphic continuum	Symptoms and other manifestations	
Hamilton Anxiety Scale	Hamilton, 1959	Inpatient/ outpatient	Trained professional	14	5	Judgment of severity	Symptoms	Sum of ratings for items contributing to each of two factors; total score
Hamilton Depression Scale	Hamilton, 1960	Inpatient/ outpatient	Trained professional	21	3–5	Observational	Symptoms	Sum of ratings for items within a factor grouping; total score
Psychotic Reaction Profile	Lorr, O'Connor, & Stafford, 1960	Mental hospital	Nurse	85	2	Checklist	Symptoms	Sum of checked items for each of 4 *a priori* classifications
Symptom and Adjustment Index	Gross, Hitchman, Reeves, Lawrence, Newell, & Clyde, 1961	Hospital outpatient	Psychiatrist or psychologist and social worker	62	4	Judgment of severity	Symptoms	Total score based on sum of ratings

Continued

Table 1. (*Continued*)

I. General use-adult

Name	Author, date	Setting for rating	Rater	Number of items	Number of steps	Anchoring basis	Content	Scoring
Health–Sickness Rating Scale	Luborsky, 1962	Inpatient/ outpatient	Therapist	1	100	Case stereotype for every five steps	Graded manifestations of mental health or illness	Rating on 100-point continuum
Brief Psychiatric Rating Scale (BPRS)	Overall & Gorham, 1962	Primarily inpatients	Clinicians	18	8	Judgment of severity	Manifestations of symptom qualities	Sum of items within each of five factors. Total score
Katz Adjustment Scale (KAS)	Katz & Lyerly, 1963	Admission procedure	Family informant	127	4	Judgment of frequency of occurrence	Symptoms and social behavior	Sum of rating for items within each of 12 groupings based on cluster analysis
Inpatient Multidimensional Psychiatric Scale (IMPS)	Lorr, Klett, McNair, & Lasky, 1963	Mental hospital	Interviewer	89	2–9	Judgments of severity or frequency and some checklist items	Symptoms	Sum of ratings of items within each of 10 factors
Wittenborn Psychiatric Rating Scale Revised (WPRS Revised)	Wittenborn, 1964	Mental hospital	Trained personnel	72	3–4	Descriptions of observed behavior arranged by severity	Symptoms rated for maximal severity	Sum of rating for items withn each of 12 factors
Wittenborn Psychiatric Rating Scale—Short Survey (WPRS–S)	Wittenborn, 1976	Outpatient/ inpatient	Trained observer	17	3–4	Descriptions of observed behavior arranged by severity	Symptoms rated for maximal severity	Sum of ratings of items within each of 6 factors
Nurse Observation Scale for Inpatient Evaluation (NOSIE)	Honigfeld & Klett, 1965	Inpatient	Nurse	80	5	Judgments of frequency	Ward behavior	Sum of ratings of items within each of 7 factors

Physicians Outpatient Psychopathology Scale	Free & Guthrie, 1969	Psychoneurotic outpatient	Physician	15	6	Judgment of severity	Symptoms	Sum of items comprising each of 4 factors
Physicians Questionnaire (PQ; revised Scale includes 13 items)	Rickels & Howard, 1970	Outpatient	Physician	10	7	Judgment of severity	Symptoms	Sum of ratings of items within each of 3 factors
Schedule for Affective Disorders and Schizophrenia (SADS) Part I	Endicott & Spitzer, 1978	Psychiatric inpatient	Trained personnel	120	Variable; 2–7	Severity of judgments or observations of behavior	Affective and schizophrenic symptoms	Sum of ratings of items within 8 groupings defined by authors
Social Adjustment Scale Revised (SAS II)	Schooler et al., 1979	Outpatient, especially schizophrenic	Trained interviewer	52	5	Judged impairment	Informant responses to specific questions	Rating for each of 8 areas of role adjustment
Social Competence	Lewinsohn et al., 1980	Depressed outpatient	Trained observer	17	7	Judged degree of characterization	Adjectives descriptive of subjects during group discussions	Total score
Observation Record of Inpatient Behavior (ORIB)	Rosen et al., 1980	Psychiatric inpatient	Trained observer	8		Observed occurrence in a series of 5-second intervals	Specific behavior in dayroom situation and gym	Average frequency of each behavior
Patient Behavior Assessment Schedule (PBAS)	Platt et al., 1981	Interview with family informant	Interviewer	34	3	Judged degree of disturbance	Social or symptomatic behavior	Mean of ratings for behavior items and for social performance items separately
Psychotherapy Competence Assessment Schedule (PCAS)	Liston et al., 1981	Psychotherapeutic interview	Experienced supervisors	85	7	Simple adjectives on graphic continuum	Statements descriptive of videotape recording of therapeutic behavior	Sum of ratings for items relating to each of four performance categories

Continued

Table 1. (*Continued*)

II. Geriatric-use

Name	Author, date	Setting for rating	Rater	Number of items	Number of steps	Anchoring basis	Content	Scoring
Ward Behavior Rating Scale	Burdock, Elliott, Hardesty, O'Neill, & Sklar, 1960	Inpatient	Nursing staff	112	2	Dichotomous	Ward behavior	A count of the items with pathological rating
Life Satisfaction Rating Scale (LSR)	Neugarten, Havighurst, & Tobin, 1961	Taped interviews with community volunteers	two raters	5	5	Descriptive judgments	Qualities-of-life satisfaction as inferred from multiple verbally reported indications	Rattings for each quality, combined for summary score
Crichton Geriatric Rating Scale (source of Plutchik, Conte, Lieberman, Bakur, Grossman, & Lehrman, 1970)	Robinson, 1964	Hospital	Trained observer	11	5	Descriptions of observable behavior	Aspects of geriatric impairment	Sum of ratings for the 11 items
Stockton Geriatric Rating Scale (SGRS)	Meer & Baker, 1966	Geriatric inpatient	Nurse	33	3	Judgment of frequency	Manifestations of geriatric impairment	Sum of ratings for items relating to each of 4 factors
Psychiatric Rating Scale for Geriatric Patients	Lennon, 1971	Geriatric inpatient	Qualified observer	118	4	Judgment of frequency	Ward behavior	Sum of ratings for items relating to each of 13 factors
Physical and Mental Impairment of Function Evaluation (PAMIE)	Gurel, Linn, & Linn, 1972	Geriatric inpatient	Nurse	77	2	Dichotomous	Manifestations of geriatric impairment	Sum of ratings for items relating to each of 10 factors
Sandoz Clinical Assessment—Geriatric (SCAG)	Shader, Harmatz, & Salzman, 1974	Geriatric inpatient/outpatient	Physician	19	7	Judged severity	Symptoms of impairment	Sum of ratings for items—FA

Name	Author, date	Setting for rating	Rater	Number of items	Number of steps	Anchoring basis	Content	Scoring
Dementia Rating Scale	Lawson, Rodenburg, & Dykes, 1977	Mental hospital	Nursing staff	27	2	Dichotomous	Manifestations of impairment	A count of items checked from each of 4 factors
London Psychogeriatric Rating Scale (LPRS Ontario)	Hersch, Kral, & Palmer, 1978	Geriatric inpatient	Ward staff	36	3	Judged severity	Mental and other manifestations of impairment	Total score and subscores for each of the 4 qualities

III. Use for children

Name	Author, date	Setting for rating	Rater	Number of items	Number of steps	Anchoring basis	Content	Scoring
Child-Rearing Practices	Wittenborn, Astrachan, DeGooyer, Grant, Janoff, Kugel, Myers, Riess, & Russell, 1956	Interview	Trained interviewer (interview with mother)	34	2	Dichotomous	Parent report of observed behavior in the home	Number of items checked "yes" for each of 5 groupings of intercorrelated items
Personality Factors in Nursery School Children	Peterson & Cattell, 1958	Home	Parent	44	3	Bipolar extremes	Child's behavior	14 interpretable factors were generated and are apparently suitable for scoring
Problem Behavior Scale	Peterson, 1961	School	Teacher	58	3	Judged severity	Problem behaviors	Sum of ratings for items contributing to each of 2 factors
Children's Behavior Inventory (CBI)	Burdock & Hardesty, 1964	Institution	Professional	139	2	Dichotomous	Observable behavior	Number of indicated items for each of 9 factors

Continued

Table 1. (*Continued*)

III. Use for children

Name	Author, date	Setting for rating	Rater	Number of items	Number of steps	Anchoring basis	Content	Scoring
Pittsburgh Adjustment Survey Scales (PASS)	Ross, Lacey, & Parton, 1965	School	Teacher	94	3	Judged applicability	Statements descriptive of behavior	Sum of ratings for items within each of 4 factors
Louisville Behavior Checklist (BCL)	Miller, 1967	Outpatient	Parent	119	2	Dichotomous	Problem behavior	Number of items contributing to each of 8 factors
Conners Teacher Questionnaire	Conners, 1969	School	Teacher	39	4	Judged severity	Problem behavior	Sum of ratings for items contributing to 5 factors
Conners Parent Questionnaire Revised and Lengthened	Conners, 1970	Home	Parent	30	4	Judged severity	Problem behavior	Sum of ratings for items contributing to factors
Classroom Behavior Rating Scale	Armentrout, 1971	School	Teacher	30	4	Judged severity	Problem behavior	Sum of ratings for each of 2 *a priori* item groupings
Sibling Deviant Behavior	Arnold, Levine, & Patterson, 1975	Home	Trained observer	14	rate per minute for each behavior	Frequency of behaviors	Noxious home behaviors	Average rate per minute based on 3–5 observational sessions

Health Resources Inventory	Gesten, 1976	School	Teacher	54	5	Judged competence	Classroom behaviors, or attributes manifested in classroom	Exact factor scores for 5 factors
Psychiatric Impairment Inventory	Langner, Gersten, McCarthy, Eisenberg, Greene, Herson, & Jameson, 1976	Interview with mother in home	Interviewer	35	2	Dichotomous (pathognomonic vs. all others)	Problem behavior	7 factor scores plus total score based on 5 items from each of 7 factors
Classroom Observation Code	Abikoff, Gittelman-Klein, & Klein, 1977	School	Trained classroom observer	14	Incidence of behavior in time sampling	Frequency of behaviors	Observed hyperactivity	Mean incidence behavior in time sampling through an observational period
Hyperkenesis Rating Scales	Zukow, Zukow, & Bentler, 1978	Home	Parent	28	2	Dichotomous	Hyperkinetic behavior	Sum of ratings for items contributing to each of 3 factors
Hyperkenesis Rating Scales	Zukow et al., 1978	School	Teacher	15	5	Judgmental	Observed behavior	Sum of ratings for items contributing to each of 2 factors

procedure is unique in the sense that it provides an assessment of therapeutic competence and is not patient-oriented.

The number of items in the adult rating scales ranges from 1 to 120; the number of steps provided ranges from 2 to 100, and the number of scores generated ranges from 1 to 12. Most of these scores represent factor clusters. The steps comprised by the continuua may be unlabeled, judgmental statements, explicit behavioral observations, or illustrative cases.

Table 1 includes 12 scales designed to reflect behavioral disturbances or impairments in geriatric patients. All but one of these procedures are concerned with the behavior of hospitalized individuals. The number of items varies from 5 to 118, and at least five of the procedures provide multiple scores derived from factor analyses.

The tabular summary includes 15 scales to be used in work with children. Eight of these describe the child's behavior in schools or in institutions, and in all but two of these the teacher was the rater. Seven of the procedures describe the child's behavior at home, and the rater is usually a parent. For one scale, however, the rater is a trained observer, and for two scales the mother is the informant. As in the other rating procedures, multiple scores, usually based on factor analyses, are available.

From the published reports, it is difficult to know how generally the rating procedures are used or how effective they are. In general, the preferred procedures are relatively short and ask that the rater make judgments rather than report specific observations. Assured familiarity of content and convenience for the rater may be more important determiners of scale usage than is precision, including degree of reliability and validity.

CURRENT ISSUES

The Optimal Number of Steps

Many of the issues which seem currently to be important and worthy of inquiry have been intermittently noted by investigators for 50 years or more. Among these may be included the question of the optimal number of steps for a rating procedure. This question has been approached from the standpoint of a diversity of considerations, and the conclusions drawn have been equally diverse. After considering both statistical and judgmental factors, Symonds (1924) concluded that seven was the optimal number of steps for rating personality, and Finn (1972) reported that 9-point scales produced less reliable ratings than those of shorter scales. The work of Bendig (1954), Komorito and Graham (1965), and Matell and Jacoby (1971) supports the hypothesis that the number of steps in a rating procedure is not important from the standpoint of reliability (cf. Miner, 1917).

Schutz and Rucker (1975) examined one aspect of the issue of the optimal number of steps by using ratings of the appropriateness of different foods in a standard set of situations. The ratings of appropriate use were prepared in 4 forms: 2-step, 3-step, 6-step, and 7-step. A different group of student raters was used for each of the respectively different forms. For each group of students, the pattern of appropriate use for one food was correlated with the pattern for every other food. Each of these 4 sets of intercorrelations was factor-analyzed separately and the factor patterns for the 4 groups of students compared. There were no appreciable differences among the factor patterns for the 4 groups, and the number of steps in the rating procedure was judged not to affect the pattern. Another aspect of the number-of-steps issue has been clarified by the work of Lissitz and Green (1975). They found that, for a given covariance, reliability increased with number of steps; but there was very little advantage in more than 5 steps.

Training Raters

Little is to be gained by asking the rater to distinguish among observations that he has not made and is unlikely to make without training in the pertinent observations. A study by Muslin, Thurnblad, and Meschel (1981) suggests that many presumably qualified and interested observers may fail to note, remember, or report pertinent observable events. Videotape recordings of medical students' interviews with patients during a psychiatric clerkship were contrasted with the videotapes of the supervisory sessions. Of the themes identified in the student's interview with his patients, 54% were not reported in the interview with the supervisor. The frequency with which a given theme was reported varied from 5% to 86%. There was no relationship between the importance of the theme and its susceptibility to being omitted from reports to the supervisor.

A study by Bernardin and Walter (1977) explored the importance of rater training by comparing teacher ratings by 5 groups of students. The groups differed in training and preparation for the use of a set of 7-step behavior expectation scales, where each step was documented by a critical incident. Each set of scales involved critical incidents relevant to 7 qualities or dimensions of instructional performance. During the first week of the term, Group 1 received training concerning rater's errors (e.g., "halo" effect, and leniency); they were given copies of the Behavior Expectation Scale, which they would use to evaluate the instructors at the end of the term, and they discussed the 7 qualities to be rated. They observed the instructor's performance and maintained an observational diary in which they recorded critical incidents each day. Group 2 received the same training as Group 1 but did not see a copy of the actual rating scale. They discussed the qualities or dimensions involved in the scale and were asked to observe the instructor's performance with respect to the 7 qualities, but they did not keep a diary. Group 3 received the same instruction as Group 2, but their training did not occur until the 10th week. Group 4 received no instruction but were told about halo and leniency effects at the time of the 10th-week rating. The amount of halo effect varied inversely with the amount of preparation and training. Group 1 was less susceptible than the other groups to leniency errors, and Group 2 was less susceptible than Group 4 to leniency. In this instance, at least one major component of rater error was diminished by training in the observation of pertinent incidents.

Later, Bernardin (1978) examined the effect of training over an extended period of time. One group of students was given a training program that provided a 1-hour training session and included definitions, illustrations, and examples of the rating errors of leniency, halo, and central tendency as they might find expression in ratings for 7 qualities of teacher performance. A second group of students was given a 5-minute training session where the 3 basic errors were defined and illustrated. A test of psychometric knowledge and either a behavioral expectation rating procedure or a summated rating scale was used to compare the effects of these two levels of training. At the 14th week of the current semester and during the 14th week of the following semester, the students rated the instructors, and the students were tested for psychometric knowledge for the second and third times, respectively. Some of the intergroup differences were significant during the current semester, but all effects were lost the following semester.

To examine the question of the accuracy of ratings, Borman (1979) used the consensus of a group of experts to provide criterion ratings of the performance of two kinds of jobs: management and recruiter. The trainee raters were assigned at random to a standard program of training in one of 5 rating procedure formats. These formats included (1) scales where the steps for the quality to be rated were defined in terms of specific behavior, (2) scales where the steps for each level of the behavioral quality were defined in general terms, (3) summational scales where the rating for a quality was the sum of simple ratings for several behaviors considered to be illustrative of the quality, (4) scales which required steps for rating a trait as contrasted with behavior, and (5) numerical scales comprising an unanchored sequence of numbers.

The results of the training were analyzed from the standpoint of halo effect, convergent and discriminant validity, and accuracy with respect to the experts' rating. The group receiving training in the numerical and in the summated format showed less halo and more discriminant validity than the groups trained in the other formats. Ratings from the numerical format group also corresponded most closely with the ratings of experts. It was found, however, that the accuracy of the ratings varied in a significant manner with both the behavioral quality being rated and the job being rated (i.e., training in the numerical format was best for the recruiter's job, and training in the summated format was worst for the management job).

Behavioral Referents

Very few of the recent studies examined the contribution of training in the observation of pertinent behavioral content, but some reports compared scales that provided a behavioral documentation of the steps with scales where behavioral referents were not used to define the steps. Borman and Dunnett (1975) compared performance ratings made on 9-step behaviorally anchored scales with performance ratings of the same 14 qualities made on 5-step judgmental scales having no behavioral anchors. Commanders and captains used these two sets of scales to rate 126 lieutenants. As expected, the interrater agreement for the behaviorally anchored scales was higher than the agreement for simple numerical scales, but the contrast was not quite significant statistically ($p < .06$). There was appreciably less "halo" effect in the anchored form than in the numerical form.

Motowidlo and Borman (1977) developed behaviorally anchored scales for assessing the morale of military units. Critical incidents of morale were classified into several categories and rated with respect to their pertinence to the category. The agreement among several different military units provided the basis for the final selection of categories and for the position of the incidents in the scaling on the respective categories. Independently, company commanders rated several platoons with respect to unit effectiveness, destructive sabotage, and drug abuse. These three criteria were found to have respective median correlation coefficients of .72, .55, and .39 with the critical incident ratings. On the basis of this report, it appears that such behaviorally anchored scales can have a useful level of validity.

Expectation Scales

Ultimately, the merit of ratings depends on the rater's interest in the events or situations being rated and on his respect for the rating procedure. Smith and Kendall (1963) recognized the validity of this principle and incorporated it in their recommendations for the construction of rating procedures which they called behavior expectation scales. Potential raters who shared the goal of the development of an assessment procedure selected and identified the qualities to be rated and proposed critical incidents of behavior relevant to each of the respective qualities. The group rated each critical incident on a numerical continuum from the standpoint of the degree to which each incident characterized or illustrated the quality to be rated. In this way, the critical incidents were placed on the continuum. At some point in this developmental sequence, the illustrative behavioral incidents are changed into the form of expectations (e.g., instead of "the child threw his pencil," the scale step would read, "the child could be expected to throw his pencil"). To assure that the statements were pertinent to the behavioral quality to be assessed, independent judges assigned the expectation statements to one of the behavioral qualities. If a statement was not successfully reassigned, it was eliminated.

Modifying the scale steps from descriptions of observed behaviors to statements describing expected behaviors changed the meaning of the rating procedure in two important respects. First, the expectation format invites the rater to infer a behavioral trait or a disposition that he can apply to a potential situation. Second, the scale no longer asks the rater to record a verifiable observation. Instead, the rater is asked to generalize on the basis of his prior observations and make predictions of behavior in potential situations. The ratings must now reflect whatever influences determine such predictions.

Several investigators have examined the possibility that rater participation in the development of rating procedures contributed to the quality of the ratings, as suggested by the Smith and Kendall procedure (1963). Friedman and Cornelius (1976) compared rater participation in scale construction with no such participation in two different formats: a behavior expectation format and a graphic format. Groups that participated in scale construction showed less random error, greater interrater agreement, and greater ratee discrimination than groups that did not participate. There were no important differences between the two formats, however.

The behavior expectation procedure for scale development has provoked an unaccountably large volume of literature. Most of this literature compares the behavior expectation format with other formats. The results have been equivocal, if not pessimistic about the power of the expectation format. Burnaska and Hollmann (1974) found no important differences between the expectation procedure and a procedure where the behavioral qualities were rated in terms of simple adjectives. DeCotiis (1977) compared the expectation format with two other formats: a numerically anchored scale where a low number meant *weak* and a high number meant *outstanding* and trait ratings that required the judgmental ascription of simple adjectives ranging from *very poor* to *very good*. The three rating procedures were indistinguishable with respect to leniency, central tendency, and halo.

In a mixed standard disguised format, steps of the various scales are presented in a random order in a single list, and the dimensions being assessed by the various scales are disguised (Blanz & Chiselli, 1972). Saal and Landy

(1977) compared expectation scales with mixed standards scales having the same content. The mixed standard procedure showed less leniency error and less halo than the expectation procedure, but interrater reliabilities of the expectation procedure tended to be higher than that of the mixed standard procedure. Finley, Osburn, Dubin, and Jeanneret (1977) compared expectation scales in two formats (obvious and disguised dimensions). The mixed standard disguised format involved the same specific behavioral items as the format in which the dimension being rated was obvious. The comparisons yielded no appreciable contrasts and supported no firm conclusion. Dickinson and Zellinger (1980) compared the expectation format with the mixed standard format. No important differences between the two procedures were found.

The numerous studies providing comparisons between the behavior expectation format of Smith and Kendall (1963) and other forms have been relatively unproductive. After a review of what they described as "all published studies," Kingstrom and Bass (1981) concluded that the behavior expectation format differs from other formats relatively little with respect to such factors as leniency, halo, interrater agreement, ratee discriminability, and accuracy.

Reliability

The distinction between interrater reliability and test–retest reliability can be important. Two raters may confirm themselves consistently in a test–retest series; nevertheless, these two raters may be rating the same patients in an entirely different fashion, so that the interrater reliability is negligible. In addition to the content of a scale and its wording, the distinctions among various approaches to the estimation of interrater reliability reflect such factors as the number of points that the scale comprises, the uniformity of the interpoint distinctions, the presumptive nature of the sequence represented by the scale steps (e.g., whether disjunctive or continuous), and whether the distribution of scale values has a Gaussian character.

When the scales are binary (e.g., yes–no checklist) and one wishes to examine the agreement between ratings provided by two raters, the kappa statistic is generally considered preferable to the familiar chi square statistic as applied to "significance of changes" (McNemar, 1949, pp. 204–207). Maxwell (1977) described several applications of kappa (Cohen, 1960) and included a model involving the assumption that the portion of responses that were doubtfully made by each rater is randomly distributed between his binary choices. On this basis he recommended the use of a random error coefficient which he called *RE*. In addition, he offered two formulas for estimating a reliability correlation coefficient *r* from the components of an analysis of variance format. Various ways of using analysis of variance components to estimate the reliability of rating procedures are described in the literature (e.g., Ebel, 1951). One of these correlation coefficients estimates the reliability of the data; another estimates the reliability of the rater.

The intraclass correlation (Fisher, 1948) may be most generally applicable to the task of estimating interrater reliability. Tinsley and Weiss (1975) explicitly recognized several interpretations for the appropriate use of the intraclass correlations. The distinctions they emphasized include whether the intraclass correlations represent the average reliability of the individual raters or the reliability of the composite rating, and they determine whether the interrater reliability can be generalized to samples other than the one used in the estimation. Later, Shrout and Fleiss (1979) analyzed 6 different forms of intraclass correlations applicable to the study of rater reliability and offered some guidelines for their use.

In their recent critical evaluation of rating scales and rating scale data, Saal, Downey, and Lahey (1980) summarized distinguishable definitions of interrater reliability in terms of respectively different approaches to the assessment of reliability. One approach examined the standard deviation of the ratings assigned to a ratee by a sample of raters. Another approach requires correlations between the assessments of a given quality provided by pairs of raters. A third approach involves the use of intraclass correlations. An additional approach is based on analysis of variance procedures (i.e., reliability is necessary to support a large ratee effect); an absence of a significant main ratee effect could indicate an absence of reliability. These writers also suggest a particularly severe criterion for interrater reliability—the correlation between initial ratings and reratings when the first and second ratings were made by different raters. Reliability can obviously be construed in many different ways, and it is fortunate that appropriate computational procedures have been proposed to serve most of these. These distinctions and the methods they suggest are helpful, but the basic problem of reliability seems to be in the data themselves.

In addition to other limitations, there is also the great likelihood that the steps on the rating continuum are unequal. Numerous concepts and procedures have been proposed for constructing scales of equal intervals. None of them has been wholly satisfactory. The investigator who devises procedures based on the assumption that the intervals are or should be equal in some respects or another (Gaito, 1980) sooner or later finds himself confronted with a situation in which the "equal" intervals are not equal. For most practical purposes, it may be sufficient to assume that the steps along a rating scale serve only to distinguish ratees along a continuum, either real or hypothetical, and that the most defensible comment about the equality of the step intervals may be in terms of the number of persons distinguished by them. Cicchetti (1976) examined the problem of weighting the step intervals in order to support estimated reliability and offered some suggestions.

Although the present reviewer is skeptical of most such schemes to increase reliability, the question of sample size appears to be subject to a defensible answer. The sample size necessary to assess reliability of a rating scale has been described as primarily a function of the number of steps that the rating scale comprises. Using a Monte Carlo method, Cicchetti and Fleiss (1975) found that 20 cases might be sufficient for a 3-step scale, while 100 cases would be appropriate for a 7-point scale (cf. Lissitz & Green, 1975).

Borman (1978) explored the upper limits of reliability by synthesizing true means, standard deviations, and intercorrelations for 13 dimensions of employees' performance (6 for recruiters and 7 for managers). He then prepared scripts and videotapes of a recruiter and of a manager in an hypothetical job situation and arranged for these tapes to be rated. The synthesized job behavior was judged to be consistent with reality and appropriate for the "true" synthetic ratings. Under conditions that Borman described as nearly ideal, the interrater reliability for the 13 dimensions rated were quite high, (i.e., 9 were above .80). The actual ratings were very close to the synthetic ratings, and validity coefficients under .80 were reported for only 6 of the 13 dimensions. Borman emphasized the importance of the rater's observing and noting relevant behavior and the ever-present danger that idiosyncrasy may affect the inferential process, particularly when the rater must generalize from his observation to a judgment concerning a quality of behavior.

Ordinarily, errors of bias may be subsumed under such categories as halo, leniency, and restriction of range. It is conceivable, however, that some unidentified forms of bias, perhaps in the form of pertinent knowledge, may detract from the accuracy of ratings. Borman (1977) prepared videotapes of hypothetical performances by 8 managers and hypothetical performances by 8 job recruiters. Criterion or "true" ratings by a panel of experts were prepared for 13 qualities or dimensions, 7 scales for managers and 6 scales for the recruiters. For the sample of 8 managers (or recruiters), the correlation between the criterion scores established by the experts for a given quality and the respective ratings provided by the rater was used as an index of the rater's accuracy for that quality. Borman calculated a correlation expressing the general accuracy of ratings for the managers' job and for the recruiters' job. In addition, it was possible to get a correlation that indicated the relationship between accuracy in rating the managers' performance and accuracy in rating the recruiters' performance. This correlation was low relative to the correlation within job categories, thereby suggesting that a person who might be an accurate rater of performance in one job may not be nearly as accurate in rating performance on some other job.

Sources of Bias

High intrarater reliability in the presence of low interrater reliability suggests the effects of bias. Some of the recent examinations of bias effects are obviously pertinent to the reliability problem and suggest the possibility of applying principles of human behavior to the task of devising standard procedures for the observation and reporting of behavior.

Kent, O'Leary, Diament, and Dietz (1974) examined the biasing effect of rater's expectations concerning treatment effects. Over a period of 16 days, raters were trained thoroughly in viewing videotapes of children's classroom behavior and in recording 9 disruptive behaviors. The average interrater reliability was .72. Half of the raters were then informed that a decrease in disruptive behavior was predicted, and the other half were informed that no change in disruptive behavior was expected. In a series of 6 additional sessions, the observers viewed videotapes and made behavioral ratings for each session. After the 6 sessions, the raters evaluated the changes. Whether the raters had received a positive or negative prediction of treatment effect was unrelated to their ratings of specific behaviors. In contrast, however, global ratings of treatment effect were largely determined by the predictive bias. Thus, ratings of specific behavior by trained raters can be sufficiently objective to be relatively independent of predictions, but the effect of such a source of bias can be a major factor in global ratings.

Later, Nisbett and Wilson (1977) compared the responses of two groups of students to videotapes of a teacher who assumed a "warm" guise for one recording and a "cold" guise for the other. They found evidence that the student raters' general impresssion of a teacher ratee affected the way in which he rated such specifics as the teacher's appearance, mannerisms, and accent. In this inquiry, however, the raters were unaware that their general impression of the ratee had affected the rating of these three qualities. The studies of Kent *et al.* (1974) and of Nisbett and Wilson (1977) differed substantially in the nature of the rating and the amount of training the raters received. This may account for the fact that the Kent *et al.* (1974) study, which used raters highly trained to observe specific behavior, showed that the general (predicted) evaluation had little influence on the ratings of specific behaviors, while the Nisbett and Wilson (1977) study, in which the training was minimal and the rating judgmental, showed that the raters' general impression influenced ratings of specific qualities.

Rozelle and Baxter (1981) studied the effect of the rater's responsibility and accountability on the quality of the rating—responsibility in the sense that the ratings would have an important effect on decisions concerning the ratee and accountability in the sense that the ratees would have access to the rating.

When the instruction to the rater informed him that he was either accountable or responsible for his rating, his ratings for a given subject tended to agree with those of other raters. When the rater was told that the ratings were strictly confidential and would not affect an important decision concerning the ratee, ratings for a given ratee by two different raters were no more likely to be in agreement than a given rater's rating for two different ratees.

CONCLUSIONS

Shortly after 1950, the availability of solid-state electronics initiated an almost explosively accelerating development of powerful computing facilities. In consequence, every investigator soon had recourse to personnel and machines that made possible complex analyses which previously had been the product of months, if not years, of effort by highly trained specialists. Multiple scoring of rating scales on the basis of factor analysis (Wittenborn, 1950) became commonplace. The availability of increasingly large amounts of grant money encouraged the amassing of large bodies of data and the preparation of rating scale procedures for the explorations of many qualities of human behavior. As a result, there was a proliferation of rating procedures; Table 1 is only a partial listing of scales that could be applicable to the assessment of psychotherapeutic efficacy.

Rating scales have come to be used with increasing frequency, confidence, and resourcefulness. Although this increased use of rating scales has been rewarding, it is not obvious that there has been any significant evolution (i.e., it is not clear that the rating scales developed in the last decade are better than their predecessors). It is possible, however, that advancements in rating procedures could be facilitated by careful distinctions between the judgmental and the observational basis for rating. Advances in the use of judgmental ratings could be implemented by explication of concepts that a judge employs as he selects and weighs the various sources of information used in formulating his judgments. Unless the various components of judgments can be known and made explicit, judgmental ratings may have an unnecessarily limited communication value.

When constructing an observational scale, one should first identify the population of behaviors to which one may wish to generalize on the basis of the observations. Ratings of observed behavior are, in essence, a sampling procedure, and planful anticipation of the use to which one expects to put the distinctions provided by the rating procedure may be necessary for the construction of scales of maximal validity. Ambiguity concerning behaviors that may be excluded from consideration and that must be included should be reduced to a minimum. The rater must be trained to observe, and he must be reconciled to the fact that his judgments concerning the patient are not to be reflected in the observations that he records. How these rated observations can best be combined to meet the goals of the assessment becomes an analytical matter and should not be confounded with the assessment *per se.*

Whether future advances will emphasize judgments or observations cannot be foretold at this time, but quantitative refinements cannot compensate for obscure or irrelevant data. The confounding of judgments and observations should be recognized as a limitation in any body of rating scale data, and insofar as possible this source of uncertainty should be minimized.

REFERENCES

Abikoff, H., Gittelman-Klein, R., & Klein, D. F. Validation of a classroom observation code for hyperactive children. *Journal of Consulting and Clinical Psychology, 1977, 45,* 772–783.

Armentrout, J. A. Parental child-rearing attitudes and preadolescents' problem behaviors. *Journal of Consulting and Clinical Psychology, 1971, 37,* 278–285.

Arnold, J. E., Levine, A. G., & Patterson, G. R. Changes in sibling behavior following family intervention. *Journal of Consulting and Clinical Psychology, 1975, 43,* 683–688.

Ayers, L. P. *A scale for measuring the quality of handwriting of school children.* New York: Russell Sage Foundation, 1912.

Bendig, A. W. Reliability and the number of rating scale categories. *Journal of Applied Psychology, 1954, 38,* 38–40.

Bernardin, H. J. Effects of rater training on leniency and halo errors in student ratings of instructors. *Journal of Applied Psychology, 1978, 63,* 301–308.

Bernardin, H. J., & Walter, C. S. Effects of rater training and diary-keeping on psychometric error in ratings. *Journal of Applied Psychology, 1977, 62,* 64–69.

Blanz, F., & Ghiselli, E. E. The mixed standard scale: A new rating system. *Personnel Psychology, 1972, 25,* 185–190.

Borman, W. C. Consistency of rating accuracy and rating errors in the judgment of human performance. *Organizational Behavior and Human Performance, 1977, 20,* 238–252.

Borman, W. C. Exploring upper limits of reliability and validity in job performance ratings. *Journal of Applied Psychology, 1978, 63,* 135–144.

Borman, W. C. Format and training effects on rating accuracy and rater errors. *Journal of Applied Psychology, 1979, 64,* 410–421.

Borman, W. C., & Dunnette, M. D. Behavior-based versus trait-oriented performance ratings: An empirical study. *Journal of Applied Psychology, 1975, 60,* 561–565.

Boyce, A. C. Methods for measuring teachers' efficiency. *14th Yearbook, National Society for the Study of Education, Part II,* 1915.

Burdock, E. I., & Hardesty, A. S. A children's behavior diagnostic inventory. *Annals of New York Academy of Sciences, 1964, 105,* 1964.

Burdock, E. I., Elliott, H. E., Hardesty, A. S., O'Neill, F. J., & Sklar, J. Biometric evaluation of an intensive treatment program in a state mental hospital. *The Journal of Nervous and Mental Disease, 1960, 130,* 271–277.

Burnaska, R. F., & Hollmann, T. D. An empirical comparison of the relative effects of rater response biases on three rating scale formats. *Journal of Applied Psychology, 1974, 59,* 307–312.

Champney, H. The variables of parent behavior. *Journal of Abnormal and Social Psychology,* 1941, *36,* 525–542.

Cicchetti, D. V. Assessing inter-rater reliability for rating scales: Resolving some basic issues. *British Journal of Psychiatry,* 1976, *129,* 452–456.

Cicchetti, D. V., & Fleiss, J. L. *A comparison of the null distributions of weighted kappa and the ordinal statistic.* Invited paper presented at the Central Regional Meeting of the American Statistical Association, St. Paul, Minn., 1975.

Cohen, J. A coefficient of agreement for nominal scales. *Educational Psychological Measurement,* 1960, *20,* 37–46.

Conners, C. K. A teacher rating scale for use in drug studies with children. *American Journal of Psychiatry,* 1969, *126,* 884–888.

Conners, C. K. Symptom patterns in hyperkinetic, neurotic, and normal children. *Child Development,* 1970, *41,* 667–682.

DeCotiis, T. A. An analysis of the external validity and applied relevance of three rating formats. *Organizational Behavior and Human Performance,* 1977, *19,* 247–266.

Dickinson, T. L., & Zellinger, P. M. A comparison of the behaviorally anchored rating and mixed standard scale formats. *Journal of Applied Psychology,* 1980, *65,* 147–154.

Ebel, R. L. Estimation of the reliability of ratings. *Psychometrika,* 1951, *16,* 407–424.

Ellson, D. G., & Ellson, E. C. Historical note on the rating scale. *Psychological Bulletin,* 1953, *50,* 383–384.

Endicott, J., & Spitzer, R. L. A diagnostic interview: The schedule for affective disorders and schizophrenia. *Archives of General Psychiatry,* 1978, *35,* 837–844.

Filer, H. A., & O'Rourke, L. J. Progress in civil service tests. *Journal of Personal Research,* 1923, *1,* 484–520.

Finley, D. M., Osburn, H. G., Dubin, J. A., & Jeanneret, P. R. Behaviorally based rating scales: Effects of specific anchors and disguised scale continua. *Personnel Psychology,* 1977, *30,* 659–669.

Finn, R. H. Effects of some variations in rating scale characteristics on the means and reliabilities of ratings. *Educational and Psychological Measurement,* 1972, *32,* 255–265.

Fisher, R. A. *Statistical methods for research workers.* New York: Hafner, 1948.

Free, S. M., & Guthrie, M. B. A rating scale for evaluating clinical response in psychoneurotic outpatients. *Journal of Clinical Pharmacoloy,* 1969, *9,* 187–194.

Friedman, B. A., & Cornelius, E. T. Effect of rater participation in scale construction on the psychometric characteristics of two rating scale formats. *Journal of Applied Psychology,* 1976, *61,* 210–216.

Gaito, J. Measurement scales and statistics: Resurgence of an old misconception. *Psychological Bulletin,* 1980, *87,* 564–567.

Galton, F. *Inquiries into human faculty and its development.* London: Macmillan, 1883.

Gesten, E. L. A health resources inventory: The development of a measure of the personal and social competence of primary-grade children. *Journal of Consulting and Clinical Psychology,* 1976, *44,* 775–786.

Gross, M., Hitchman, I. L., Reeves, W. P., Lawrence, J., Newell, P. C., & Clyde, D. J. Objective evaluation of psychotic patients under drug therapy: A symptom and adjustment index. *Journal of Nervous and Mental Disease,* 1961, *133,* 399–409.

Gurel, L., Linn, M. W., & Linn, B. S. Physical and mental impairment-of-function evaluation in the aged: The PAMIE scale. *Journal of Gerontology,* 1972, *27,* 83–90.

Hamilton, M. The assessment of anxiety states by rating. *British Journal of Medical Psychology,* 1959, *32,* 50–55.

Hamilton, M. A rating scale for depression. *Journal of Neurology, Neurosurgery, and Psychiatry,* 1960, *23,* 56–61.

Hartshorne, H., & May, M. A. *Studies in service and self-control.* New York: Macmillan, 1929.

Hepner, T. W. Better judgments of men. *Industrial Psychology,* 1926, *1,* 19–24.

Hersch, E. L., Kral, V. A., & Palmer, R. B. Clinical value of the London psychogeriatric rating scale. *Journal of the American Geriatrics Society,* 1978, *26,* 348–354.

Hollingworth, H. L. Judgments of the comic. *Psychological Review,* 1911, *18,* 132–156.

Holzinger, K. J. *Preliminary report on Spearman-Holzinger unitary trait study.* Chicago: University of Chicago Press, 1935.

Honigfeld, G., & Klett, C. The Nurses' observation scale for patient evaluation (NOSIE): A new scale for measuring improvement in chronic schizophrenia. *Journal of Clinical Psychology,* 1965, *21,* 65–71.

Hull, C. L., & Montgomery, R. B. An experimental investigation of certain alleged relations between character and handwriting. *Psychological Review,* 1919, *26,* 63–75.

Katz, M. M., & Lyerly, S. B. Methods for measuring adjustment and social behavior in the community: 1. Rationale, description, discriminative validity and scale development. *Psychological Reports,* 1963, *1,* 503–535.

Kelley, T. L. *Statistical method.* New York: Macmillan, 1923.

Kent, R. N., O'Leary, K. D., Diament, C., & Dietz, A. Expectation biases in observational evaluation of therapeutic change. *Journal of Consulting and Clinical Psychology,* 1974, *42,* 774–780.

Kingstrom, P. O., & Bass, A. R. A critical analysis of studies comparing behaviorally anchored rating scales (BARS) and other rating formats. *Personnel Psychology,* 1981, *34,* 263–289.

Knight, F. B., & Franzen, R. Pitfalls in rating schemes. *Journal of Educational Psychology,* 1922, *13,* 204–213.

Komorita, S. S., & Graham, W. K. Number of scale points and the reliability of scales. *Educational and Psychological Measurement,* 1965, *25,* 987–995.

Langner, T. S., Gersten, J. C., McCarthy, E. D., Eisenberg, J. G., Greene, E. L., Herson, J. H., & Jameson, J. D. A screening inventory for assessing psychiatric impairment in children 6 to 18. *Journal of Consulting and Clinical Psychology,* 1976, *44,* 286–296.

Lawson, J. S., Rodenburg, M., & Dykes, J. A. A dementia rating scale for use with psychogeriatric patients. *Journal of Gerontology,* 1977, *32,* 153–159.

Lennon, W. J. *Development of a behavior rating scale for geriatric psychiatric patients.* Unpublished doctoral dissertaion, Rutgers University, 1971.

Lewinsohn, P. M., Mischel, W., Chaplin, W., & Barton, R. Social competence and depression: The role of illusory self-perceptions. *Journal of Abnormal Psychology,* 1980, *89,* 203–212.

Lissitz, R. W., & Green, S. B. Effect of the number of scale points on reliability: A Monte Carlo approach. *Journal of Applied Psychology,* 1975, *60,* 10–13.

Liston, E. H., Yager, J. & Strauss, G. D. Assessment of psychotherapy skills: The problem of interrater agreement. *American Journal of Psychiatry,* 1981, *138,* 1069–1074.

Lorr, M., Schaefer, E., Rubinstein, E. A., & Jenkins, R. An analysis of an outpatient rating scale. *Journal of Clinical Psychology,* 1953, *9,* 296–299.

Lorr, M., O'Connor, J. P., & Stafford, J. W. The psychotic reaction profile. *Journal of Clinical Psychology,* 1960, *16,* 241–245.

Lorr, M., Klett, C. J., McNair, D. M., & Lasky, J. J. *Manual: Inpatient multidimensional psychiatric scale.* Palo Alto, Calif.: Consulting Psychologists Press, 1963.

Luborsky, L. Clinicians' judgments of mental health. *Archives of General Psychiatry,* 1962, *7,* 407–417.

Lucero, R. J., & Meyer, B. T. A behavior rating scale suitable for use in mental hospitals. *Journal of Clinical Psychology,* 1951, *7,* 250–254.

Malamud, W., Hoagland, H., & Kaufmann, I. C. A new psychiatric rating scale. *Psychosomatic Medicine,* 1946, *8,* 243–245.

Matell, M. S., & Jacoby, J. Is there an optimal number of alternatives for Likert scale items? Study I: Reliability and validity. *Educational and Psychological Measurement*, 1971, *31*, 657–674.

Maxwell, A. E. Coefficients of agreement between observers and their interpretation. *British Journal of Psychiatry*, 1977, *130*, 79–83.

McNemar, Q. *Psychological statistics*. New York: Wiley, 1949.

Meer, B., & Baker, J. A. The Stockton geriatric rating scale. *Journal of Gerontology*, 1966, 21, 392–403.

Miller, L. C. Louisville behavior check list for males, 6–12 years of age. *Psychological Reports*, 1967, *21*, 885–896.

Miner, J. B. Evaluation of a method for finely graduated estimates of abilities. *Journal of Applied Psychology*, 1917, *1*, 123–133.

Moore, T. V. The essential psychoses and their fundamental syndromes. *Studies in Psychology and Psychiatry*, 1933, *3*, 1–128.

Motowidlo, S. J., & Borman, W. C. Behaviorally anchored scales for measuring morale in military units. *Journal of Applied Psychology*, 1977, *62*, 177–183.

Muslin, H. L., Thurnblad, R. J., & Meschel, G. The fate of the clinical interview: An observational study. *American Journal of Psychiatry*, 1981, *138*, 822–825.

Neugarten, B. I., Havighurst, R. J., & Tobin, S. S. The measurement of life satisfaction. *Journal of Gerontology*, 1961, *16*, 134–143.

Nisbett, R. E. & Wilson, T. D. The halo effect: Evidence for unconscious alteration of judgments. *Journal of Personality andSocial Psychology*, 1977, *35*, 250–256.

Osbourne, J. W. Determinations of subjective temperature. *Proceedings of the 25th Meeting of AAAS*, 1877.

Overall, J. E., & Gorham, D. R. The brief psychiatric rating scale. *Psychological Reports*, 1962, *10*, 799–812.

Pearson, K. On the relationship of intelligence to size and shape of head, and to other physical and mental characters. *Biometrica*, 1906, *5*, 105–146.

Peterson, D. R. Behavior problem of middle childhood. *Journal of Consulting Psychology*, 1961, *25*, 205–209.

Peterson, D. R., & Cattell, R. B. Personality factors in nursery school children as derived from parent ratings. *Journal of Clinical Psychology*, 1958, *14*, 346–355.

Platt, S. D., Hirsch, S. R. & Knights, A. C. Effects of brief hospitalization on psychiatric patients' behaviour and social functioning. *Acta Psychiatrica Scandinavia*, 1981, *63*, 117–128.

Plutchik, R., Conte, H., Lieberman, M., Bakur, M., Grossman, J., & Lehrman, N. Reliability and validity of a scale for assessing the functioning of geriatric patients. *Journal of the American Geriatrics Society*, 1970, *18*, 491–500.

Rickels, K., & Howard, K. The physician questionnaire: A useful tool in psychiatric drug research. *Psychopharmacologia*, 1970, *17*, 338–344.

Robinson, R. A. The diagnosis and prognosis of dementia. In W. F. Anderson (Ed.), *Current achievements in geriatrics*. London, Cassell, 1964.

Rogers, C. R., & Dymond, R. F. (Eds.), *Psychotherapy and personality change*. Chicago: University of Chicago Press, 1954.

Rosen, A. J., Tureff, S. E., Daruna, J. H., Johnson, P. B., Lyons, J. S., & Davis, J. M. Pharmacotherapy of schizophrenia and affective disorders: Behavioral correlates of diagnostic and demographic variables. *Journal of Abnormal Psychology*, 1980, *89*, 378–389.

Ross, A. O., Lacey, H. M., & Parton, D. A. The development of a behavior checklist for boys. *Child Development*, 1965, *36*, 1013–1027.

Rozelle, R. M., & Baxter, J. C. Influence of role pressures on the perceiver: Judgments of vid-
     eotaped interviews varying judge accountability and responsibility. *Journal of Applied Psy-
     chology,* 1981, *66,* 437–441.
Saal, F. E., & Landy, F. J. The mixed standard rating scale: An evaluation. *Organizational
     Behavior and Human Performance,* 1977, *18,* 19–35.
Saal, F. E., Downey, R. B., & Lahey, M. A. Rating the ratings: Assessing the psychometric
     quality of rating data. *Psychological Bulletin,* 1980, *88,* 413–428.
Schooler, N., Hogarty, G., & Weissman, M. Social adjustment scale II (SAS II). In W. A.
     Hargreaves, C. C. Attkisson, & J. E. Sorenson (Eds.), *Resource materials for community
     mental health program evaluators,* Publication No. (ADM) 79328, United States Depart-
     ment of Health, Education, and Welfare, 1979.
Schutz, H. G., & Rucker, M. H. A comparison of variable configurations across scale lengths:
     An empirical study. *Educational and Psychological Measurement,* 1975, *35,* 319–324.
Scott, W. D. *The army rating scale.* Pittsburgh: Carnegie Institute of Technology, 1917.
Scott, W. D., & Clothier, R. C. *Personnel management.* New York: Shaw, 1923.
Shader, R. I., Harmatz, J. S., & Salzman, C. A new scale for clinical assessment in geriatric
     populations: Sandoz Clinical Assessment Geriatric (SCAG). *Journal of American Geria-
     trics Society,* 1974, *22,* 107–113.
Shrout, P. E., & Fleiss, J. L. Intraclass correlations: Uses in assessing rater reliability. *Psycho-
     logical Bulletin,* 1979, *86,* 420–428.
Smith, P., & Kendall, L. M. Retranslation of expectations: An approach to the construction of
     unambiguous anchors for rating scales. *Journal of Applied Psychology,* 1963, *47,* 149–155.
Spearman, C. The proof and measurement of association between two things. *American Journal
     of Psychology,* 1904, *15,* 72–101.
Sweet, L. The measurement of personal attitudes in younger boys. *Young Men's Christian Asso-
     ciation,* Occasional Studies, No. 9, 1929.
Symonds, P. M. On the loss of reliability due to coarseness of the scale. *Journal of Experimental
     Psychology,* 1924, *7,* 456–460.
Symonds, P. M. *Diagnosing personality and conduct.* New York: Century, 1931.
Thorndike, E. L. Handwriting. *Teachers College Record,* 1910, *2,* 1–93.
Thurstone, L. L. A law of comparative judgment. *Psychological Review,* 1927, *34,* 273–286.
Thurstone, L. L. *The vectors of mind.* Chicago: University of Chicago Press, 1935.
Tinsley, H. E. A., & Weiss, D. J. Interrater reliability and agreement of subjective judgments.
     Journal of Consulting Psychology, 1975, *22,* 358–375.
Voelker, P. F. The function of ideals and attitudes in social education: An experimental study.
     Teachers College Contributions to Education, 1921, *112,* 126.
Webb, E. Character and intelligence. *British Journal of Psychology Monographs,* 1915, *3,* 1–
     99.
Webster's third new international dictionary. Springfield, Mass.: Merriam, 1966.
Willoughby, R. R. A scale of emotional maturity. *Journal of Social Psychology,* 1932, *3,* 3–35.
Wittenborn, J. R. A new procedure for evaluating mental hospital patients. *Journal of Consult-
     ing Psychology,* 1950, *14,* 500–501.
Wittenborn, J. R. Symptom patterns in a group of mental hospital patients. *Journal of Con-
     sulting Psychology,* 1951, *15,* 290–302.
Wittenborn, J. R. The dimensions of psychosis. *Journal of Nervous and Mental Disease,* 1962,
     134, 117–128.
Wittenborn, J. R. *Wittenborn psychiatric rating scale* (revised, 1964). Formerly published by
     The Psychological Corporation, 304 East 45th Street, New York, New York. (Available
     from author.)

Wittenborn, J. R. Reliability, validity, and objectivity of symptom-rating scales. *Journal of Nervous and Mental Disease*, 1972, *154*, 79–87.

Wittenborn, J. R. Wittenborn psychiatric rating scale (short form), 1964, In W. Guy (Ed.), *ECDEU Assessment manual for psychopharmacology, revised*. U.S. Department of Health, Education and Welfare, 1976.

Wittenborn, J. R., & Bailey, C. The symptoms of involutional psychosis. *Journal of Consulting Psychology*, 1952, *16*, 104–106.

Wittenborn, J. R., & Holzberg, J. D. The generality of psychiatric syndromes. *Journal of Consulting Psychology*, 1951, *15*, 372–380.

Wittenborn, J. R., & Mettler, F. A. Some psychological changes following psychosurgery. *Journal of Abnormal and Social Psychology*, 1951, *46*, 548–556.

Wittenborn, J. R., & Weiss, W. Patients diagnosed manic depressive psychosis—manic state. *Journal of Consulting Psychology*, 1952, *16*, 193–198.

Wittenborn, J. R., Mandler, G., & Waterhouse, I. K. Symptom patterns in youthful mental hospital patients. *Journal of Clinical Psychology*, 1951, *7*, 323–327.

Wittenborn, J. R., Bell, E. G., & Lesser, G. S. Symptom patterns among organic patients of advanced age. *Journal of Clinical Psychology*, 1951, *7*, 328–331.

Wittenborn, J. R., Herz, M. I., Kurtz, K. H., Mandell, W., & Tatz, S. The effect of rater differences on symptom rating scale clusters. *Journal of Consulting Psychology*, 1952, *16*, 107–109.

Wittenborn, J. R., Holzberg, J. D., & Simon, B. Symptom correlates for descriptive diagnosis. *Genetic Psychological Monographs*, 1953, *47*, 237–301.

Wittenborn, J. R., Astrachan, M. A., DeGooyer, M. W., Grant, W. W., Janoff, I. Z., Kugel, R. B., Myers, B. J., Riess, A., & Russell, E. C. A study of adoptive children: I. Interviews as a source of scores for children and their homes. *Psychological Monographs*, 1956, *408*, 1–58.

Wittenborn, J. R., Plante, M., Burgess, F., & Livermore, N. The efficacy of electroconvulsive therapy, iproniazid and placebo in the treatment of young depressed women, *Journal of Nervous and Mental Disease*, 1961, *133*, 316–332.

Wittenborn, J. R., Plante, M., Burgess, F., & Maurer, H. A comparison of imipramine, electroconvulsive therapy and placebo in the treatment of depressions. *Journal of Nervous and Mental Disease*, 1962, *135*, 131–137.

Wittenborn, J. R., Weber, E. S. P., & Brown, M. Niacin in the long-term treatment of schizophrenia. *Archives of General Psychiatry*, 1973, *28*, 308–315.

Wittman, P., & Sheldon, W. A proposed classification of psychotic behavior reactions. *American Journal of Psychiatry*, 1948, *105*, 124–128.

Zukow, P. G., Zukow, A. H., & Bentler, P. M. Rating scales for the identification and treatment of hyperkinesis. *Journal of Consulting and Clinical Psychology*, 1978, *46*, 213–222.

5

Self-report Ratings and Inventories

D. CARTWRIGHT and GIDEON WEISZ

INTRODUCTION

This chapter reviews self-report methods in psychotherapy research since the second edition of *The Handbook of Psychotherapy and Behavior Change* (Garfield & Bergin, 1978). Its topic is limited to studies either using self-report ratings or inventories or doing research and development (R & D) on self-report devices. It is also mainly limited to studies of adult psychotherapy, behavior therapy, or cognitive therapy: individual rather than group, chiefly outpatient. In the space allotted to this chapter, the aim has had to be reasonable sampling rather than comprehensiveness.

Studies using self-report devices in psychotherapy research are most often concerned with the issue of effectiveness, either in its general form (Does psychotherapy in general produce results superior to spontaneous remission of placebo administrations?) or a specific form (Does this type of psychotherapy have these effects on given patients over and above controls?). A weaker version of effectiveness research is simple demonstration that changes do occur in patients concomitantly with treatment or thereafter in follow-up periods.

In organizing the materials within each section, we draw on several authors' work to formulate a suitable framework. Luborsky (1976) has proposed some major divisions of the process of psychotherapy. A major distinction is between the goals to be achieved and the means for achieving them. For example, a patient may say that he or she wants to be rid of depression and the therapist may suggest cognitive therapy as the means. Means may be divided into techniques and relationships. Techniques seem to be of two main

---

D. CARTWRIGHT and GIDEON WEISZ   ●   Department of Psychology, University of Colorado, Boulder, Colorado 80309.

kinds: the structure of treatment and specific interventions. *Structure* refers to matters such as degree of supportiveness versus expressiveness or degree of focus on understanding and degree of focus on specific attainment of clearly defined goals. *Interventions* include such matters as interpretations. Relationships are again divided into transference types, or the nature of core conflictual themes as reenacted with the therapist, and helping alliances of two types. In a Type 1 helping alliance the patient develops confidence that the therapist is supportive and will provide the help the patient needs. In a Type 2 helping alliance the patient and therapist form a working partnership in tackling the problems that confront the patient. We may call these respectively *supportive* and *working alliances*.

Kanfer (1979) says that psychotherapy has gone through a dialectic process over the last 25 years and is now emerging as a new synthesis of earlier forms. The new synthesis is neither psychodynamic nor behavioral. Primarily focusing on cognitive processes, it integrates elements of earlier insight and behavior modification approaches. The new approach may be broadly labeled as cognitive-behavioral or cognitive-mediational. Covert behaviors such as thinking, planning, fantasizing, and decision making are now incorporated both as goals of treatment and as means of treatment.

Kanfer notes that behavioral clinicians are increasingly concerned with self-control, imitative processes, and self-regulation. Research suggests that behavior change is maintained to a greater extent if the patient attributes the change to his or her own efforts, at least in part. Gains from therapy are most long-lasting when therapists help clients to have confidence that they are actively participating in the treatment and can make their own behavior changes eventually, without supervision (Kanfer, 1979, p. 199). In Luborsky's terms, therapist and patient must establish a working (as opposed to a supportive) alliance.

The working alliance is crucial to Kanfer's model of self-regulation, in which the aim is to increase patients' self-esteem by having them formulate self-generated criteria that they can actually achieve. The self-regulatory methods consist of three stages in a process that is cued by any event which breaks up the smooth flow of behavior. The first stage is self-monitoring of performance and emotional reactions. The second stage is self-evaluation, in which patients compare the information obtained by self-monitoring with performance criteria previously generated for situations of the given kind. In the third stage, if performance has met or exceeded criteria, patients give themselves positive reinforcement; if performance was poor, they opt for repetition or self-criticism. Thus, self-monitoring, self-evaluation, and self-reinforcement are the component sets of skills in self-regulation. It is believed that correction of deficits in any of these components will improve the self-regulatory process as a whole (Kanfer, 1979, p. 213).

It seems possible that many different forms of psychotherapy implicitly or explicitly seek, through technique or relationship, an improvement in self-regulatory skills. Our framework will include self-monitoring as the basis of self-regulation; it will omit interventions and transference relationships, since these are probably not well assessed by self-report techniques.

We may summarize the framework and state it in terms suitable for casting in structural equation models (Bentler, 1980)—that is, in classes of dependent, independent, and mediating variables.

Goal variables of symptom reduction or personality change qualify as dependent variables.

Since treatment structure is chosen by the therapist, variables of the structure seem best considered as independent variables. Also, variables of patients' initial status may be viewed as independent, subject to free variation (for prediction studies) or to selection (for homogeneous patient samples).

Variables of the helping alliance are less subject to control, since such alliances may or may not emerge in interaction with a given patient. These variables are perhaps best conceived as mediating variables. Likewise, variables of self-monitoring (which reflect the patient's work or homework) are mediating.

REVIEW OF RECENT RESEARCH

Studies are reviewed under two categories, effectiveness and R and D. Within each category, materials are organized into sections on symptom reduction, personality change, initial status, treatment structure, helping alliances, and self-monitoring.

Effectiveness

Symptom Reduction

Epstein and Vlok (1981) have examined the evidence for effectiveness in psychotherapy and conclude that (outside of the work of Smith & Glass, 1977) there is little evidence for effectiveness except for anxiety and nonpsychotic depression. Indeed, these are the two main variables that have been studied recently. We include phobias and fears in the same category as anxiety and also add a miscellaneous section.

Anxiety. Foa and Chambless (1978) used a 100-point scale of current anxiety to study variations within and across sessions of imagery-flooding therapy for eight obsessive-compulsive and six agoraphobic patients. Significant differences were found between groups, sessions, and time segments within sessions. It was concluded that subjective anxiety is habituated (reduced) during flooding within and across sessions. Agoraphobics seemed to show the most reduction in anxiety.

Marks and Mathews (1979) developed a one-page questionnaire covering the patient's main phobia, a global phobia rating, items about the 15 most common phobias, and five related anxiety or depression items. Test–retest correlations of about .8 were obtained. In a preliminary study of validity, it was found that agoraphobic patients declined 11.6 points on the agoraphobia items; social-phobic patients declined 5.7 points on the social-phobic items. This scale deserves special mention because, although not standardized in the usual sense, it was developed explicitly so that several clinics might have an "agreed measure of change," and three hospitals in England currently use this scale.

In the period covered, only one study using the Ipat Anxiety Scale was found (Krug, Scheier, & Cattell, 1979). The study by Sookman and Solyom (1978) used this scale and others in an evaluation of two cases of depersonalization treated by flooding. The scale has 40 items measuring the five primary factors found in the second-order anxiety factor of the Sixteen Personality Factor Questionnaire (16 PF; Cattell, Eber, & Tatsuoka, 1970).

Agras and Jacob (1981) carried out an interesting meta-analysis of 48 studies of the effects of treatment on phobia. They used indices of sensitivity, including especially the percentages of studies in which a given measuring instrument (or type of instrument) had shown significant differences between pretreatment and posttreatment means (within-group = WG) between treatment and control groups (between groups = BG). Out of 48 methodologically suitable studies found between 1960 and 1978, the percentages of each kind (WG and BG) were tabulated in a bar graph (Agras & Jacob, 1981, p. 50). Behavioral measures had the highest percentages (WG = 74, BG = 53). Next on WG (64) was the Fear Survey Schedule (FSS; Wolpe & Lang, 1964). This is a 76-item list of specific fears (e.g., of harmless snakes) which the patient checks, rating each item on a scale of 0 to 4 for intensity of fear ("none" to "very much"). But next on BG was self-report of the patient's main phobia (BG = 45). Measures of heart rate were next in sensitivity (WG = BG = 39). Next in WG (38) was again the self-report of main phobia. Ratings by therapist and others came next, and then the BG percentage (29) for the FSS.

Agras and Jacob (1981, p. 51) comment that the results show equally adequate sensitivity for both client ratings and behavioral measures, but physiological measures do less well (GSR: BG = 19). However, the tabulation of personality inventory data showed these instruments to be quite insensitive. Personality scales had WG = 16, BG = 3! We discuss these results below.

Depression. Substantial progress has been made in studying effects of treatment for depression. Beck, Rush, Shaw, and Emery (1979) describe three studies using the Beck Depression Inventory (BDI; Beck, 1978, 1979) and other measures. In one study, cognitive therapy was compared with pharmacotherapy. Patients were assigned randomly to treatment groups ($n = 18$ and $n = 17$ respectively). Treatment lasted up to 12 weeks. Patients who completed treatment showed significant decreases in BDI scores exceeding "reported ranges for placebo response in depressed outpatients" (Beck *et al.*, 1979, p. 391). The gains were maintained through 3-month, 6-month, and 12-month follow-ups for the cognitive therapy group, who scored significantly lower on the BDI than the pharmacotherapy group at the 12-month point. Beck *et al.* (1979) report on a study of 11 hospitalized depressive patients. Following an average of 8 weeks of cognitive therapy, the patients dropped from a mean score of approximately 30 at intake to a mean score of 16 at discharge and a mean of 14 at follow-up 4 weeks later. In a third study reported by Beck *et al.* (1979), patients were randomly assigned to cognitive or cognitive plus pharmacotherapy. Both groups showed roughly equal and significant drops in mean BDI score (from 30 to 10, approximately).

The BDI has rapidly gained widespread acceptance as a measure of depression. Each of its 21 items has just four alternative responses scaled 0–3. Instructions ask the patient to report on feelings during the past week.

Lewinsohn and Arconad (1981) of the Oregon Depression Unit report on two samples studied by means of the BDI. The patients were treated by behavioral methods based on social-learning theory. In both samples ($n = 21$ and $n = 12$), significant decreases were obtained on mean BDI scores by termination and at follow-up (6 months). Also, in two samples ($n = 21$ and $n = 14$), these authors used the MMPI depression scale (MMPI-D). Significant mean reductions by termination and at follow-up were found.

Lewinsohn and Arconad (1981) hold that depression is a function of a low rate of positive reinforcement and a high rate of aversive experience. Therapy aims to increase the positive reinforcements experienced by patients and/or to decrease the aversive experiences. Accordingly, two self-report scales are used: the Pleasant Events Schedule (PES) and the Unpleasant Events Schedule (UES; Lewinsohn & Amenson, 1978). An 80-item general-purpose form of each schedule is given in Appendix A of the volume by Clarkin and Glazer (1981). In studies at the Oregon Depression Unit mentioned above, the PES and UES were given early in treatment and at termination. Expected changes were found. However, the authors do not claim specific efficacy because, in related studies (cf. Zeiss, Lewinsohn, & Munoz, 1979), no differences have been found between various treatments such as cognitive, interpersonal, and pleasant-events-focused treatments. All methods reduced depression as measured by the BDI.

McLean (1981) takes the position that if a treatment works for depression, it does so across all modalities; therefore, if patients can be helped to improve their performance levels, benefits will spread to cognitive, motivational, and even physiological functioning. Accordingly, treatment aims to correct skill and performance deficits. Studying 196 clinically depressed patients and 161 normal subjects, McLean showed that self-reported performances are substantially less frequent among patients prior to treatment. They do not go out much and they do not do much at home except watch television. After 3 months of treatment, however, the patients substantially increase their social activities and their personal activities (hobby, letters, paperwork, etc.) are, on average, as frequent as those of normals.

Rehm (1981) reports on a self-control therapy program for depressed patients ($n = 8$). This program was compared with nonspecific form of treatment and waiting list controls. Using the BDI, it was found that both therapy groups improved significantly and were significantly less depressed than the controls on the posttest. Similar results were obtained on the MMPI-D. A version of the PES was also used, and the self-control group showed the greatest improvement. The therapy lasted 6 weeks and results were maintained over a further 6-week follow-up.

In another study reported by Rehm (1981), patients were given either the self-control program ($n = 24$) or assertion training ($n = 14$). Both groups improved significantly, as measured by the BDI and MMPI-D, over the treatment period and there were no significant differences between them at termination or at follow-up 1 year later.

Rehm introduced the valuable idea of disassembly studies. Patients were randomly assigned to treatment programs having one or more components of the program removed. All patients were nonpsychotic, with MMPI-D over 69. Five treatment conditions were (1) the full program, including self-monitoring, self-evaluation, and self-reinforcement modules; (2) self-monitoring and self-evaluation only; (3) self-monitoring and self-reinforcement only; (4) self-monitoring only; (5) waiting-list control. Sample sizes varied from 9 to 15. The BDI and MMPI-D were major dependent variables. A version of the UES was used as well as the PES. Results on the BDI and PES showed a general treatment effect, with all treatment groups doing better than the waiting-list controls. No differences were found between the treatment groups, however.

It seemed possible that the self-monitoring component by itself instigated higher levels of activity and thereby improvement in depression. Rehm (1981) reports a kind of second-stage disassembly study having four conditions: (1) full self-control program; (2) full program less explicit homework assignments (although the principles of self-monitoring were explained); (3) self-monitoring and self-evaluation only; (4) a comparison group of patients receiving interpersonal therapy. Samples had 11, 11, 12, and 5 patients respectively. Therapy was conducted for 12 weeks in this study. Overall effects of treatment were found, but no significant differences were obtained between the groups. Doubling the length of treatment (from 6 to 12 weeks) apparently made no difference; improvement was comparable in both studies.

These results surprised Rehm and his colleagues. Could it be that the samples were too small? Could it be that the therapists had different competence, which worked in opposite directions from the relative efficacies of the several treatment programs? Could it be that simply explaining the principles of self-monitoring is enough to help people reduce depression? Could it be a Hawthorne effect? Is it really conceivable that a self-reinforcing component adds absolutely nothing to a treatment program? The questions must fly around in everyone's head.

Another study of differences between treatments is the Vanderbilt Psychotherapy Research Project, directed by Hans Strupp. This project had the "primary aim . . . to investigate the relative contribution of the therapist's technical skills and the qualities inherent in any good human relationship to outcome in time-limited psychotherapy" (Strupp & Hadley, 1979, p. 1125). Only male college students were studied, each having T scores of 60 or more on MMPI-D and the MMPI psychasthenia and social introversion scales. Following an interview to establish three target complaints, patients were assigned (at random except for clinical constraints) to one of three groups: (1) professional therapist ($n = 16$); (2) alternate therapist (college professors; $n = 15$); (3) delayed, minimal treatment group ($n = 14$). Also a "silent control" group was studied ($n = 19$). These were students who had difficulties similar to the patient group but had not applied for treatment.

The Vanderbilt study used several instruments, including an 11-point rating scale of degree of improvement. Ratings were made at termination (averaging 17 sessions) and at follow-up.

While there were significant decrements in MMPI scores from intake to termination for the treatment groups, there were also significant decrements for the silent control group. No difference in the amount of change (residual gain scores) between treatment and control groups could be detected. On the global rating of improvement from intake to termination, the three treatment groups all had a mean of approximately 8.3. Thus, once again evidence is found for the idea that specifically different treatment programs do not result in specifically different outcomes, at least in brief therapy and as measured by patient self-report.

A study with somewhat different results as regards specific treatment effects is that of LaPointe and Rimm (1980). Cognitive, assertive, or insight-oriented therapy was given to three groups of depressed women (scoring over 49 on the BDI). All treatments reduced mean BDI scores significantly, but there were significant differences in the degree of improvement on the BDI according to groupings by initial level of depression and by treatment group.

Miscellaneous. Strupp and Hadley (1979) used the Target Complaints (TC; Battle, Imber, Hoehn-Saric, Stone, Nash, & Frank, 1966). The TC had to be further processed before suitable psychometric study could be done, because each patient's complaints were different. Strupp and Hadley report that the complaints fell into three content categories: dating and other interpersonal skills, academic performance, and self-identity. However, patients differed in the number of complaints mentioned and in the distribution of those complaints across the content categories. Severity ratings were assigned to each complaint; therefore the investigators decided to average the severity ratings of a given patient's complaints across whatever categories they were placed. What this seems to mean is that the target complaints score represents average severity of complaint, regardless of exact content (although we know the three general content areas in which the complaints could be categorized). The measures then analyzed were difference scores: complaint score at termination minus complaint score at intake. There were no significant differences between the three treatment groups, and the average change across all groups was about 3.1 (reduction in average severity rating).

Karle, Corriere, Hart, and Klein (1981), using the Eysenck Personality Inventory (EPI; Eysenck & Eysenck, 1968), found that the EPI measure of neuroticism decreased significantly during the course of feeling therapy with $n = 211$ outpatients. This measure is highly correlated with measures of anxiety and depression (Eysenck & Eysenck, 1968).

Personality Change

Although the evidence comes from only a few sources, it nevertheless suggests very strongly that extraversion (vs. introversion) is a variable of prime importance. Reports indicate that patients become more extraverted during the course of treatment. Other variables of importance are self-actualization, self-control, and assertiveness.

Extraversion. Karle *et al.* (1981), in the study of feeling therapy mentioned earlier, also found a significant increase in the EPI extraversion mean score from pre- to posttherapy.

Strupp and Hadley (1979) found a significant decrease in social introversion, as measured by the MMPI, for all treatment and control groups from intake to termination assessment points. A significant decrease was also found, from intake to follow-up points, for all groups except minimal therapy controls. The professional therapy group dropped roughly 15 T-score points, while the silent control group dropped roughly 6 points over the same period (intake to follow-up).

Wilson and Kennard (1978) studied 114 residents of a therapeutic community for drug abusers. Marked changes, as measured by the second-order extraversion factor of the 16PF, were found in those who stayed 6 months or more. The longer the patients stayed in the community, the more they increased in extraversion.

In all the studies mentioned above, the patients had mean scores on intake that on the average were somewhat introverted. Thus, the changes toward extraversion should perhaps be interpreted as movements away from introversion and toward a middle point on the bipolar scale.

Self-actualization. Karle *et al.* (1981) used the Personal Orientation Inventory (POI; Shostrom, 1974). This inventory was developed on the basis of personality theory, especially as conceived by Maslow and Perls. It focuses on variables important for personal growth and self-actualization. It was found that patients did increase in self-actualizing tendencies, the more so with greater time in treatment. Dosamentes-Alperson, Smokler, and Merrill (1980) also used the POI to assess effects of experiential movement therapy. The treatment group showed greater improvement in self-actualization than did control subjects.

Knapp, Shostrom, and Knapp (1978) summarize several studies of personality change, as measured by the POI, during individual or group therapy. In general, self-actualization increases during psychotherapy, the more so with greater time in treatment.

Knapp *et al.* (1978, p. 109) note that the spontaneity scale of the POI correlates highly ($-.67$) with social introversion as measured by the MMPI. So it is possible that the results on self-actualization and the results on extraversion reported above are closely related. For patients, becoming less introverted is associated with becoming more spontaneous and more self-actualizing generally.

Self-control. Rehm (1981) used a Self-control Concepts Test (SCT) developed specially for the program of self-control therapy. It consists of 25 statements such as "I have high standards for my own performance" (paraphrased). Respondents indicate their agreement with each statement on a 5-point scale. In three studies it was found that the SCT yielded higher scores for treatment groups than for control subjects (measured at termination only). There was mixed evidence concerning differences between treatment groups on the SCT.

Assertiveness. One of the control or comparison groups in Rehm's studies (1981) received assertion training. Outcomes were measured in part by the Wolpe–Lazarus Assertion Scales (Wolpe & Lazarus, 1966). No significant differences were found, but there was evidence of improvement in assertion skills in both the assertion training and self-control program groups.

LaPointe and Rimm (1980), using the Rathus Assertiveness Schedule (Rathus, 1973), found that patients in cognitive, assertiveness, and insight-oriented group therapies showed improvements in assertiveness.

It may be noted that the 16PF second-order factor of extraversion includes an important primary factor of assertiveness (H-factor). While there is no direct evidence, it seems possible that, like self-actualization, improvements in assertiveness are associated with changes in the broader dimension of extraversion.

Patient's Initial Status

Initial status refers to the patient as described on intake. Two uses may be distinguished: selection and prediction. We have already seen numerous instances above of patient selection, as, for example, in the Vanderbilt project, where all patients met the 2–7–0 syndrome criterion (T scores above 59 on MMPI-D, psychasthenia, and MMPI-Si). This section will, therefore, concentrate on the few prediction studies that have appeared recently.

Prediction. Luborsky, Todd, and Katcher (1973) report on a Social Assets Scale (SAS) that has not previously been reviewed. The SAS is currently used in the University of Pennsylvania Psychotherapy Project (Luborsky, 1981). It is self-administered, and items refer to education, occupation, marital status, birthplace, parents' marital status, childhood experiences, physical health, and so on. Alternative responses were selected by agreement among judges rating the extent to which a response constitutes either a social asset (+1, +2, or +3) or liability (−1, −2, −3) or is neutral (0). Modal ratings were used as weights to be assigned to each response. As an example, one item asks about parents' quarrels during the respondent's childhood, with alternatives and scores as follows: "All the time" (−2.0), "Frequently" (−1.0), "Rarely" (0.0), "Never" (0.5).

Several forms of the SAS are available (for students, nonstudents, men, women, special groups such as nurses, and so on). The form given in the cited articles has 31 scored items with alternative phrasings for different groups. Internal consistency is about .70. In one study, the SAS correlated with clinicians' ratings of predicted improvement in psychotherapy ($n = 75$; $r = .35$). In another study, the SAS correlated with improvement in brief psychiatric hospitalization as judged from ratings by several sources ($n = 65$, $r = .30$, point-biserial).

A related scale is the Addiction Severity Index (ASI) (McLellan, Luborsky, Woody, & O'Brien, 1980). By contrast with the SAS, the ASI has numerous components explicitly designed for the study of change as well as prediction; the relevant items are given a time frame of "during the past 30 days."

The ASI derives from an interview in which objective data are collected and the patient rates how much he or she has been bothered by problems and feels the need of treatment for them. The problems fall into six areas: chemical abuse, medical, psychiatric, legal, family/social, and employment/support. Objective data and patient ratings are integrated to arrive at a judgment of severity in each area, using a 10-point scale. Reliability estimates are high and validity evidence is good. The ratings for the six areas have relatively low intercorrelations, the median based on 524 subjects being .17. A comparison of alcohol and drug abuse patients shows the former significantly higher in severity of medical problems and the latter significantly higher in legal problems. Cluster analysis shows six distinct homogeneous patient subgroups, indicating differential treatment needs. The Psychiatric Severity Scale correlates .3 to .4 with outcome in psychotherapy (Luborsky, 1981). The interview can be administered by a technician in 30 minutes. The ASI deserves to become standard in all projects dealing with substance abuse.

An important study related psychoticism as measured by the Eysenck Personality Questionnaire (EPQ; Eysenck & Eysenck, 1975) to outcome of treatment for neurotic patients (Rachman & Eysenck, 1978). It was hypothesized that patients diagnosed as neurotic would prove relatively less responsive to psychiatric treament if they also had higher scores on psychoticism. They would take longer in treatment and would eventually give the therapist problems by becoming quarrelsome, hostile, and argumentative. Male ($n = 188$) and female ($n = 116$) samples were studied. Each was divided into high and low groups on the scale of psychoticism in the EPQ. Some were on antidepressant drug treatment as well as in psychotherapy. The average number of days in treatment was indeed found to be greater for those high in psychoticism regardless of sex or drug treatment. Also, psychiatrists' ratings of the patients on a 6-point scale of prognosis at the beginning of treatment and of assessed status at the end yielded an improvement score. Significant correlations for both male and female samples indicated that higher psychoticism was associated with smaller degrees of improvement.

Zivich (1981) reports on the use of the 16PF and the Personality Research Form (PRF; Jackson, 1974) in studying types of alcoholics. It was found that treatment was most effective with types described as less disturbed.

Treatment Structure

In Luborsky's (1976) analysis, treatment structure refers to one aspect of therapeutic technique. It includes supportiveness versus expressiveness, focus on understanding versus focus on attaining specific, defined goals, and so on.

It is not clear that patient report data are specifically useful in appraising the structure of treatment. Ordinarily, such appraisal would be made by an observer and would presumably focus on the behavior of the therapist. Nevertheless, two recent studies have provided evidence pertinent to treatment structure, even though they did not have this particular formulation of the topic in mind.

Hanfmann (1978) studied some 400 responses to a posttherapy questionnaire mailed routinely to students who had received counseling during the given school year. The author compared the responses of those who had "found counseling helpful" with others (75% vs. 25%). Most positive comments referred to the readiness of the counselor to "take your problems to heart." Most frequent complaints referred to inactivity on the part of the counselor. Unproductive counseling was reported by students who complained about detachment on the counselor's part. Such counselors were described as distant, impersonal, rigid in observing rules, stilted in language, withholding of opinions and information, and so on. Conversely, most productive counseling came from counselors described as being there, talking as well as listening, engaging in a joint effort toward problem solving, making the student think, and confronting.

Gomes-Schwartz (1978) and Gomes-Schwartz and Schwartz (1978), working in the Vanderbilt project, found that analytically oriented therapists were characterized as engaged in listening, questioning, and interpreting while maintaining interpersonal distance. Experiential therapists, on the other hand, were more like the college professors: more relaxed, more free in exchanging views, more apt to give advice and guidance.

Putting together the results so far, it might be expected that the analytically oriented therapists in the Vanderbilt study would have less productive outcomes in Hanfmann's terms. But that is not so, as demonstrated most clearly in the study by Waterhouse (1978), also of the Vanderbilt project. She used a modified version of the Relationship Inventory (MRI; Barrett-Lennard, 1962). The MRI has 16 forced-choice items, each with contrasting descriptions. Although dealing with relationships, some of these items focus on the therapist's behavior and thus are suitable for an assessment of treatment structure. For example, one item indicates the following alternatives: (1) therapist gives suggestions and advice or (2) therapist avoids answering questions and wants patient to make all decisions. Data were obtained after the third session and at termination.

Waterhouse reports on 9 patients in the analytic therapy group and 7 in the experiential therapy group as well as 15 in the alternate therapy (college professor) group. Significant differences were found on four items. On three of these, the alternates were contrasted with the professionals, with alternates being more active (doing their share of the talking), disclosing their own feelings more, and being judged as friends more than the professionals. On one item the analytic therapists were judged as different from the experiential and alternate therapists: analytically oriented therapists were more likely to be stiff and formal.

But were these differences related to outcome? They were not. Three other items differentiated high- from low-success groups. As of the third interview, those who would later be in the high-success group experienced themselves as liked by the therapist, feeling free to talk about anything, and feeling at ease in the relationship. These matters seem unrelated to treatment structure.

Helping Alliance

In Luborsky's (1976) account, relationship factors are divided into two kinds: transference and helping alliance. The latter are further divided into supportive and working alliances.

Hanfmann's (1978) study reported that students who had experienced productive counseling described their counselor as one who "engaged in a joint effort toward problem solving." This seems to be an essential mark of the working alliance. Since all were presumably ambulatory, outpatient, actively enrolled students, it is quite likely that few needed a supportive alliance and possible that those who did were the ones who failed to find their counseling experience productive.

Nowhere in the MRI used in the Vanderbilt project are there items suggesting a truly supportive alliance, although genuine interest in helping the patient was characteristic of all therapists. But there are no items indicating, for example, that the patient really needs the therapy in order to stay out of trouble or hospital.

It seems possible that certain items in the MRI do refer to characteristics of a working alliance, however. These are precisely the items related to outcome. Using outside clinicians' ratings of change in overall disturbance, psychic stress, performance, and global state, Waterhouse (1978, p. 23) found that patients' third interview ratings (of feeling liked, at ease, and free or open) were highly positively related to outcome. One way to interpret these findings is to infer that ease and freedom were experienced by those patients who were engaged in a more collaborative or working alliance: they were doing more in a joint effort. Another piece of evidence for this interpretation also comes from the Vanderbilt project, from investigators (McDaniel, Stiles, & McGaughey, 1981) who found that improvement was associated with number of patient utterances during the entire course of therapy. This result again suggests a greater degree of collaborative work on the part of the patient.

Luborsky's own approach to the assessment of helping alliances (1976) uses content analysis of the patient's speech to count signs of the two types of helping alliance. A comparison of most and least improved patients in this project shows that the former had mainly developed a positive Type 1 (supportive) alliance early in treatment whereas the latter had developed a negative Type 1 alliance.

Self-monitoring

It was suggested above that self-monitoring data are of two kinds: those referring to self-help activities and those describing current symptom conditions. The latter are used in one way as input to the therapist in modulating treatment approaches and in another way as materials for evaluation of change from intake period (a baseline) to termination period (a goal line).

Current Symptoms. Lewinsohn and Arconad (1981) report on four instruments used in daily monitoring of symptoms. The Depression Adjective Checklists (Lubin, 1965) have seven alternative lists of adjectives that the patient checks each evening to record the day's feelings. It is pointed out that these data give the therapist an opportunity to find out whether certain days are associated with increased depression and then to explore possible reasons for this. The second instrument is a 9-point scale ranging from very happy to very depressed. The scale is shown as Appendix B in Clarkin and Glazer (1981, p. 373). The final pair of instruments consists of the PES and the UES, rated on a daily basis, with $0 =$ no occurrence, $1 =$ occurred but neutral, and $2 =$ occurred and pleasant (or unpleasant). An example of one patient's daily logs for 48 days is given (Lewinsohn & Arconad, 1981, p. 54). It provided impressive visual evidence of fluctuations day by day and gradual improvement beginning about the tenth day. The authors point out that the data soon begin to convince the patient of the relationship between events and depressive moods and of the value of pleasant events in reducing depression. The patient then becomes able to diagnose his or her own depression.

Self-help Activities. An example of self-help activities monitoring is also available from Lewinsohn and Arconad (1981). Part of their program consists of classical relaxation training. Patients are asked to monitor each day the amount of time they spend in practicing relaxation techniques, also rating how relaxed they were before the practice and afterward. The form is given as Appendix C in Clarkin and Glazer (1981, p. 374).

Study of Change. Lewinsohn and Arconad (1981) also provide an example of the use of self-monitoring in studying change. The same form for recording relaxation practice also asks the patient to record an average tension score for the day and scores for the least and most relaxed occasions. The patient also indicates when, where, and what the situation was. Such daily logs provide a record of symptom change and also a baseline from which to assess progress.

Research and Development

We turn now to our second main division of studies, R and D. The same organization of subdivisions will be followed.

Symptom Reduction

Anxiety. The Fear Survey Schedule (FSS) has been given extensive consideration in recent years. Tasto (1977, pp. 156–174) provides a survey and points out that there have been many versions, dating back to 1956, using between 50 and 122 items and either 5-point or 7-point rating scales. Different versions have had relatively little item overlap. Factor analysis of a 122-item version yielded five factors, but only 40 of the items correlated .35 or better with one or more factors, suggesting that most items have a good deal of unique variance. The five factors (with between 5 and 10 items per factor) were fear of small animals, of incipient hostility, of moral infringements, of isolation and loneliness, and of bodily harm. Within each factor, items are somewhat homogeneous in content. For example, the small animals include bats and worms and harmless spiders.

Bellack and Hersen (1980, p. 170) mention two of the FSS versions and stress that, in general, individual item scores or factor scores are preferable to total scores. This is certainly borne out in the survey by Agras and Jacob (1981) concerning sensitivity, as discussed earlier. But why would this be so? And how is it that the FSS total scores did so much better than the single-item self-reports of main phobia (PSRMP) in within-group (WG) comparisons? For that matter, why would the FSS total scores do less well than the PSRMP in BG comparisons? There is a curious inversion of results here.

Perhaps the answer is that WG effects with total scores can be inflated. Suinn (1969) showed that over a period of 5 weeks, college students ($n = 82$), under conditions of anonymity, reduced their FSS scores an average of 10 points, or roughly 0.10 point per item. Presumably, this means that a measure of general fearfulness is likely to show reductions in a normal or untreated population on repeated testing. Such a component would inflate apparent change in a WG design but would decrease the net gain for the treatment group in a BG design.

It is not clear why PSRMP scores should be more sensitive in BG than in WG comparisons. On the face of it, such a result is counterintuitive. Surely patients have to reduce their main symptom in order to have an improvement exceeding that of a control group? Perhaps the answer is that control groups are apt to increase their one-item PSRMP score if they are not treated properly—a complaint. It must also be true that only patient control groups could have nonzero scores on PSRMP. Obviously, this one item is not just more specific but also more important. Bellack and Hersen (1980) report on this matter and reproduce the Geer (1965) version of the FSS. They point to results showing that as an item becomes increasingly specific (e.g., by indicating the situation in which one might encounter a snake and making it closely descriptive of the actual laboratory test situation), the correlations between item score and *in vivo* test behavior increase. For snakes, the correlations go from .37 to .68; for rats, from .53 to .81.

Allen (1980) reports on numerous studies involving six different self-report instruments for measuring test anxiety. Only 45% of interventions yielded improvements exceeding the nonspecific response to placebo. Here again it would likely be useful to improve the specifity of items.

On a related topic, Marzillier and Winter (1978) showed that changes on a 12-point self-rating scale of anxiety were highly individualized. In one 49-week treatment, the patient showed reduction in anxiety only during weeks 9 through 16. In another patient, no anxiety reduction was found during treatment (12 weeks) but did appear during an 8-week follow-up period. Other idiosyncrasies suggest that evaluations need to be tailored more carefully.

Levy and Guttman (1982) use carefully designed national survey data to show some interesting differences between fear, worry, and concern. Fear is found to be a sufficient condition for worry but not a necessary one. Worry shows the statistical features of a "semiattitude" in that frequency of discussing a given topic (such as terrorist activity) has a J-shaped regression on amount of worry about that topic. The relevant items were all designed according to facet theory and a mapping sentence. This allows subtle conclusions, such as that fear may be either a cognitive or an instrumental reaction but worry is an affective reaction.

A related distinction has been made under the title of *desynchrony*. Agras and Jacob (1981) indicate that physiological changes, avoidance behaviors, and self-reports of fear do not always correspond. Therefore measures of each should be included in a thorough study of the effects of treatment for phobia. Indeed, Bellack and Hersen (1980, p. 175) believe that concordance among the three response systems is rare. Two recent studies again show the lack of concordance (Andrasik, Turner, & Ollendick, 1980; Marzillier, Carroll, & Newland, 1979).

Depression. New volumes have appeared on the MMPI (cf. Butcher, 1979; Dahlstrom & Dahlstrom, 1980). Faschingbauer and Newmark (1978) produced a volume specializing in short forms of the MMPI; they recommend one in particular—the FAM, having only 166 items. This form takes only 35 minutes to complete. Among the scales of immediate interest, those for depression, psychasthenia, and social introversion all have 1-week test–retest reliabilities above .83 for 50 patients. For another short form, the MMPI-168, Overall and Eiland (1982) give norms for bright young college graduates.

An especially important problem with self-report devices is the potential for ethnic bias. Gynther, Lachar, and Dahlstrom (1978) reported that the MMPI standard F scale did not have the same implications for pathology among blacks that it does among whites. They devised a new scale that was expected to have the same implications for both populations. Smith and Graham (1981) show that neither the old scale nor the new scale is associated with pathology among black patients. Thus, caution is advised in use of MMPI scales with black patient samples. Similar issues for Native American subjects are addressed by Pollack and Shore (1980). Publication of a large project on this topic is due in 1982 (Dahlstrom, 1981).

A new tool for monitoring and evaluating treatment of depression has been described by Hudson and Proctor (1977). With 25 very simply worded items, the Generalized Contentment Scale (GCS) has reliability of alpha = .96 in two studies (Hudson, Hamada, Keech, & Harlan, 1980). Construct, convergent, and discriminant validities have been strong, notably in distinguishing groups of depressed and nondepressed patients, and in correlations with a 13-item version of the BDI (.85 and .76 in two studies). The GCS is part of measurement package to be described below.

A final comment in this section has been reserved for the Clinical Analysis Questionnaire (CAQ; Delhees & Cattell, 1971). This well-constructed test measures, among other things, seven different kinds of depression. However, it is not even mentioned in recent volumes on depression or behavioral assessment.

Miscellaneous. Measures of psychotic symptoms are probably less satisfactory than measures of other variables. Indeed, Last (1980) believes that mania, in particular, is extremely difficult to measure by self-report techniques, although the MMPI mania scale has some potential. Neale and Oltmanns (1981) praise the Present State Exam (PSE; Wing, Cooper, & Sartorius, 1974) as having high diagnostic reliability. The PSE has 500 questions for an interview, allowing the rating of 107 symptoms on the basis of patient self-report (another portion is based on clinician's direct observations). Neale and Oltmanns (1981) provide a valuable appraisal of data on the 12 symptoms that best discriminate schizophrenics from other groups. Conceivably, these could provide the basis for a new brief rating or self-rating.

Beech and Vaughn (1978) and Mavissakalian and Barlow (1981) both recommend the Maudsley Obsessional-Compulsive Inventory (MOC) (Hodgson & Rachman, 1977) for study of changes in obsessional symptoms.

Measures of mood states appear to be particularly appropriate for detection of symptom change provided that enough observations are made to establish baseline and postline. The Eight State Questionnaire (8SQ; Curran & Cattell, 1974) offers measures of anxiety, stress, depression, regression, fatigue, guilt, extraversion, and arousal, all as mood states. Two parallel forms are available, which is always a useful feature in repeated testing. The two forms correlate .93 on the average. An example of using both forms is given by Rohner and Miller (1980) in a demonstration of the effects of brief "sedative" music on high-anxious subjects.

Lorr and McNair (1980) have recently published a 72-item mood scale (POMS) in which items are adjectives or adjectival phrases; the respondent marks a range 0 to 3 for feelings ranging from much *unlike* this to much *like* this right now. This form scores all moods as bipolar: composed–anxious, agreeable–hostile, energetic–fatigued, elated–depressed, clear-thinking–confused (Lorr, McNair, & Fisher, 1982). This work culminates several years of research into the question of whether unipolarity of mood states is a function of response sets such as acquiescence. It is. Accordingly, corrections are introduced into format and analysis in order to reveal the underlying bipolarity (Lorr & Shea, 1979; Lorr & Wunderlich, 1980).

Hudson (1982; in press) reports on a Clinical Measurement Package (CMP). The package contains 9 scales, each consisting of 25 simply worded items and each item scored 0 to 4 on a frequency basis (rarely to most of the time). Thus, each scale is easily scored, with a range from 0 to 100. In addition to the GCS mentioned earlier, the CMP contains scales for self-esteem, marital satisfaction, sexual satisfaction, child's attitude toward mother, child's attitude toward father, parental attitudes, family relations stress, and peer relations. These scales were originally developed for use in repeated measurement of individuals, couples, and families in treatment. Reliabilities average .94; validities range from .52 to .92. Each scale has a clinical cutting score of 30, above which a significant problem in the given area may be assumed. Complete scales and a field manual are contained in the volume by Hudson (1982).

Personality Change

We saw above that variables of extraversion, self-actualization, self-control, and assertiveness have recently been found to show change in the course of psychotherapy. Also, we have noted that some researchers have rejected personality inventories as not being useful for evaluating effects of behavior therapy, specifically therapy of phobias. It is our conviction that inventories have been maligned for failing to serve a purpose for which they were not designed. In many other kinds of therapy programs, it is reasonable to expect that personality changes as well as symptom reduction will be found. Moreover, inventory construction has not stopped. Sharply improved technologies have already been used to create new inventories. Other new technologies (for example, facet theory) will no doubt yield further refinements in self-report inventory construction.

Among inventories already mentioned in this report, several have established their utility. These include the EPI (Eysenck & Eysenck, 1968), measuring extraversion and neuroticism; the EPQ (Eysenck & Eysenck, 1975), which measures psychoticism as well as extraversion and neuroticism; the POI (Knapp, 1976; Shostrom, 1974, 1979); and the 16 PF (Cattell, Eber, & Tatsuoka, 1970). The POI has recently been shown to reflect significant changes in every one of its 12 scales over a 6-month period of an educational marathon group (Watson, Janes, & Shostrom, 1981). The scales measure time-competence, inner-directedness, acceptance of aggression, and capacity for intimate contact as well as other factors directly related to self-actualization. The 16PF has many parallel and several special forms (as for low literacy), comes in many languages, has computer scoring available with profile printout and various predictive interpretations, and covers a wide range of variables (including ego strength, superego strength, guilt-proneness, suspiciousness, ergic tension, self-sufficiency, and self-sentiment strength). A recent study using an Israeli-language version of the 16PF (Zak, 1982) showed substantial changes in personality patterns of 18-year-old youth as a result of the Yom Kippur war. Wilderman (1981) recommends including the 16PF in a battery for cost-effective assessment of outcome in community mental health centers.

A widely used inventory deserving further exploration in psychotherapy research is the California Psychological Inventory (CPI) (Gough, 1957). The explicit purpose in constructing this inventory was to yield predictions of how individuals will behave in specific situations and how they will be described by others (Gough, 1977). Megargee (1977) describes the 18 scales of the CPI as falling into four groups: interpersonal adequacy (e.g., self acceptance or sociability, which should be pertinent to social skills training), socialization (e.g., responsibility or self-control, both of which should be pertinent to outcomes of self-regulation treatment), achievement, and interests. The scales have good reliability and validity (self-control, for instance, has internal consistency averaging .82 over several studies). Gynther (1978) provides a useful review of the CPI.

The Revised Adjective Check List (ACL; Gough & Heilbrun, 1980) offers measures of interest to psychotherapy researchers, especially those studying transactional analysis. The ACL measures 15 needs (e.g., affiliation, exhibition, aggression) and several other variables, including self-control, counseling readiness, critical parent, nurturing parent, adult, free child, and adapted child. Uses for diagnosis, comparison of self and ideal self, and studies of addictions are listed in "minibibliographies" (Gough, 1979, 1980) that will be found very useful. Use of the counseling readiness scale in predicting early dropouts from psychotherapy is reported by Cartwright, Lloyd, and Wicklund (1980).

The Personality Research Form (PRF; Jackson, 1974) was designed mainly for assessments of normal personality but can be used in clinics and guidance centers. A very carefully developed inventory, the PRF comes in several forms: two parallel shorter and two parallel longer. The short forms have 300 true–false items measuring 15 variables, the long forms have 440 items measuring 22 variables. The PRF was developed using Murray's (1938) theory of personality, and each trait to be measured was first given explicit theoretical definition. Items were selected to maximize homogeneity within scales, and response bias (such as social desirability) was suppressed as far as possible at the item level. Convergent and discriminant validation studies were carried out.

A recent example of research use of the PRF is given by Jackson and Chan (1979). Among the measured variables are aggression, affiliation, exhibition, dominance, and impulsivity. Many substantial correlations with CPI scales are reported. For example, PRF impulsivity correlates $-.53$ with CPI self-control; PRF dominance correlates .78 with CPI dominance. Very small changes in mean scores are found over a 1-week period (little spontaneous change), and test–retest reliabilities are high (approximately .81 on the average).

A new inventory with specially advanced technology of test construction is available in the Comrey Personality Scales (CPS; Comrey, 1970, 1980). The CPS gives scores on eight personality dimensions: trust, orderliness, conformity, activity, emotional stability, extraversion, masculinity, and empathy. Each variable is measured by 20 items organized into five sets. Each set consists of four highly correlated items, two positively and two negatively worded (to control for acquiescence response set). These four items are all highly loaded on a lower-order factor. Each set is called a Factor Homogeneous Item Dimension (FHID). As an example, the variable of trust (vs. defensiveness) is measured by the following five FHIDs: lack of cynicism, lack of paranoia, lack of defensiveness, belief in human worth, trust in human nature. Thus, each scale represents a higher-order factor. Factor structures have been confirmed over a large number of studies.

Subjects respond to each CPS item on a 7-point scale, either of frequency (never to always) or of certainty (definitely not to definitely). The complete test has 180 items, including 20 that check for random or biased responding. Scale internal consistency reliabilities average .93. Extensive work on validity has been carried out (cf. Comrey, 1970, 1981; Comrey, Soufi, & Backer, 1978; Comrey, Wong, & Backer, 1978; Montag & Comrey, 1982). The scales are found useful in detecting emotional disturbance and in psychiatric screening. The CPS has been found factorially stable across cultural and ethnic groups (Montag & Comrey, in press; Zamudio, Padilla, & Comrey, 1983). The recent *Handbook of Interpretations for the CPS* (Comrey, 1980) gives extensive theoretical discussion of the CPS scales in relation to psychoanalytic and Jungian theory as well as a clinical interpretation for each of 29 individual profiles.

An even more recently developed inventory is the Howarth Personality Questionnaire (HPQ; Howarth, 1973, 1980b). Following extensive study and empirical analyses of other major inventories (Howarth, 1980a), the HPQ was designed to measure 10 of the most replicable factors previously established, using just 12 simply worded true–false items per factor. Factor variables measured include persistence, conscience, and impulsiveness (all of which seem likely to be related to self-control). Preliminary norms are available (Howarth, 1980c). Reliabilities average .73.

Lorr and his associates (Lorr & More, 1980; Lorr, More, & Mansueto, 1981) show that assertiveness as measured by self-report is not a unitary variable. Rather, it has four distinct though correlated components: directiveness, social assertiveness, defense of rights and interests, and independence. The four factors are well replicated, and balanced 8-item scales are given in Lorr and More (1980, p. 132). Internal consistency reliabilities average .85; test–retest reliabilities average .82.

Rosenbaum (1980) reports on a schedule for assessing self-control behaviors (SCB). The scale has 36 items covering various aspects of control, including use of cognitions to control emotion and perceived ability to delay gratification and employ problem-solving strategies. Test–retest reliability over a 4-week interval is .86; internal consistency reliabilities average .80 in several samples (in English and Hebrew versions of the scale). Various estimates of validity are available, including low but significant correlations as predicted with the Rotter I-E scale and the 16PF scale of superego strength.

Cartwright and associates report a new technology for assessment of personality variables using self-reported vividness of imagery (SVI; Cartwright, 1980; Cartwright, Jenkins, Chavez, & Peckar, 1982; Jenkins & Cartwright, 1981). Items theoretically relevant to ego strength (adaptive, not defensive), superego strength, id derivatives, and identity are rated by subjects for the extent to which they evoke vivid images. The resulting scales have internal consistency reliabilities averaging .80. Construct validation studies and external validity studies (against the 16PF and other scales) suggest that the measures are valid (except for the measure of ego strength, which so far has not been checked against an adequate alternative measure of adaptive functions of the ego). Diagnostic and treatment implications of individual profiles based on these scales are suggested by Jenkins and Cartwright (1981).

Initial Status

Woody (1980) presents an encyclopedia of clinical assessment techniques, many of which are likely to be useful in assessing aspects of initial status, including assessments relevant to psychosomatic disorder and Types A and B behavior.

New procedures for identifying ideal types of personality and pathology are presented by Jackson, using the Jackson Personality Inventory (Jackson, 1978) and the Differential Personality Inventory (Jackson & Messick, 1971; Paquin & Jackson, 1981). Individuals are classified in one of seven major types by correlation between the individual's profile of test scores and the profile of the ideal type. On cross-validation, roughly 72% of subjects can be clearly classified in this way. The procedure should be useful in selecting patients for psychotherapy research studies.

Sarason, Johnson, and Siegel (1979) report on refinements in the assessment of life changes. They differentiate between positive and negative stress effects (i.e., subjects rate whether a given change event was, in fact, positive or negative for them). It is shown that only the negative events show significant correlations with measured anxiety.

Promising kinds of measures of initial status include some that might not be measures of personality. For example, the ACL Counseling Readiness scale predicts premature termination of therapy for males, and this is important to know for selecting a homogeneous study sample. Likewise, the SAS gives an indication of life supports that could be pertinent to treatment progress.

Treatment Structure

Strupp (1980a,b,c) reports on the close analysis of matched pairs of cases in time-limited dynamic therapy. Among the data are patient reports obtained at follow-up. These retrospective reports—using a questionnaire developed by Strupp, Fox, and Lessler (1969)—include descriptions of the therapist's manner, tone, language use, activity, topical emphases, and so on; hence they are pertinent to assessment of treatment structure.

Helping Alliances

Luborsky's "Manual for Scoring Signs of Helping Relationships" (Luborsky, 1976, Appendix A) should be consulted as a very promising form of a self-report technique. A new publication on the method will appear shortly (Luborsky, in press).

A new version of the Relationship Inventory, using a Likert-scale format, has been developed by Strupp (1981) and will be used in research in the Vanderbilt Project.

Scales to assess supportive and working alliances by direct self-report are still needed. The methods developed by Orlinsky and Howard (1975) deserve wider application.

Self-monitoring

Lewinsohn and Chaplin (1979) present a manual for the Unpleasant Events Schedule. Miller (1981, p. 281) describes forms for self-report of drinking activities, including quantities, place, time, and so on. Wincze and Lange (1981, pp. 309–310) describe questionnaires for assessing sexual behavior.

An excellent review of self-monitoring procedures is given by Ciminero, Nelson, and Lipinski (1977). The reader is referred to that review for coverage of methods for self-recording of target behaviors, antecedents and consequents, self-monitoring devices (including mechanical counters, etc.), reactive effects of self-monitoring, and the important distinction between self-monitoring for assessment (requiring maximum accuracy) and self-monitoring for therapeutic purposes (requiring maximum reactivity).

CURRENT STATE OF THE FIELD

Psychotherapy research has had a few additions to its bibliography, a couple of movements forward, and one or two failures to realize expectations. Otherwise, things are much the same now as they were in 1978.

The general question of effectiveness lies where it was 30 years ago: unanswered, since there is no acceptable evidence. Then the demand was for change exceeding rates and quantities associated with spontaneous remission (Eysenck, 1952). Now it is for improvements exceeding those obtainable with a placebo (Eysenck, 1978). But little evidence is in hand. How can one create a placebo for psychotherapy? Perhaps the minimal treatment group in the Vanderbilt Project comes closest.

As to the special effectiveness of particular treatments, little in the way of positive information has been gained. The disassembly studies of Rehm (1981) are promising as methods but yielded the same absence of difference among treatments that has become standard. Special effectiveness will have to be shown for each kind of treatment. Garfield (1978, p. 191) pointed out that over 100 different kinds of psychotherapy are being practiced. The number has probably become larger since then. Scientific evaluation will have much to do.

The last decade has seen the fruition and dissolution of a great idea in psychotherapy research. Waskow and Parloff (1975) reported results of a national committee's deliberations on how to maximize comparability among results obtained by different researchers across the country. A "core battery" was recommended, including the MMPI, the TC, and others. All were aimed at assessing symptom reduction, since it was expected that such would be acceptable across disciplines and theories. To date, only one project (Vanderbilt) has used the core battery. Earlier, the notion was subject to discussion. Garfield (1978, p. 225) appeared to endorse it. Gottman and Markman (1978, p. 43) thought it was a step backward; the market seems to have agreed with them.

Yet some instruments are emerging as standard across limited domains. For example, the BDI is typically used in studies of cognitive and other treatments for depression. Some version of the FSS is becoming standard for phobia research.

Relatively little programmed research on psychotherapy is currently being done or even planned outside of a few strongholds. Federal funds seem to be diminishing.

Yet one senses an unrest among clinicians and researchers. It shows in efforts to provide better conceptual frameworks (cf. Kanfer, 1979), better means of analysis (cf. Agras & Jacob, 1981), and better measurement devices. It suggests determination "to really get to the bottom of the problem" by searching anew for measurable variables of relationships in treatment (cf. Luborsky, 1976), by disassembly studies (Rehm, 1981), and by studying the interaction of selected patient–therapist pairs (Strupp, 1980a,b,c). One has a sense of building drama and expectancy despite the currently prevailing silence.

CURRENT ISSUES

The question of effectiveness is still an issue. Despite the meta-analyses of Smith and Glass (1977), Eysenck (1978) and others still maintain that there is no acceptable evidence for effectiveness, largely because the individual studies included for meta-analysis had inadequate controls. Moreover, Eysenck also points to the fact reported by Smith and Glass that the largest effect sizes were obtained from self-report data, presumably the most subjective.

Perspectives

Bergin and Lambert (1978) labeled the problem of "perspectives": the fact that, in general, measures of improvement taken from the patient, the therapist, and clinician observers usually have, at best, low correlations. It is often recommended that researchers include measures from several perspectives. One might conclude that the patient's perspective is one that would be hard to exclude from consideration and difficult to assess without using self-report techniques. Given the problem of perspectives, it seems unwise to urge that validation must necessarily be sought from agreements between the patient and others. Each perspective has its own vantage point and meanings even when ostensibly the "same" variables and scales are being used (cf. Cartwright, Kirtner, & Fiske, 1963). Validity, then, must be sought in other ways.

Utility of Self-report

We are surely between a rock and a hard place. If self-report data cannot be trusted yet the patient perspective must be included, some sleight of hand is needed to resolve the impasse. What are the views of technical specialists?

Ericsson and Simon (1980), speaking chiefly on research in cognitive psychology, say that self-report can be trusted when the subject is given careful instructions and is reporting on immediately occurring cognitive events. Much less confidence can be placed in reports that call upon long-term memory and require inference and interpretation. They propose that the process of self-report is itself in need of theoretical understanding. Fiske (1980) states that the essential problem with self-reports is doubtful veridicality. Where the subject reports on external events, other observers can be used to check on veridicality; when the subject is reporting inner experience, no such check is available. Likewise, when the subject is averaging over instances in response to a questionnaire item (e.g., "Are you usually able to assert your rights?"), long-term memory must be accessed and summarized. Interpretations of items and approaches to responding vary over subjects and add to the lack of veridicality.

The whole issue of faking, response sets, and other sources of invalidity in questionnaires and self-ratings is one that has a long history and a long list of attempts at correction. The writing of "subtle" versus "obvious" items has been expected to increase validity by making it more difficult for subjects to fake. But Holden and Jackson (1979) showed that less subtle items had greater empirical validity; Holden and Jackson (1981) found that even when subjects are motivated toward distortion, subtle scales do not have superior validity.

What then is the relation between veridicality and validity and trustworthiness? A veridical answer is true if it correctly represents its referent—an inner experience or average of remembered behaviors or whatever. A valid item is one that correlates as theoretically required (positive, negative, or zero) with specified external variables, typically some criterion measure, or an "accepted" test. Obviously these are quite different. An item may be answered untruthfully by all subjects and still correlate significantly with a criterion. Which, then, shall be the basis of our trust?

In our judgment, validity should be the criterion. Veridicality is not quite irrelevant. The same conclusion is reached by Fiske (1981) as far as "applied studies" (such as guidance, counseling, or assessing initial status) are concerned.

There is work to do. Reliability values set upper limits to validity. Sources of invalidity must be controlled or measured. But the recent suspicions of self-report data as a generic type are not well based. They probably refer to possible lack of veridicality or the inability to check up on it. If we stick to validity as our touchstone, and remember that good measurement of any kind requires care and precision, then we have every reason for confidence in many modern devices for obtaining self-report data. Moreover, we might do well to remember that suitable treatment of such data provides the foundation of scientific psychology. Fechner's law, for example, states a relation between a magnitude of subjective sensation (S) and a logarithmic function of an amount of stimulation (R) measured in units of absolute threshold (RL) where even this unit is determined by self-report techniques:

$$S = k \log R$$

As Boring says (1957, pp. 275–296), this expression represents a fundamental law of psychophysics.

Relevance of Measured Variables

Agras and Jacob (1981) showed that the EPI and the Rotter scale of internal-external control (I-E; Rotter, 1966), among other "personality" scales, were less "sensitive" than self-report phobia scales. But why would anyone expect a general measure of neuroticism (EPI) to detect changes in specific phobias? The EPI has not a single item that asks "Are you afraid of X"? Likewise, the I-E scale is based on a theory of expectancies, with I-E tapping the highest order of generalization of expectancy that one's own behavior will result in desired reinforcements. Its items are drawn from earlier scales on alienation, powerlessness, meaninglessness, and so on. Any reasoning that links the I-E to treatment of phobia (or any other treatment, for that matter) must be tenuous at best. Measures have got to be measuring variables relevant to the patients and the treatment goals.

On a related matter, Agras and Jacob (1981) found the patient's self-report of main phobia (PSRMP) to be superior to the Fear Survey Schedule (FSS) in sensitivity for between-group comparisons (BG) but inferior in within-group design (WG). Ordinarily, a single-item device like PSRMP would have to have lower reliability than a full scale like FSS; hence, have greater proportion of error variance; hence, have less ability to detect any differences at all. The trouble is that PSRMP is not even a single-item device. The actual item is presumably different for each patient. Thus, all patients have their own tests. Despite the fact that all fears are rated on a 0 to 4 scale, these measures are not comparable, so that you can add them up and take a mean, compare groups, and so on. It would be easy to argue that the PSRMP is not a measuring device at all. Its status is moot.

What Is the Role of Self-monitoring?

Ordinarily, no claim is made that self-monitoring instruments are reliable, valid, or have other psychometric desiderata. They are used to help homework and provide treatment guides to the therapist. Yet, they also have therapeutic effects, apparently. Such effects are well documented (e.g., McFall, 1970). Kalish (1981, pp. 299–301) suggests that the information from self-monitoring serves as a reward for behavior changes.

Glazer, Clarkin, and Hunt (1981, p. 12) point out that self-monitoring tools yield three kinds of information: baseline data on how the patient was before treatment, data input into the process of treatment interventions, and progress data by which effectiveness may be estimated. There appear to be two kinds of materials in these self-monitoring logs, one describing symptom states and one reporting on the results of assigned activities. For example, in the Daily Record of Dysfunctional Thoughts (DRDT; Beck *et al.*, 1979, p. 403), the patient describes each relevant event or train of thoughts and also states what the emotion was (e.g., sadness, anger) and gives it a rating from 1 to 100 for intensity. By contrast, the same authors also give a form entitled "Possible Reasons for Not Doing Self-help Assignments" in which one item, for example, indicates that the patient is too busy to do them. This second form clearly refers to the patient's contribution to the means of treatment (or lack of contribution).

Burgeoning interest in self-monitoring and its effects brings to light an important issue in all psychotherapy research: What does the patient do? What does the patient's work contribute to the treatment process?

Initial Status

Selection and prediction uses of initial status work at cross purposes. To select, you reduce heterogeneity of patients and hence reduce variance. To predict, you need to maximize variance in the predictor (and criterion) measures.

Even within a selection strategy there are problematic issues. The choice of a syndrome (cf. Strupp & Bloxom, 1975) does not mean that all patients will be homogeneous on all possible important variables. Not all persons with scores above 60 or 70, or whatever, on the MMPI-D are going to be equally motivated for treatment or equally free of psychoticism or other relevant matters. Just how wide should we spread the net in selection?

Future Directions

Effectiveness Research

In addition to placebo studies it seems likely that dissasembly studies will become increasingly desirable as a means to understanding questions of specific effectiveness. The stronger such studies become, the more reliance can then be placed on studies using principles of meta-analysis. For in the end, the general question of effectiveness can be answered only in terms of collected studies of specific effects.

A formulation in terms of independent (initial status, structure) and mediating (alliance, self-monitoring) variables further suggests that attention should be paid to the cumulations and interactions of these several sources of variance in dependent variables (symptom reduction, personality change). This formulation also suggests that suitable techniques of analysis should be employed, notably structural equation models (Bentler, 1980; Bentler & Bonett, 1980). In many cases there will also, however, be interest in individual differences. Suitable techniques include multidimensional individual differences scaling (Carroll & Chang, 1970). Facet theory designs and smallest-space analysis (Levy & Guttman, 1982) should also be profitable in psychotherapy research. Effectiveness research will benefit from improved self-report criterion measures. These should include self-monitoring devices for a wide range of purposes, client–therapist relationship measures articulating the full "nonspecific" therapeutic context, and measures of initial motivation, defensiveness, and so on. One may hope for discussions and research on the interrelationships between such qualitatively different measures, both simultaneously and in the evolution of the therapy.

Research and Development

Psychometric research proceeds apace. Increasingly, attention should be paid to the results of investigators like Goldberg (1981), who shows that the format of self-report devices substantially affects results. He also urges that dimensions of frequency and intensity be distinguished. What does *severity* mean, for example? Does it mean more frequent or more intense?

There is obviously a need for psychometric research devoted solely to the interest of clinical assessment in psychotherapy research. For example, instead of having every investigator marshal his or her own placebo control groups, one central institute or some other collective effort should provide normative data on the effects of placebo treatments on selected measures. For example, the BDI or the phobia questionnaire developed by Marks and Mathews (1979) should be studied on samples initially high and initially medium high on these measures. Suitable lengths of time should be chosen—for example, 6, 12, 26, and 52 weeks. Normative data on changes by each kind of sample, subdivided for each length of time, should then be made available to researchers. Their actual changes shown in treatment could then be compared against the standards. Implementation of such a program could be quite simple and would not require gigantic financial support.

CONCLUSIONS

This review of recent research and current issues suggests that substantial advances have been made in the technology of self-report ratings and inventories. However, more widespread adoption of the new technology is needed, and continued research and development are imperative if psychotherapy is to fulfill its promise.

ACKNOWLEDGMENTS

We wish to express our deep appreciation to numerous colleagues who kindly sent reprints and references, which greatly facilitated our review.

REFERENCES

Agras, W. S., & Jacob, R. Phobia: Nature and measurement. In M. Mavissakalian & D. H. Barlow (Eds.), *Phobia: Psychological and pharmacological treatment*. New York: Guilford, 1981.

Allen, G. J. The behavioral treatment of test anxiety: Therapeutic innovations and emerging conceptual challenges. In M. Hersen, R. M. Eisler, & P. M. Miller (Eds.), *Progress in behavior modification* (Vol. 9). New York: Academic Press, 1980.

Andrasik, F., Turner, S. M., & Ollendick, T. H. Self-report and physiologic responding during *in vivo* flooding. *Behaviour Research and Therapy*, 1980, *18*, 593–594.

Barrett-Lennard, G. T. Dimensions of therapy response as causal factors in therapeutic change. *Psychological Monographs*, 1962, *76* (43, Whole No. 562).

Battle, C. C., Imber, S. D., Hoehn-Saric, R., Stone, A. R., Nash, E. R., & Frank, J. D. Target complaints as criteria of improvement. *American Journal of Psychotherapy*, 1966, *20*, 184–192.

Beck, A. T. *Depression inventory*. Philadelphia: Center for Cognitive Therapy, 1978. Also in A. T. Beck, A. J. Rush, B. F. Shaw, & G. Emery, *Cognitive therapy of depression*. New York: Guilford, 1979.

Beck, A. T., Rush, A. J., Shaw, B. F., & Emery, G. *Cognitive therapy of depression*. New York: Guilford, 1979.

Beech, H. R., & Vaughn, M. *Behavioral treatment of obsessional states*. New York: Wiley, 1978.

Bellack, A. S., & Hersen, M. *Introduction to clinical psychology*. New York: Oxford University Press, 1980.

Bentler, P. M. Multivariate analysis with latent variables: Causal modelling. *Annual Review of Psychology*, 1980, *31*, 419–456.

Bentler, P. M., & Bonett, D. G. Significance tests and goodness of fit in the analysis of covariance structures. *Psychological Bulletin*, 1980, *88*, 588–606.

Bergin, A. E., & Lambert, M. J. The evaluation of therapeutic outcomes. In S. L. Garfield, & A. E. Bergin (Eds.), *Handbook of psychotherapy and behavior change: An empirical analysis* (2nd ed.). New York: Wiley, 1978.

Boring, E. G. *A history of experimental psychology*. New York: Appleton Century Crofts, 1957.

Butcher, J. N. (Ed.). *New developments in the use of the MMPI*. Minneapolis: University of Minnesota Press, 1979.

Carroll, J. D., & Chang, J. J. Analysis of individual differences in multidimensional scaling via an N-way generalization of "Eckart-Young" decomposition. *Psychometrika*, 1979, *35*, 283–319.

Cartwright, D. Exploratory analyses of verbally stimulated imagery of the self. *Journal of Mental Imagery*, 1980, *4*, 1–21.

Cartwright, D., Kirtner, W. L., & Fiske, D. W. Method factors in changes associated with psychotherapy. *Journal of Abnormal and Social Psychology*, 1963, *66*, 164–175.

Cartwright, D., Jenkins, J. L., Chavez, R., & Peckar, H. *Studies in imagery and identity*. Unpublished manuscript. University of Colorado, 1980.

Cartwright, R., Lloyd, S., & Wicklund, J. Identifying early dropouts from psychotherapy. *Psychotherapy: Theory, Research and Practice*, 1980, *17*, 263–267.

Cattell, R. B., Eber, H. W., & Tatsuoka, M. *Handbook for the 16PF*. Champaign, Ill.: Institute for Personality and Ability Testing, 1970.

Ciminero, A. R., Nelson, R. O., & Lipinski, D. P. Self-monitoring procedures. In A. R. Ciminero, K. S. Calhoun, & H. E. Adams (Eds.), *Handbook of behavioral assessment*. New York: Wiley, 1977.

Clarkin, J. F., & Glazer, H. I. (Eds.), *Depression: Behavioral and directive intervention strategies*. New York: Garland STPM Press, 1981.

Comrey, A. L. *EITS manual for the Comrey Personality Scales*. San Diego, Calif.: Educational and Industrial Testing Service, 1970.

Comrey, A. L. *Handbook of interpretations for the Comrey personality scales*. San Diego, Calif.: Educational and Industrial Testing Service, 1980.

Comrey, A. L. Detecting emotional disturbance with the Comrey personality scales. *Psychological Reports*, 1981, *48*, 702–710.

Comrey, A. L., Soufi, A., & Backer, T. E. Psychiatric screening with the Comrey personality scales. *Psychological Reports*, 1978, *42*, 1127–1130.

Comrey, A. L., Wong, C., & Backer, T. E. Further validation of the social conformity scale of the Comrey personality scales. *Psychological Reports*, 1978, *43*, 165–166.

Curran, J. P., & Cattell, R. B. *The eight state questionnaire battery*. San Diego, Calif.: Educational and Industrial Testing Service, 1974.

Dahlstrom, W. G. Personal communication, 1981.

Dahlstrom, W. G., & Dahlstrom, L. (Eds.), *Basic readings on the MMPI: A new selection of personality measurement*, Minneapolis: University of Minnesota Press, 1980.

Delhees, K. H., & Cattell, R. B. *Manual for the clinical analysis questionnaire (CAQ)*. Champaign, Ill.: Institute for Personality and Ability Testing, 1971.

Dosamentes-Alperson, A. E., Smokler, I., & Merrill, N. Growth effects of experiential movement psychotherapy. *Psychotherapy: Theory, Research and Practice*, 1980, *17*, 63–88.

Epstein, N. B., & Vlok, L. A. Research on the results of psychotherapy: A summary of evidence. *American Journal of Psychiatry*, 1981, *138*, 1027–1032.

Ericsson, K. A., & Simon, H. H. Verbal reports as data. *Psychological Review*, 1980, *87*, 215–251.

Eysenck, H. J. The effects of psychotherapy: An evaluation. *Journal of Consulting Psychology*, 1952, *16*, 319–324.

Eysenck, H. J. An exercise in mega-silliness. *American Psychologist*, 1978, *33*, 517.

Eysenck, H. J., & Eysenck, S. B. G. *Manual for the Eysenck personality inventory*. San Diego, Calif.: Educational and Industrial Testing Service, 1968.

Eysenck, H. J., & Eysenck, S. B. G. *Manual for the Eysenck personality questionnaire*. San Diego., Calif.: Educational and Industrial Testing Service, 1975.

Faschingbauer, T. R., & Newmark, C. S. *Short forms of the MMPI.* Lexington, Mass.: Heath, 1978.

Fiske, D. W. When are verbal reports veridical? In R. Schweder (Ed.), *New directions for methodology of social and behavioral science: Fallible judgments in behavioral research* (No. 4). San Francisco: Jossey-Bass, 1980.

Fiske, D. W. *What can we do with personality questionnaires?* Unpublished manuscript. Chicago: University of Chicago, 1981.

Foa, E. B., & Chambless, D. L. Habituation of subjective anxiety during flooding in imagery. *Behaviour Research and Therapy,* 1978, *16,* 391–399.

Garfield, S. L. Research on client variables in psychotherapy. In S. L. Garfield & A. E. Bergin (Eds.), *Handbook of psychotherapy and behavior change* (2nd ed.). New York: Wiley, 1978.

Garfield, S. L., & Bergin, A. E. (Eds.) *Handbook of psychotherapy and behavior change: An empirical analysis* (2nd ed.). New York: Wiley, 1978.

Geer, J. H. The development of a scale to measure fear. *Behaviour Research and Therapy,* 1965, *3,* 45–53.

Glazer, H. I., Clarkin, J. F., & Hunt, H. F. Assessment of depression. In J. F. Clarkin & H. I. Glazer (Eds.), *Depression: Behavioral and directive intervention strategies.* New York: Garland STPM Press, 1981.

Goldberg, L. R. Language and personality: Developing a taxonomy of trait-descriptive terms. In D. W. Fiske (Ed.), *New directions for methodology of social and behavioral science* (No. 9). *Words: What you choose them to mean.* San Francisco: Jossey-Bass, 1981.

Gomes-Schwartz, B. Effective ingredients in psychotherapy: Prediction of outcome from process variables. *Journal of Consulting and Clinical Psychology,* 1978, *46,* 1023–1035.

Gomes-Schwartz, B., & Schwartz, J. M. Psychotherapy process variables distinguishing the "inherently helpful" person from the professional psychotherapist. *Journal of Consulting and Clinical Psychology,* 1978, *46,* 196–197.

Gottman, J., & Markman, H. J. Experimental designs in psychotherapy research. In S. L. Garfield & A. E. Bergin, *Handbook of psychotherapy and behavior change: An empirical analysis.* New York: Wiley, 1978.

Gough, H. G. *Manual for the California psychological inventory.* Palo Alto, Calif.: Consulting Psychologists Press, 1957.

Gough, H. G. Foreword to Megargee, E. I., *The California psychological inventory handbook.* San Francisco: Jossey-Bass, 1977.

Gough, H. G. *Selected and classified list of ACL references.* Palo Alto, Calif.: Consulting Psychologists Press, 1979.

Gough, H. G. *ACL minibibliography of studies on alcoholism and drug abuse.* Palo Alto, Calif.: Consulting Psychologists Press, 1980.

Gough, H. G., & Heilbrun, A. B. *Manual for the revised adjective check list.* Palo Alto, Calif.: Consulting Psychologists Press, 1980.

Gynther, M. D. Review of the California psychological inventory. In O. K. Buros (Ed.), *The eighth mental measurements yearbook.* Highland Park, N.J.: Gryphon Press, 1978.

Gynther, M. D., Lachar, D., & Dahlstrom, W. G. Are special norms for minorities needed? Development of an MMPI *F*-scale for blacks. *Journal of Consulting and Clinical Psychology,* 1978, *46,* 1403–1408.

Hanfmann, E. *Effective therapy for college students.* San Francisco: Jossey-Bass, 1978.

Hodgson, R. J., & Rachman, S. Obsessional-compulsive complaints. *Behaviour Research and Therapy,* 1977, *15,* 389–396.

Holden, R. R., & Jackson, D. N. Item subtlety and face validity in personality assessment. *Journal of Consulting and Clinical Psychology,* 1979, *47,* 459–468.

Holden, R. R., & Jackson, D. N. Subtlety, information and faking effects in personality assessment. *Journal of Clinical Psychology,* 1981, *37,* 380–386.

Howarth, E. *Howarth personality questionnaire.* Edmonton, Canada: University of Alberta, 1973.

Howarth, E. Major factors of personality. *Journal of Psychology,* 1980, *104,* 171–183. (a)

Howarth, E. Interrelations between state and trait: Some new evidence. *Perceptual and Motor Skills,* 1980, *51,* 613–614. (b)

Howarth, E. *Scale descriptions, norm tables and bibliography.* Edmonton, Canada: University of Alberta, 1980. (c)

Hudson, W. W. *The clinical measurement package: A field manual.* Homewood, Ill.: Dorsey Press, 1982.

Hudson, W. W. A measurement package for clinical workers. *Journal of Applied Behavioral Science,* in press.

Hudson, W. W., & Proctor, E. K. Assessment of clinical affect in clinical practice. *Journal of Consulting Psychology,* 1977, *45,* 1206–1207.

Hudson, W. W., Hamada, R. S., Keech, R., & Harlan, J. *A comparison and revalidation of three measures of depression.* Unpublished manuscript. Tallahassee, Fla.: School of Social Work, Florida State University, 1980.

Jackson, D. N. *Personality research form manual* (2nd ed.). Port Huron, Mich.: Research Psychologists Press, 1974.

Jackson, D. N. Interpreter's guide to the Jackson personality inventory. In P. McReynolds (Ed.), *Advances in psychological assessment* (Vol. 4). San Francisco: Jossey-Bass, 1978.

Jackson, D. N., & Chan, D. W. Implicit personality theory: Is it illusory? *Journal of Personality,* 1979, *47,* 1–10.

Jackson, D. N., & Messick, S. *Differential personality inventory.* London, Ontario, Canada: Authors, 1971.

Jenkins, J. L., & Cartwright, D. A mental imagery approach to id, ego, and superego evaluation: Diagnostic and treatment implications. *Academic Psychology Bulletin,* 1981, *3,* 47–60.

Kalish, H. I. *From behavioral science to behavior modification.* New York: McGraw-Hill, 1981.

Kanfer, F. H. Self-management: Strategies and tactics. In A. P. Goldstein & F. H. Kanfer (Eds.), *Maximizing treatment gains: Transfer enhancement in psychotherapy.* New York: Academic Press, 1979.

Karle, W., Corriere, R., Hart, J., & Klein, J. The personal orientation inventory and the Eysenck personality inventory as outcome measures in a private outpatient clinic. *Psychotherapy: Theory, Research and Practice,* 1981, *18,* 117–122.

Knapp, R. R. *Handbook for the personal orientation inventory.* San Diego, Calif.: Educational and Industrial Testing Service, 1976.

Knapp, R. R., Shostrom, E. L., & Knapp, L. Assessment of the actualizing person. In P. McReynolds (Ed.), *Advances in psychological assessment* (Vol. 4). San Francisco: Jossey-Bass, 1978.

Krug, S. E., Scheier, I. H., & Cattell, R. B. *IPAT anxiety scale.* Champaign, Ill.: Institute for Personality and Ability Testing, 1979.

LaPointe, K., & Rimm, D. C. Cognitive, assertive and insight-oriented group therapies in the treatment of reactive depression in women. *Psychotherapy: Theory, Research and Practice,* 1980, *17,* 312–321.

Last, U. Psychodiagnostic assessment and psychological function in mania. In R. H. Belmaker & H. M. Van Praag (Eds.), *Mania: An evolving concept.* New York: Spectrum Publications, 1980.

Levy, S., & Guttman, L . Worry, fear and concern differentiated. In C. D. Spielberger, I. G. Sarason, & N. A. Milgram (Eds.), *Stress and anxiety* (Vol. 8). New York: McGraw-Hill, 1982.

Lewinsohn, P. M., & Amenson, C. Some relations between pleasant and unpleasant mood related activities and depression. *Journal of Abnormal and Social Psychology,* 1978, *87,* 644–654.

Lewinsohn, P. M., & Arconad M. Behavioral treatment of depression: A social learning approach. In J. R. Clarkin & H. I. Glazer (Eds.), *Depression: Behavioral and directive intervention strategies.* New York: Garland STPM Press, 1981.

Lewinsohn, P. M., & Chaplin, W. *Manual for the unpleasant events schedule.* Department of Psychology, University of Oregon, 1979. (Mimeographed)

Lorr, M., & McNair, D. *POMS-BI.* San Diego, Calif.: Educational and Industrial Testing Service, 1980.

Lorr, M., McNair, D. M., & Fisher, S. Evidence for bipolar mood states. *Journal of Personality Assessment,* 1982, *46,* 432–436.

Lorr, M., & More, W. W. Four dimensions of assertiveness. *Multivariate Behavioral Research,* 1980, *15,* 127–133.

Lorr, M., & Shea, T. M. Are mood states bipolar? *Journal of Personality Assessment,* 1979, *5,* 468–472.

Lorr, M., & Wunderlich, R. A. Mood states and acquiescence. *Psychological Reports,* 1980, *46,* 191–195.

Lorr, M., More, W. W., & Mansueto, C. S. The structure of assertiveness: A confirmation study. *Behaviour Research and Therapy,* 1981, *19,* 153–156.

Lubin, B. Adjective checklists for the measurement of depression. *Archives of General Psychiatry,* 1965, *12,* 57–62.

Luborsky, L. Helping alliances in psychotherapy. In J. L. Claghorn (Ed.), *Successful psychotherapy.* New York: Brunner/Mazel, 1976.

Luborsky, L. Personal communiction, 1981.

Luborsky, L. The helping alliance rating method. *Archives of General Psychiatry,* in press.

Luborsky, L., Todd, T. C., & Katcher, A. H. A self-administered social assessment scale for predicting physical and psychological illness and health. *Journal of Psychosomatic Research,* 1973, *17,* 109–120.

Marks, I. M., & Mathews, A. M. Brief standard self-rating for phobic patients. *Behaviour Research and Therapy,* 1979, *17,* 263–267.

Marzillier, J. S., & Winter, K. Success and failure in social skills training: Individual differences. *Behaviour Research and Therapy,* 1978, *16,* 67–84.

Marzillier, J. S., Carroll, D., & Newland, J. R. Self-report and physiological changes accompanying repeated imagining of a phobic scene. *Behaviour Research and Therapy,* 1979, *17,* 71–77.

Mavissakalian, M. R., & Barlow, D. H. Assessment of obsessive-compulsive disorders. In D. H. Barlow (Ed.), *Behavioral assessment of adult disorders,* New York: Guilford; 1981.

McDaniel, S., Stiles, W., & McGaughey, K. Correlates of male college students' verbal response mode use in psychotherapy with measures of psychological disturbance and psychotherapy outcomes. *Journal of Consulting and Clinical Psychology,* 1981, *49,* 571–582.

McFall, R. M. Effects of self-monitoring on normal smoking behavior. *Journal of Consulting and Clinical Psychology,* 1970, *35,* 135–142.

McLean, P. Remediation of skills and performance deficits in depression: Clinical steps and research findings. In J. F. Clarkin & H. I. Glazer (Eds.), *Depression: Behavioral and directive intervention strategies.* New York: Garland STPM Press, 1981.

McLellan, A. T., Luborsky, L.,Woody, G. E., & O'Brien, C. P. An improved diagnostic evaluation instrument for substance abuse patients: The addition severity index. *Journal of Nervous and Mental Disease,* 1980, *168,* 26–33.

Megargee, E. I. *The California psychological inventory handbook.* San Francisco: Jossey-Bass, 1977.

Miller, P. M. Assessment of alcohol abuse. In D. H. Barlow (Ed.), *Behavioral assessment of adult disorders.* New York: Guilford, 1981.

Montag, I., & Comrey, A. C. Comparison of certain MMPI, Eysenck and Comrey personality constructs. *Multivariate Behavioral Research,* 1982, *17,* 93–98.

Montag, I., & Comrey, A. C. Personality construct similarity in Israel and the United States. *Applied Psychological Measurement,* in press.

Murray, H. A. *Explorations in personality.* New York: Oxford, 1938.

Neale, J. M., & Oltmanns, T. F. Assessment of schizophrenia. In D. H. Barlow (Ed.), *Behavioral assessment of adult disorders.* New York: Guilford, 1981.

Orlinsky, D. E., & Howard, K. I. *Varieties of psychotherapeutic experience.* New York: Teachers College Press, 1975.

Overall, J. E., & Eiland, D. C. MMPI-168 norms and profile sheets for bright young college graduates. *Journal of Clinical Psychology,* in press.

Paquin, M. J., & Jackson, D. N. The dimensionality of psychopathology descriptors. *Multivariate Behavioral Research,* 1981, *16,* 291–308.

Pollack, D., & Shore, J. H. Validity of the MMPI with Native Americans. *American Journal of Psychiatry,* 1980, *137,* 946–980.

Rachman, M. A., & Eysenck, S. B. G. Psychoticism and response to treatment in neurotic patients. *Behaviour Research and Therapy,* 1978, *16,* 183–189.

Rathus, S. A. A 39-item schedule for assessing assertive behavior. *Behavior Therapy,* 1973, *4,* 398–406.

Rehm, L. P. A self-control therapy program for treatment of depression. In J. F. Clarkin & H. I. Glazer (Eds.), *Depression: Behavioral and directive intervention strategies.* New York: Garland STPM Press, 1981.

Rohner, S. J., & Miller, R. Degrees of familiar and affective music and their effects on state anxiety. *Journal of Music Therapy,* 1980, *17,* 2–15.

Rosenbaum, M. A schedule for assessing self-control behaviors: Preliminary findings. *Behavior Therapy,* 1980, *11,* 109–121.

Rotter, J. B. Generalized expectancies for internal versus external control of reinforcement. *Psychological Monographs,* 1966, *80* (1, Whole no. 609).

Sarason, I. G., Johnson, J. H., & Siegel, J. M. Assessing the impact of life changes: Development of the life experiences survey. In I. G. Sarason & C. D. Spielberger, *Stress and anxiety* (Vol. 6). New York: Wiley, 1979.

Shostrom, E. L. *Manual for the personal orientation inventory.* San Diego, Calif.: Educational and Industrial Testing Service, 1974.

Shostrom, E. L. *Bibliography for the POI.* San Diego, California: Educational and Industrial Testing Service, 1979.

Smith, C. P., & Graham, J. R. Behavioral correlates for the MMPI standard *F*-scale and for a modified *F*-scale for black and white psychiatric patients. *Journal of Consulting and Clinical Psychology,* 1981, *49,* 455–459.

Smith, M. L., & Glass, G. V. Meta-analysis of psychotherapy outcome studies. *American Psychologist,* 1977, *32,* 752–760.

Sookman, D., & Solyom, L. Severe depersonalization treated by behavior therapy. *American Journal of Psychiatry,* 1978, *135* 1543–1545.

Strupp, H. H. Success and failure in time-limited psychotherapy: A systematic comparison of two cases: Comparison 1. *Archives of General Psychiatry,* 1980, *37,* 595–603. (a)

Strupp, H. H. Success and failure of time-limited psychotherapy: A systematic comparison of two cases: Comparison 2. *Archives of General Psychiatry,* 1980, *37,* 708–716. (b)

Strupp, H. H. Success and failure in time-limited psychotherapy: With special reference to the performance of a lay counselor. *Archives of General Psychiatry,* 1980, *37,* 831–848. (c)

Strupp, H. H. Personal communication, 1981.

Strupp, H. H., & Bloxom, A. L. An approach to the problem of defining a patient population in psychotherapy research. *Journal of Counseling Psychology,* 1975, *22,* 231–237.

Strupp, H. H., & Hadley, S. W. Specific vs. nonspecific factors in psychotherapy. *Archives of General Psychiatry,* 1979, *36,* 1125–1136.

Strupp, H. H., Fox, R. E., & Lessler, K. *Patients view their psychotherapy.* Baltimore: John Hopkins Press, 1969.

Suinn, R. M. Changes in non-treated subjects over time: Data on a fear survey schedule and test anxiety scale. *Behaviour Research and Therapy,* 1969, *7,* 205–206.

Tasto, D. L. Self-report schedules and inventories. In A. R. Ciminero, K. S. Calhoun, & H. E. Adams (Eds.), *Handbook of behavioral assessment.* New York: Wiley, 1977.

Waskow, I. F., & Parloff, M. E. (Eds.), *Psychotherapy change measures.* Rockville, Md.: National Institute of Mental Health, 1975.

Waterhouse, G. J. *Perceptions of facilitative therapeutic conditions as predictors of outcome in brief therapy.* Unpublished master's thesis, Department of Psychology, Vanderbilt University, Nashville, Tenn., 1978.

Watson, G. W., Janes, T. L., & Shostrom, E. L. *The effect of an educational marathon group as measured by the POI.* Unpublished manuscript. Fullerton, Calif.: Department of Psychology, California State University, 1981.

Wilderman, R. Psychotherapy in a community mental health facility: A practical and inexpensive approach to outcome assessment. *Evaluation and the Health Professions,* 1981, *4,* 189–205.

Wilson, S., & Kennard, D. The extraverting effect of treatment in a therapeutic community for drug abusers. *British Journal of Psychiatry,* 1978, *132,* 296–299.

Wincze, J. P., & Lange, J. D. Assessment of sexual behavior. In D. H. Barlow (Ed.), *Behavioral assessment of adult disorders.* New York: Guilford, 1981.

Wing, J. K., Cooper, J. E., & Sartorius, N. *The measurement and classification of psychiatric symptoms.* Cambridge, England: Cambridge University Press, 1974.

Wolpe, J., & Lang, P. J. A fear survey schedule for use in behavior therapy. *Behaviour Research and Therapy,* 1964, *2,* 27–30.

Wolpe, J., & Lazarus, A. A. *Behavior therapy techniques.* Oxford: Pergamon, 1966.

Woody, R. H. *Encyclopedia of clinical assessment* (Vol. 1). San Francisco: Jossey-Bass, 1980.

Zak, I. Stability and change of personality traits. In C. D. S. Spielberger, I. G. Sarason, & N. A. Milgram (Eds.) *Stress and anxiety* (Vol. 8). New York: McGraw-Hill, 1982.

Zamudio, A., Padilla, A. M., & Comrey, A. L. Personality structure of Mexican-Americans using the Comrey personality scales. *Journal of Personality Assessment,* 1983, *47,* 100–106.

Zeiss, A. M., Lewinsohn, P. M., & Munoz, R. F. Nonspecific improvement effects in depression using interpersonal, cognitive and pleasant events focused treatments. *Journal of Consulting and Clinical Psychology,* 1979, *47,* 427–439.

Zivich, J. M. Alcoholic subtypes and treatment effectiveness. *Journal of Consulting and Clinical Psychology,* 1981, *49,* 72–80.

III

Design Issues

6

Time-Series Research in Psychotherapy

THOMAS R. KRATOCHWILL and F. CHARLES MACE

INTRODUCTION

Time-series methods continue to be an area of rapidly expanding research methodology in the psychotherapy field. Many significant developments in time-series methodology have occurred over the past 10 years. Developments in time-series methodology are worthy of close examination by individuals in the psychotherapy field for a number of reasons. To begin with, time-series research provides an important alternative to traditional large-n between-group designs in psychotherapy research. Frequent concerns have been raised over the efficacy of group designs in answering certain types of questions in therapeutic studies (Barlow, 1981; Bergin & Strupp, 1972; Hersen & Barlow, 1976; Kazdin, 1980; Kiesler, 1971; Kratochwill, 1978; Myers, 1979). Concerns have centered around five main areas: (1) ethical objections (e.g., withholding the treatment from a client who participates in a no-treatment control group), (2) practical problems in collecting large numbers of clients who are homogeneous with respect to some variable (e.g., type of disorder), (3) averaging results over a group so as to obscure individual response to treatment, (4) generality of the findings from a group study to individual clients, and (5) a focus on intersubject variability without adequate focus on intrasubject variability.

A second reason that time-series methodology should be considered for

THOMAS R. KRATOCHWILL ● School Psychology Program, Department of Educational Psychology, The University of Wisconsin-Madison, Madison, Wisconsin 53706.   F. CHARLES MACE ● Department of Human Development, Lehigh University, Bethlehem, Pennsylvania 18015.

psychotherapy research is that these designs are uniquely suited to evaluation of treatments applied to a single client. Sometimes conducting a group study may simply be prohibitive due to time and cost associated with running a large number of clients. Moreover, the available numbers of subjects for a study may simply be limited. This is sometimes the case with certain behavioral or personality disorders (e.g., cases of multiple personality, selective mutism, clinical fears, etc.). Indeed, the importance of this feature of time-series methodology has not been stressed enough. Across many areas of psychotherapy research, investigators have tended to conduct studies that are an analogue to the more clinical problems faced by the practitioner. For example, in the area of children's fears and phobias, many investigators have studied clients who may not have a clinical fear (Morris & Kratochwill, 1983; Ross, 1981). Subjects who can be recruited easily and are not likely to drop out of therapy are sometimes used instead of those who actually experience a rather severe problem. Thus, time-series methodology provides an alternative for credible experimentation in this area.

A third reason for careful study of time-series methodology is that these procedures usually improve on the more traditional case study methods used across diverse theoretical approaches in psychotherapy research. Generally, case study methods have provided a rich foundation in the psychotherapy field because they have generated hypotheses about the basis of personality or behavior. Typically, many psychotherapeutic techniques have been developed and applied through case study methods (e.g., Bolgar, 1965; Kazdin, 1980, 1981; Lazarus & Davidson, 1971). Yet the researcher is able to draw few valid conclusions from the typical case study, making this a nonpreferred form of investigation most of the time. As an alternative to case study methods, the researcher can employ time-series methodology to evaluate the results. Various sources of internal and external validity can be addressed in time-series research, making these procedures more acceptable in the scientific community. This is not to imply that case study methods should never be used. There are good rationales for continuation of case study methods in psychotherapy research, particularly when attempts are made to improve the internal validity of these methods (see Kazdin, 1981). However, in such instances where validity can be improved, time-series might be considered as an alternative (see later discussions).

Fourth, associated with the developments in time-series methodology have been a number of major improvements in assessment measures of outcome for adult and child disorders. Various behavioral measures such as self-monitoring, direct observation, and psychophysiological recordings across cognitive, behavioral, and physiological content areas have advanced our knowledge of what treatments are effective with certain types of disorders. Particularly important in this domain of contribution has been the repeated measurement strategy associated with time-series designs. The repeated measures taken on a client (or clients) have allowed study of individual variability as a function of environmental influences (Hersen, 1982; Hersen & Barlow, 1976; Sidman, 1960). A major task for the applied researcher is to discover sources of variability and exert therapeutic control. In this regard, the repeated-measures feature of time-series methodology also allows ongoing evaluation of the treatment. Flexibility in changing the treatment is also available, since data provide feedback on the efficacy of treatment.

A final reason for considering work based on time-series methodology is that these strategies have frequently been recommended as a model for practitioner evaluation of therapeutic work (e.g., Barlow, 1981; Hayes, 1981; Hersen & Barlow, 1976; Jayaratne & Levy, 1980). It is also suggested that single-case time-series strategies will promote research among practitioners, possibly much better than traditional research approaches. However, whether or not these strategies will prove useful in this regard is still subject to considerable debate (see Kratochwill & Piersel, 1983). Specifically, the degree to which this task can be accomplished depends on several considerations, including the unit of analysis in the research, the ease with which assessment allows analysis of the individual case, design employed, and ethical and legal considerations (see later discussion in the chapter).

The aforementioned areas of time-series methodology show why enthusiasm has been high for use of these procedures in psychotherapy research. Nevertheless, there are many considerations associated with the use of these procedures and several unresolved issues surrounded by controversy. In the remainder of the chapter we provide an overview of time-series research and the accompanying methodological and conceptual features associated with their use in the psychotherapy field. We emphasize that although many methodological advances have been made in time-series investigation, major issues also need to be addressed for future developments to occur. In many cases, this will involve empirical investigation of certain issues, several of which are reviewed.

TIME-SERIES RESEARCH: GENERAL ISSUES

Time-series research has emerged as a major method for investigating the theraputic impact of psychotherapy. As a research tool, it shares common objectives with the more traditional research strategies. Both methods provide a description of the problem and treatment, measures used to evaluate change, and a degree of insight into the variables responsible for change or the lack of it. Although similarities do exist, the various research methods are best characterized by their differences, relative advantages and limitations, and the degree to which validity threats are controlled. In this section of the chapter, we provide a brief historical perspective of single-case and time-series designs, contrast group and case study methods with time-series designs, discuss the methodological characteristics of time-series research, and review some applications of these procedures.

Historical Perspectives

Time-series designs have many and diverse historical origins with major contributions from medicine, physiology, and experimental psychology (Hersen & Barlow, 1976; Robinson & Foster, 1979). Many of these early contributions involved single-subject case studies that formed the basis for hypotheses about human behavior and physiological responses. For example, such individuals as Johannes Muller, Claude Bernard, and Paul Broca made significant contributions to the study of the individual during the 1800s. Bernard's writings stressed careful analysis of individual data, and at least some writers (e.g., Yates, 1976) have suggested that his methods can be regarded as a general statement on experimental study of the single case. Broca's work on mapping speech functions can be regarded as critical in establishing that findings with a single case have wide-scale generality (Hersen & Barlow, 1976).

Early developments in experimental psychology frequently involved study of one or a few cases (Robinson & Foster, 1979). For example, Fechner's studies in sensory thresholds provided many basic principles in psychophysics; Wundt developed new concepts in sensation and perception; Ebbinghaus pioneered the study of learning and memory; and Pavlov established basic laws of learning and conditioning. Contributions from these individuals were made from the study of single cases. Moreover, a whole new area of developmental psychology and cognition was created by Piaget in his studies of single cases, including his own children as "subjects."

Another major contribution to contemporary time-series investigation came from case study methods discussed in the works of Allport, Lewin, Shapiro, and Chassan. For example, Shapiro emphasized the use of well-defined clinical measures taken over time in one client, with some early work in this area serving as a prototype of an A-B-A design (Shapiro & Ravenette, 1959). Shapiro's contributions extended beyond design and reached into other conceptual and methodological areas. Yates (1976) notes:

> for Shapiro, explanation involved experimental control, that is, the demonstration that the client's behavior could be systematically manipulated and changed. But, if this proved to be feasible, then it was a very short step to the further expectation that it should also be possible to use the same approach to modify the client's behavior to a degree which would enable him to return to society (or improve his adjustment to it). Behavior modification in Great Britain was thus spawned from Shapiro's original aim of explaining the abnormality of the presenting client by bringing it under experimental control. (p. 299)

Interestingly, the approach emphasized an empirical perspective for dealing with a client, independent of theoretical perspective. In this regard it bears similarity to more current perspectives on the use of single-case designs in clinical practice (Barlow, 1981; Hayes, 1981).

Case study methods were not limited to this area. Some authors reported innovative psychotherapeutic techniques in professional journals through case study methods (e.g., Bolgar, 1965; Paul, 1969), and Dukes (1965) reviewed over 200 single-case studies over a 25-year period from diverse areas of psychology which provided instances of critical findings. Nevertheless, despite the writings emphasizing the importance of single-case studies (e.g., Lazarus & Davison, 1971; Shontz, 1965), these methods were not regarded as credible research strategies in many areas of psychotherapy research. To begin with, these methods were characterized by numerous sources of uncontrolled variation (i.e., little internal validity) and inadequate description of variables; they were virtually impossible to replicate. Second, the appearance of Eysenck's (1952) psychotherapy review paper encouraged disillusionment with case studies and encouraged evaluation of psychotherapy through group designs (Hersen & Barlow, 1976). Third, even earlier, group designs were becoming the methodology of choice with rapidly evolving developments in inferential statistics. For example, Sir R. A. Fisher developed properties of statistical tests that helped researchers make generalizations to target populations (Heerman & Braskamp, 1970). With an emphasis on group variabilitiy and average performance, single-case methods were not regarded as credible by many researchers.

Despite these trends, single-case research was kept alive over the latter half of the 20th century. Three major reasons can account for this. First of all, as group designs were used with increasing frequency, the limitations for clinical practice become more salient. For example, Bergin (1966) suggested that in many group psychotherapy studies, some clients improved while others did not improve or even got worse. Such findings led individuals in the psychotherapy field to question the homogeneity of classification schemes and also the group methodology in advancing scientific knowledge in therapeutic research. Also, as noted in our introduction, a whole series of limitations of group designs were raised. These influences led to consideration of alternative methodologies, of which single case research was one.

Another major area where single-case time-series designs have a strong tradition and contemporary support is in the behavior therapy field. Specifically, early work in the experimental analysis of behavior (Skinner, 1938, 1953) and publication of Sidman's (1960) work on experimental methodology promoted single-case research. Several areas of contribution can be identified in this area. To begin with, journals (e.g., the *Journal of Applied Behavior Analysis*) were created specifically to publish work relating to the experimental analysis of behavior. Researchers with applied interests were given an outlet for their work, whereas in the past, other journals had been closed to empirical investigation of the single case. Many other behaviorally oriented journals have developed and provided numerous options for the behaviorally oriented psychotherapy researcher. Also, design and methodology became an active area within the field of behavior modification or behavior therapy, with numerous works appearing that pioneered new design and assessment strategies. Generally, the major advances in time-series design have appeared in the behavior therapy literature.

A final area where advances have been made is in the mainstream of social science research. Since Campbell and Stanley (1963) introduced time-series methods as quasi-experimental, researchers have been given an expanded context within which to evaluate the internal and external validity of time-series designs (see Cook & Campbell, 1979). Also there has been an emphasis on the development of statistical techniques for the analysis of time-series experiments (Kratochwill, 1978). Advances here have raised consciousness over methods that are inappropriate as well as suitable for therapeutic time-series research. Such developments have been important in the perceptions of single-case time-series methods as credible research strategies.

Traditional Methods of Research

Historically, a vast amount of psychotherapy research has been conducted using group designs or case study methods. Understanding how time-series research differs from these traditional approaches requires a closer look at both group and case study procedures.

Group Designs

Research designs commonly used to evaluate psychotherapy include between-group and within-subject strategies. Each approach has purposes to which it is best suited and advantages and limitations to consider. Between-group strategies typically involve the selection of a single group of subjects from a common population. Subjects from this single group are then randomly assigned to groups receiving the experimental and control conditions. Therefore, the number of groups needed is equal to the number of conditions being compared. If certain subject characteristics are thought to interact with treatments to influence outcome measures, then subjects are matched according to the salient characteristics with the aim being to achieve equivalent groups, (Myers, 1979). Following exposure to the experimental and control conditions, subjects within the groups are assessed on the dependent variable at approximately the same point in time. Because the groups were considered equivalent prior to exposure, any between-group differences obtained afterward may be attributed to the effects of treatment.

Between-group designs have a number of advantages that make them useful for psychotherapy research. First, they provide a simple means of comparing an experimental treatment with no treatment or evaluating the relative effectiveness of two or more different treatments. By assigning subjects to a single condition, treatment effects may be studied independent of multiple treatment interference or temporal ordering effects (Kazdin, 1980). Second, between-group designs are useful in assessing therapeutic impact on the entire group. This information can assist program planners in selecting, from among available group treatments (e.g., milieu therapy vs. token economy), the procedure most effective for most group members. Finally, these designs permit the evaluation of interaction effects, including subject characteristics $\times$ treatment and treatment $\times$ treatment interactions (Myers, 1979).

By contrast, within-subject designs involve (1) exposing subjects within a group to each of the different treatments under study or (2) providing a single treatment to all clients and evaluating their performance over time. This requires fewer subjects than between-group designs because each subject is assessed under all conditions in the study, not merely under a single condition. This arrangement is also advantageous from a statistical standpoint. Variability due to individual differences can be isolated and removed from the error term in the F test. With less error, statistical power is increased even though fewer subjects are used. The primary disadvantage of the within-subject strategy is the internal validity threat, multiple-treatment interference. Since each subject is exposed to all treatments, there is a good likelihood that the effects of Treatment A may carry over or add to the effects of Treatment B. The result is that the separate contribution of each treatment is impossible to discern.

The utility of group designs in applied research must be weighed against their potential drawbacks. Sample characteristics needed for group designs include large numbers of clients who are homogeneous on important variables. Unfortunately, it often is impossible in clinical settings to obtain the sample size needed when clients are to be matched on personality, behavior, background, and demographic variables. The more these factors vary, the less representative are the findings of any given client within the group. Furthermore, the averaging of results obscures individual performance and is often responsible for clinically "weak" results. That is, some clients improve only slightly or not at all, while others make clinically significant progress. When performance is averaged across subjects, the net improvement will appear weak. Data analysis procedures used in group designs have been criticized for their overreliance on statistics to determine treatment impact (Meehl, 1978: Michael, 1974). Statistically significant findings often lack the magnitude necessary for psychotherapy to be considered successful for clinical problems (Barlow, 1981). At the same time, effects that do not reach rejection levels are often assumed to be nonexistent, thereby overlooking variables with therapeutic potential for inclusion in treatment packages. Finally, between-group designs have come under fire for withholding treatment from no-treatment control groups, raising ethical objections over their use. Clients suffering from various disorders have been reported to decline during the course of the study, violating their right to treatment. Similarly, the inflexibility of group research regarding the duration and nature of treatment means that some clients will receive insufficient treatment while others are possibly treated in "excess."

Case Study

The case study has played an important role in the investigation of psychotherapy over the years. It is synonymous with the intense study of the individual that is the hallmark of applied psychology and psychiatry. Traditionally, case study refers to detailing the course of psychotherapy for an individual or group of individuals in the absence of experimental controls (Kazdin, 1980). Case study investigation can take various forms, ranging from anecdotal impressions of the investigator, to extensive narrative accounts of procedures and outcomes, to well-defined treatments and dependent measures within the context of experimental design. Yet regardless of the form, interpretations are marked by considerable ambiguity that prevents the identification of causal relationships in therapy.

The value of the case study must be viewed with appreciation for the unique role it can play in research. Case reports can be an important means of generating hypotheses about the variables operating to maintain a particular disorder. Detailed cases may provide insights that might otherwise never surface and may eventually lead to the development of experimental hypotheses. Case studies can also provide a forum for communicating the development of therapeutic techniques and unique technique applications. This is a particularly attractive alternative to the practicing clinician, who can offer unique contributions to the growth of therapeutic procedures. An especially important function of case studies is the examination of rare phenomena. Infrequent disorders encountered by clinicians may offer descriptive data that may characterize the problem—its course and potential therapeutic strategies. Along similar lines, case reports can provide a counterinstance to widely accepted assumptions, procedures, or theories.

Although case study methods provide a vehicle for clinicians to participate in the process of psychotherapy, information gleaned from the research is inherently ambiguous. Valid inferences regarding the relationship between treatment and outcome are not possible because numerous factors are left uncontrolled. Typically, procedures are unspecified and client progress is assessed via self-report. Both therapist and client biases are reflected in interpretations of the findings (Mischel, 1968). Because adequate controls are lacking, most of the validity threats (to be discussed in a later section) may be assumed to be operating to undermine scientific credibility (Leitenberg, 1973).

As implied above, an important function of the case study is its interplay with experimental research (Kazdin, 1980). Each method generates unique information that contributes to the development of effective psychotherapy. The case study provides a forum for cataloguing and describing the numerous variables that converge on an individual problem. Examining the case as it occurs under natural conditions facilitates appreciation of its full complexity without the artificial control involved in experiments. Hypotheses advanced in case reports may, in turn, stimulate experimental research that may have important implications for the practice of therapy (i.e., whether the treatment works and, if so, under what conditions). Conversely, information about a particular treatment derived from experimental methods may be limited to controlled conditions. Clinicians' efforts to generalize research findings to their clinical practice may be communicated via case studies. Any discrepancies that arise in this process may be the source of still further investigation. Thus, the interplay between case study and controlled research is an interdependent process that contributes significantly to the growth of psychological knowledge and offers the potential of narrowing the longstanding gap between practice and research (Kazdin, 1980).

In recognition of both the assets and liabilities of case study methods, Kazdin (1981) has identified steps to take toward drawing valid inferences from case reports. These include: (1) use of objective measurement strategies, (2) frequent assessment occasions, (3) demonstration of the stability of the problem and the treatment effect over extended periods, (4) development of treatments that have an immediate and large magnitude of effect, and (5) replication of findings with a number of heterogenous subjects. These recommendations are one attempt to bridge the gap between the scientific rigor of group designs and the flexibility of case studies. Time-series research offers a viable alternative to both methods.

Methodological Characteristics of Time-Series Research

Time-series research has become an important method for evaluating psychotherapy systematically. It is first useful to examine the terminology surrounding this area of research. Writers have tended to use a variety of terms to refer to the same or different aspects of empirical work in this area. For example, these designs have been labeled single-subject or case, $n = 1$ (Hersen,

1982; Barlow, 1976); small *n* (Robinson & Foster, 1979); intensive (Chassan, 1967, 1979; Kratochwill, 1979); intrasubject replication (Kazdin, 1980); time-series (Campbell & Stanley, 1963; Glass, Wilson, & Gottman, 1975; Kratochwill, 1978); and case study, among numerous others. We will use the term *time-series* to refer to a class of designs that involve repeated measures on the dependent variable and introduction of one or more interventions into single or multiple subjects. Time-series designs are usually distinguished from the more traditional case study on several dimensions. Inference for an intervention effect in these designs is made both within (intrasubject) and between subjects, depending on the type of design used. Finally, we should emphasize that there are research strategies that do not involve the usual characteristics of time-series designs as outlined above (e.g., repeated measures), and some of these may be based on a single case (e.g., case study, Markov models, factor analysis through the p technique). In some time-series research, decisions regarding treatment effects are made by comparing repeated measures of the dependent variables prior to and after the onset of treatment. Numerous design options have been developed for this purpose, each suited for investigating a range of clinical problems (Hersen & Barlow, 1976; Kratochwill, 1978). Despite the rich diversity in time-series designs, they share common characteristics that serve to distinguish this methodology from the more traditional methods discussed earlier.

Repeated Measurements

The assessment of client status or behavior on multiple occasions over time is fundamental to time-series work (Cook & Campbell, 1979; Hersen & Barlow, 1976; Kratochwill, 1978). Data collected prior to intervention represents the client's baseline performance (A phase) and are compared to data during therapy (B phase) to determine its effectiveness. Single or multiple measures may be taken from each of the client's motoric, cognitive, or physiological response domains. Concurrent samples from all three domains permit evaluation of treatment impact on each area of client functioning. Consider a client with severe fear of public speaking. Repeated measurement prior to and during a cognitive therapy may suggest that treatment was effective in reducing fearful thoughts during speaking occasions. However, data on the frequency of speaking occasions and heart rate during speaking were unaffected by the treatment. In view of the client's desire to also increase public speaking and reduce physical arousal when doing so, treatment can be considered only partially successful. This information, however, can be very useful in developing additional therapeutic strategies to alter the client's motoric and physiological responses. Repeated measurement may be further extended to include the independent variable. Periodic assessment of the treatment variable can help determine whether it is being administered consistently and as planned, both of which are prerequisite for the establishment of construct validity (see later discussion).

Establishing Interclient Variability

As client behavior is repeatedly measured, fluctuations or variability will be apparent over time. The level and trend of the dependent measure are estimated and inferences drawn about the future course of behavior. Performance that departs significantly from these predictions provides evidence for the therapist. Stable measures are often considered necessary in order to attribute change to the introduction of the treatment (Hersen & Barlow, 1976). Parsonson and Baer (1978) noted that either stable measures or those that trend in the opposite direction of anticipated change are necessary preconditions for applying or withdrawing intervention. However, Hayes (1981) points out that measures need only be stable enough to see effects in the event they occur. For example, highly variable rates of aggression would suffice for baseline measures given the goal of treatment is to eliminate the problematic behavior.

Specification of Conditions

Specification of the independent variable, setting, therapist and client characteristics, and assessment procedures can provide important information regarding the therapeutic value of each variable (see Kazdin & Wilson, 1978, for treatment evaluation strategies). Manipulation of any of these variables while holding the others constant can expose its unique contribution to the client's performance. As a result, the clinical researcher can combine effective variables to produce maximally effective treatments (Hersen & Barlow, 1976). Moreover, specification of conditions is essential for the replication of findings and for establishing validity.

Replication

The logic of time-series research is predicated on the replication of treatment effects. Internal validity is achieved by reproducing effects either within an individual or group (e.g., A-B-A-B design) or across clients, client measures, or settings (e.g., multiple-baseline across-behaviors design). Each instance in which changes in data patterns coincide with random phase or treatment changes reduces the plausibility of rival hypotheses (i.e., validity threats) being responsible for observed effects. For example, the effectiveness of an antidepressant drug in relieving a client's depression may be evaluated with the following phase sequence: (1) no drug, (2) inactive drug (placebo), (3) experimental drug, (4) inactive drug (placebo), and (5) experimental drug. Data patterns that indicate improvement in depression measures under the experimental drug condition relative to the no-drug and placebo conditions suggest some therapeutic value in the psychoactive drug. Evidence for this hypothesis becomes convincing, however, when performance patterns are reproduced during Phases 4 and 5. Replication is similarly critical if results are to be generalized beyond the single case or experiment. External validity is achieved by systematic replication of effects across key variables in the study (see section in this chapter on external validity).

Design Flexibility

Time-series designs have been characterized as dynamic in their ability to change in response to ongoing conditions (Hayes, 1981; Hersen & Barlow, 1976). Because individual responses to treatment are often variable, evaluation of effects should parallel the natural course of therapy. Time-series designs provide the applied researcher with this type of flexibility and represent a significant contrast to the planning involved in traditional group research. In assessing the impact of a brief psychodynamic therapy on marital discord, individual couples will likely respond to treatment at different rates. Analyzing the effects of therapy after a predetermined period of time (e.g., 6 or 8 weeks) overlooks the possibility that some couples will have progressed faster than others. Consequently, the analysis may underestimate the therapy's potential to effect change. More importantly, differential responses to treatment may be indicative of the need to alter the components of therapy in order to achieve the desired outcome. With time-series designs, repeated measurement permits the clinician to assess on an ongoing basis the need to adjust the course of treatment. In each instance, information gleaned from such investigation may spark interest in examining the issues raised under more rigorous experimental conditions.

TIME-SERIES RESEARCH: APPLICATIONS

Types of Time-Series Designs

Numerous designs and design variations have been developed in the past two decades. Each utilizes a group of core elements, combined in a creative manner, to investigate a range of clinical problems. These core design elements serve as building blocks from which all time-series designs are derived. Time-series designs may be classified according to three types: within, between, and combined series (Hayes, 1981). Within each of these types, a single subject or a group ($n > 1$) can be evaluated.

Within-Series Designs: Simple Phase Change

This type of design is characterized by examining changes in the level and trend of client-repeated measures across phases of the study. The logic of the simple phase-change design is in comparing measures at the point of phase change for abrupt differences in level or trend. The A-B design represents this strategy in its simplest form. If stable baseline measures (A) change suddenly following the administration of some therapy (B), we are led to believe that the therapy was responsible. Our confidence in the causal relationship between treatment and outcome increases when withdrawal of the treatment results in a return or approximation of initial baseline patterns (the A-B-A design). If treatment effects can be demonstrated a second time by reinstating the B phase (the A-B-A-B design), we can be reasonably confident that therapy is the change agent and that alternative explanations (i.e., validity threats) are less plausible. This strategy of replicating treatment effects within subjects or groups of subjects can similarly be used to compare differing treatments, B and C (e.g., B-C-B; B-C-B-C designs).

Within-Series Designs: Complex Phase Change, Interaction Designs

Interaction designs operate under the same logic as the simple phase-change designs. The primary difference is that interaction designs examine the effects of multiple treatment components. In its reduced form, the interaction design compares the effect of adding or subtracting a treatment component from another treatment component. For example, suppose a researcher is interested in assessing the combined effect of a reinforcement program (B) plus a cognitive restructuring strategy (C) compared with the reinforcement program only. Two phase-change options are available: B-BC or BC-B. As with the simple phase-change designs, greater confidence in the findings can be achieved by repeating the phase changes in order to replicate the initial data patterns (e.g., B-BC-B; B-BC-B-BC; BC-B-BC; BC-B-BC-B).

Interaction designs are particularly suited for examining treatment interactions and developing effective and parsimonious treatment packages. The term *treatment interactions* refers to the effect of combining treatment components on the independent variable. Analysis of interaction effects requires the manipulation of two or more variables, separately and in combination (Hersen & Barlow, 1976). Often treatment effects will be other than additive. This means that some treatment components will contribute more therapeutically than others. The task of the applied researcher is to determine the optimal combination of treatment components to produce the best treatment package.

A recent example of within-series designs, both simple and complex phase changes, is seen in a study by Wells, Conners, Imber, and Delamater (1981). The effect of various psychoactive medications alone and in combination with a self-control program were evaluated for a 9-year-old male identified as hyperactive. Observational measures taken on the child's classroom behavior on a psychiatric inpatient unit included excessive gross motor behavior, inappropriate vocalizations, off-task behavior, and on-task with no deviant behavior. The design and treatment sequence consisted of an A-B-A-C-CD-A_1D-CD design, where A is baseline, A_1, is placebo medication, B is dextroamphetamine (Dexedrine) treatment, C is methylphenidate (Ritalin) treatment, and D is a behavioral self-control program. As presented in Figure 1, the data indicating the Dexedrine was unsuccessful in improving subject performance led the researchers to introduce Ritalin following a second baseline period. With the exception of Sessions 26 and 32, the subject's response to Ritalin appeared positive across all client measures. During the CD phase, Ritalin was combined with a self-control program involving self-monitoring of on-task behavior and a self-administered token system. As the data in the C-CD-A_1D-CD sequence indicate, the Ritalin plus self-control treatment combination appeared most effective. Neither Ritalin (C phase) nor behavior self-control (A_1D phase) alone were as powerful as their interactive effects.

There are a number of limitations to consider regarding the within-series withdrawal-type designs. First, the withdrawal of treatment may not result in anticipated changes in the dependent measure. When this occurs, the logic of the design and experimental control are compromised. A second limitation is that some measures may not deteriorate rapidly, especially when learning is presumed to occur. Consequently, the design is usually inappropriate to evaluate academic, motor, or social skill training. A final issue concerns the withdrawal of effective treatment when clients are likely to suffer in the absence of therapy. In such cases, ethical concerns should take priority and designs providing for continuous treatment should be used.

Within-Series Designs: Complex Phase Changes, Changing-Criterion Design

The changing-criterion design establishes experimental control by bringing the level of the dependent measure under the control of arbitrarily set criteria (Hall & Fox, 1977; Hartmann & Hall, 1976). Following a series of baseline observations, an intervention is implemented and maintained throughout the treatment phase. At various points in the treatment phase, stepwise changes in the level of the dependent measure are arbitrarily set as criteria.

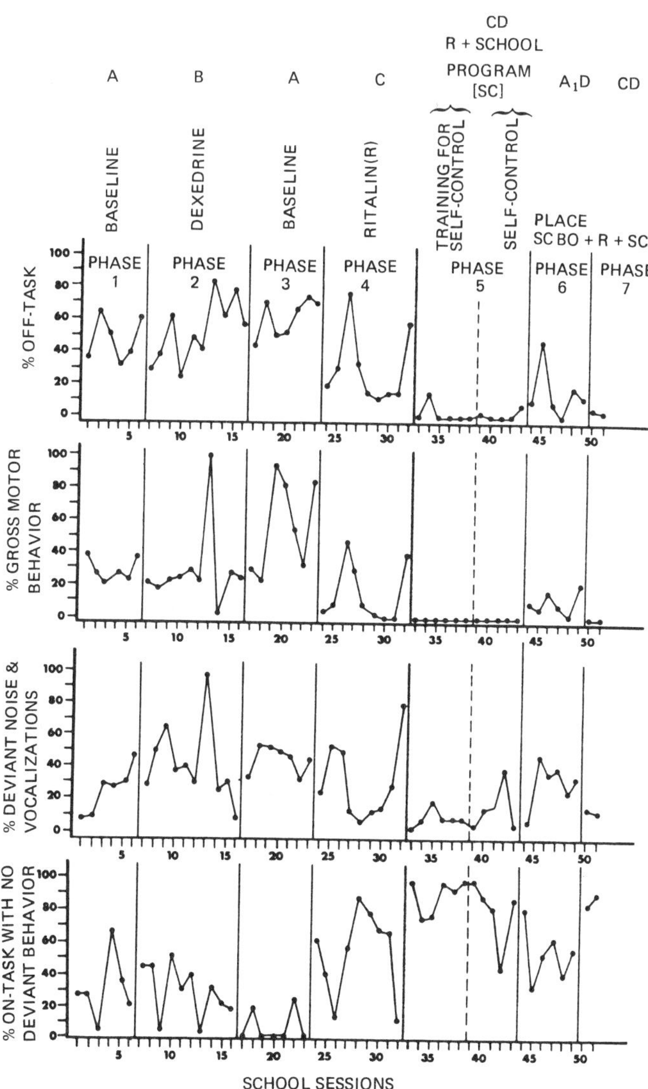

Figure 1. Percent occurrence in the classroom of off-task behavior, gross motor behavior, deviant noise and vocalizations, and on-task behavior with no other deviant behavior recorded, measured across baseline, medication, and placebo phases. (From ``Use of Single-Subject Methodology in Clinical Decision Making with a Hyperactive Child on the Psychiatric Inpatient Unit'' by K. C. Wells, C. K. Conners, L. Imber, and A. Delamater, *Behavioral Assessment*, 1981, *3*, 359–369. Copyright 1981 by Association for Advancement of Behavior Therapy. Reproduced by permission.)

Typically, criteria are linked with treatment contingencies (e.g., an anorexic client may receive tokens for meeting or exceeding the criteria for caloric consumption on a given day). If the dependent measure reliably tracks the stepwise changes in criteria, internal validity is strengthened and rival hypotheses concerning behavior change may be discounted.

There are a number of considerations in using the changing-criterion design. The key ingredient in the design is the demonstration of parallel changes in the dependent measure and the arbitrarily set criteria. In order to demonstrate parallel changes, it is important that each criterion phase have sufficient length to allow the dependent measure to stabilize before proceeding to the next stepwise level (Kratochwill, Schnaps, & Bissel, in press). Similarly, the size of the stepwise criteria changes must be large enough to distinguish treatment effects from the stochastic variability of the time series. Treatment effects can be further illuminated by randomly varying the length, depth, and direction of the criterion shifts (Hayes, 1981). This tactic serves to exaggerate the control of the criteria (and the contingencies associated with them) on the target measure.

Bernard, Dennehy, and Keefauver (1981) employed a changing-criterion design in the treatment of a woman's excessive coffee and tea drinking. Intervention consisted of contingent social praise, goal setting, a response cost procedure (payment of $2 for each cup beyond the criterion), and self-monitoring coffee or tea consumption. Following a 13-day baseline period, the subject and therapist established a terminal treatment goal of 6 cups per day by the end of a 30-day period. An initial criterion level was set at 11 cups per day (the nearest whole number below the baseline mean). The criterion was then reduced by 1 cup in a stepwise fashion on the 18th, 24th, 27th, 34th, and 37th day. As is apparent in Figure 2, the client's coffee or tea consumption consistently traced the criterion changes. Moreover, once treatment was terminated, the criterion level of coffee or tea drinking remained constant during maintenance and follow-up assessment.

Between Series: Alternating and Simultaneous Treatments Designs

The alternating treatments design (ATD) and the simultaneous treatments design (STD) are the basic types of between-series elements (Hayes, 1981). Both offer an alternative to group designs for the comparison of different therapies in the same subject. Between-series strategies compare two or more data series across time. Comparisons made between the series take into account level and trend differences for the same dependent measure as the client or clients progress through the independent treatments.

The ATD exposes the client to separate treatment conditions for equal periods of time (Barlow & Hayes, 1979). Treatments are alternated within a

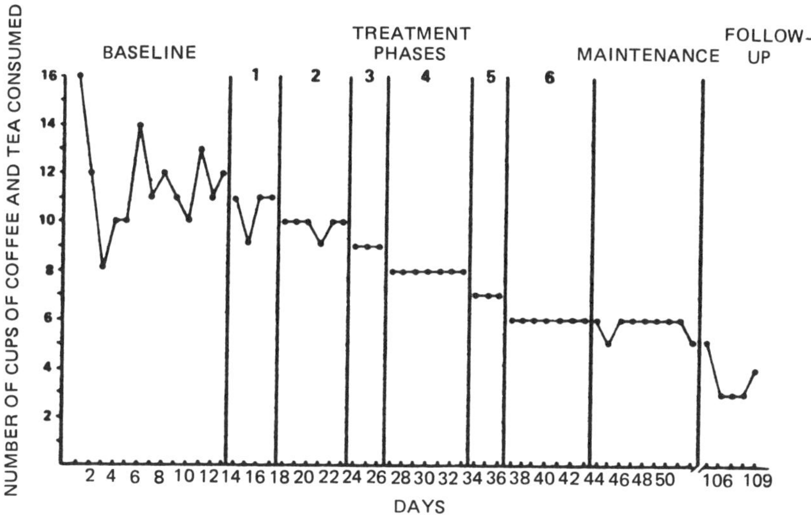

Figure 2. Subject's daily coffee and tea intake during baseline treatment, maintenance, and follow-up. The criterion level for each treatment phase was one cup less then the previous treatment phase. (From ''Behavioral Treatment of Excessive Coffee and Tea Drinking: A Case Study and Partial Replication'' by M. E. Bernard, S. Denneby, & L. W. Keefauver, *Behavior Therapy*, 1981, *12*, 543–548. Copyright 1981 by Association for Advancement of Behavior Therapy. Reproduced by permission.)

short time period (e.g., Treatment B in the morning, Treatment C in the afternoon). The sequence of times of treatment exposure should be determined randomly or by counterbalancing. This arrangement ensures that the client receives equal exposure to the interventions while providing control for the effects of time and setting differences. The primary advantages of the ATD are that differing treatments can be compared within a relatively short time period while avoiding some of the disadvantages of within-series withdrawal designs (e.g., need for stable baselines, treatment withdrawal, and history threats to internal validity). Limitations of the ATD include possible multiple-treatment interference and logistical considerations involved in administering two or more treatments (see Barlow & Hayes, 1979; Kratochwill *et al.*, in press, for suggestions on minimizing these problems). Overall, the ATD is a useful strategy for comparing distinct and independent treatments when large treatment effects are expected and time and number of subjects are limited.

An alternating treatments design was used by Ollendick, Shapiro, and Barrett (1981), comparing positive practice overcorrection, physical restraint, and a no-treatment condition in the treatment of stereotypic behaviors. Three mentally retarded children (ages 7 and 8) engaging in repetitive hand posturing or repetitive hair twisting served as subjects. A single therapist administered the treatments during three 15-minute sessions per day in which subjects were instructed to work on visual–motor tasks. Baseline observations of stereotypic behavior were taken during all three time periods. Following baseline, the two-treatment and the no-treatment conditions were administered in a counterbalanced sequence across daily sessions. One stereotypic behavior was reduced to a near-zero rate under one treatment condition. The other treatments were discontinued and replaced by the most effective procedure. Figure 3 illustrates individual subject responses to the various treatments. For Tim and Jane, positive practice overcorrection produced greater reductions in the target behavior than physical restraint, which, in turn, was more effective than no treatment. John, on the other hand, showed his greatest improvement with physical restraint followed by overcorrection and no treatment. Concurrent data taken on task preformance showed gradual improvement under all treatment conditions over the course of the study. The results illustrate how independent treatments can be evaluated simultaneously to identify the most effective procedure for the individual client.

The STD differs from the ATD in that multiple treatments are available to the subject simultaneously (Kazdin & Hartman, 1978). However, Kratochwill and Levin (1980) have pointed out that simultaneous availability does not ensure that the client will be exposed equally to the treatments under study. Rather than comparing the relative effectiveness of different treatments, the STD evaluates client "preference" among treatments. For this reason, the STD may serve as a valuable tool for establishing hierarchical orderings of treatments on the basis of restrictiveness (Hayes, 1981). Future work in this area may offer an empirical basis to ethical issues and guidelines.

Combined Series: Multiple Baseline Design

Combined series designs permit comparisons within series and between series. A familiar example of these combined elements is the multiple baseline design (MBD). Typically, the MBD employs a single within-series simple-phase element (i.e., A-B). The A-B element is replicated across two or more clients, settings, or therapists/experimenters allowing comparison between the A-B series. Control for common internal validity threats is achieved by staggering the administration of treatment at different points in time. If changes are observed at the first A-B shift and the remaining baselines remain unaffected, alternative explanations for the behavior change (e.g., history or maturation) are less likely. Confidence in the treatment effects increases each time the pattern of change is replicated.

Multiple-baseline designs represent a useful method in applied research for a variety of reasons. First, the number of data series required for the design is not fixed (Hayes, 1981), although some researchers recommend at least

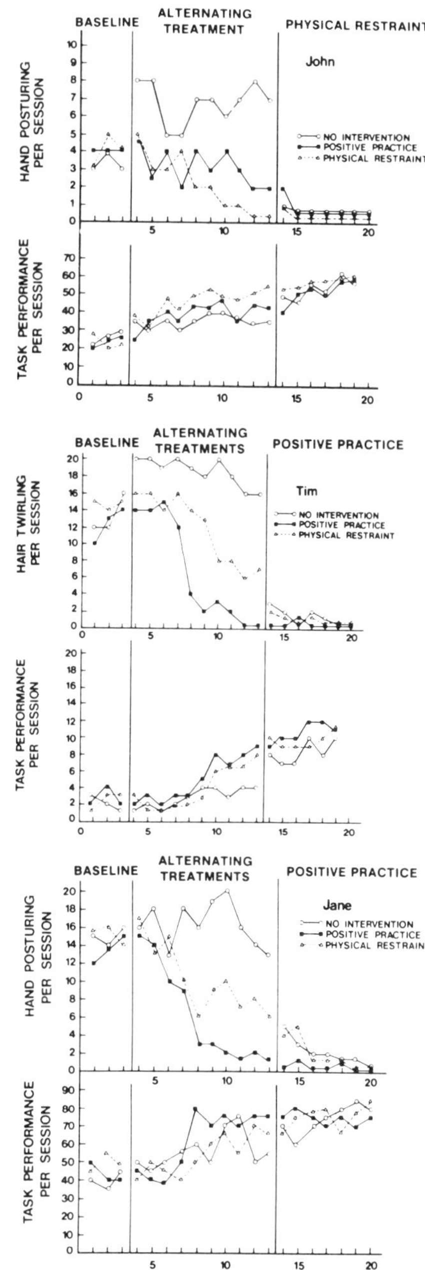

Figure 3. Stereotypic hair twirling or hand posturing and accurate task performance for John, Tim, and Jane across experimental conditions. The data are plotted across the three alternating time periods according to the schedule that the treatments were in effect. The three treatments, however, were presented only during the alternating-treatments phase. During the last phase, physical restraint or positive practice was used during all three time periods. (From ''Reducing Stereotypic Behaviors: An Analysis of Treatment Procedures Utilizing an Alternating Treatments Design'' by T. H. Ollendick, E. S. Shapiro, & R. P. Barrett, *Behavior Therapy,* 1981, *12,* 570–577. Copyright 1981 by Association for Advancement of Behavior Therapy. Reproduced by permission.)

three series to establish adequate internal validity (Hersen & Barlow, 1976). As noted earlier, validity lies on a continuum; thus, confidence in the results can be improved by simply including additional series. Second, reservations have been raised concerning the use of the MBD when correlated data series exist (Kazdin & Kopel, 1975; e.g., across behaviors or settings within single subject). However, when generalized treatment effects do occur across series, the opportunity arises to study generalization effects. At that point, the researcher may shift to a withdrawal design to demonstrate treatment effects and examine generalization issues separately (see Kendall, 1981, for design options). Finally, the MBD may at times be compatible with clinical practice. The A-B element fits the natural course of treatment. Findings from a single client may be strengthened by replicating effects with subsequent similar cases while maintaining the same treatment and therapist.

Kolko, Dorsett, and Milsan (1981) evaluated the effectiveness of a social skills training package using a multiple-baseline design. Three adolescent males described as verbally and physically aggressive received training (behavioral rehearsal plus corrective feedback) in the following skills: moderate response latency, sustained eye contact, neutral facial expressions, moderate voice loudness, and passive-compliant verbal responses. Observational data were taken on each of these measures across various role-play and word-simulation situations. Treatment effects were assessed both across behaviors within a client and across clients (see Figure 4). With few exceptions, shifts in the level and/or trend of performance coincide with the administration of training staggered across series. Since baseline measures remain relatively stable prior to the introduction of treatment, training effects are clearly demonstrated.

Special Design Issues

Length of Phases

Several factors need to be considered when deciding on the length of experimental phases in time-series designs. Among these are the data patterns in the preceding phase, the relative length of adjacent phases, requirements of data analysis procedures, staff concerns, and ethical considerations. Often, these factors are in opposition to one another, in which case the clinical researcher must strike a balance between providing effective intervention and establishing experimental control.

Many authors have advocated that the pattern of data in the preceding phase be stable (i.e., without trend or excessive variability) before changing experimental conditions (Hersen & Barlow, 1976; Johnston, 1972; Kazdin, 1980). The logic of this requirement is that apparent treatment effects may be

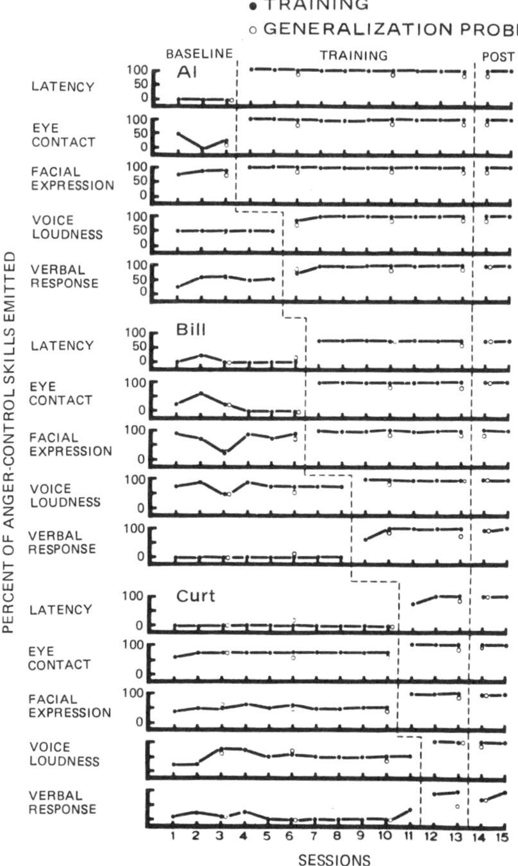

Figure 4. Percentage of anger control skills emitted by the three clients to single-response training and generalization probe scenarios during the baseline, training, and posttraining phases of the study. (From ''A Total-Assessment Approach to the Evaluation of Social Skills Training: The Effectiveness of an Anger Control Program for Adolescent Phychiatric Patients'' by D. J. Kolko, P. G. Dorsett, & M. A. Milan, *Behavioral Assessment,* 1981, *3,* 383–402. Copyright 1981 by the Association for Advancement of Behavior Therapy. Reproduced by permission.)

due to trends or high variability in the data rather than to the influence of the independent variable. Others, however, have regarded the prerequisite of data stability as too rigid. Performance trends in the opposite direction of expected treatment effects provide the clear contrast that is needed to determine effects (Parsonson & Baer, 1978). Similarly, highly variable or trending data are acceptable baseline measures when treatment is expected to produce dramatic results (Hayes, 1981). Generally, any pattern of data in the preceding phase that contrasts significantly with the data patterns in the subsequent phase is adequate for establishing experimental effects.

A related factor to consider is the relative length of adjacent phases. As a general rule, it is desirable to have experimental phases approximately equal in length (Hersen, 1982). The purpose of this is to avoid the possibility that treatment effects are, at least in part, due to differences in the amount of time they were administered. Also important in determining phase length is the problem of carryover effects across adjacent phases. Factors to consider with behavioral intervention are the development of new conditioned reinforcers (Bijou, Peterson, Harris, Allen, & Johnston, 1969) and the schedule of reinforcement used. Lengthy phases increase the probability that new conditioned reinforcers will maintain treatment effects even after the intervention is withdrawn. Similarly, effects may be maintained, temporarily at least, under their schedules of reinforcement.

The data analysis used may also influence the determination of phase length. For example, time-series data analysis procedures may require 50 or more data points for accurate model building (McCleary, Hay, Meidinger, & McDowell, 1980). This requirement may preclude the use of data analysis for more short-term evaluations. However, this series may be divided into multiple experimental phases.

Phase length may also be influenced by staff concerns in carrying out treatment. Often persons working with the client may be unsupportive of attempts to delay treatment for experimental purposes. Annoying client behavior may prompt staff to request immediate intervention. Moreover, when maladaptive behavior is potentially injurious to the client or others, there is an ethical responsibility to provide treatment as soon as possible and to shelve concern for experimental rigor (Hersen, 1982).

Evaluation of Psychoactive Medication

Clinicians frequently encounter cases that may benefit from psychoactive medication alone or in combination with some psychological intervention (e.g., Ritalin for hyperactivity, Stelazine plus social skills training for asocial behavior). Since idiographic responses to drug therapy are common, the clinical researcher may wish to evaluate drug effects systematically for the individual case or group of individuals. A number of time-series designs, each of which can be a valuable tool in determining whether medication is producing its desired effects, have been developed specifically for this purpose (Hersen, 1982; Hersen & Barlow, 1976).

Hersen and Barlow (1976) have identified specific issues to be addressed in medication evaluations; these include carry over effects, placebo effects, and blind assessments. Unlike some psychological treatments, medication may produce residual effects that remain evident after the drug is discontinued. Biochemical changes may carry over into subsequent experimental phases and influence their outcome. One means of dealing with this potential problem is to monitor drug levels in the body continuously. Initiation of the next phase may be postponed until drug levels reach an acceptable point (Herson & Barlow, 1976). For example, a child receiving Dexedrine for hyperactive behavior (B phase) may be placed on a placebo medication (A_1 phase) and then undergo a behavioral treatment package consisting of token reinforcement, praise, and activity feedback (C phase). In order to avoid carry over effects of the Dexedrine into the A_1 or C phases, active drug and placebo treatment can be separated by several days in which no treatment is given.

Another important issue is that the controlling for possible expectancy or placebo effects accompanying the administration of medication.[1] This usually entails administering an inert medication (placebo) and evaluating it is a separate phase (A_1). This strategy permits the effect of the experimental drug to be differentiated from change due to client expectancies. A common sequence illustrating this procedure is the A-A_1-B-A_1-B sequence. Note that a single variable is manipulated, with each phase change allowing its independent effects to be assessed. Clear differences between A_1 and B provide evidence for the therapeutic value of the psychoactive drug.

A final issue to be considered in the evaluation of medication is conducting blind assessments. Single-blind experiments are those in which either the service provider or the clients are unaware of whether they are receiving the active drug or a placebo. Double-blind studies, on the other hand, keep both parties naive about the nature of the medication administered to the client. As a general rule, it is desirable to achieve double-blind conditions to control for possible experimenter bias and subject expectancies. However, as Hersen and Barlow (1976) have pointed out, obtaining "true" double-blind conditions can be difficult in applied settings. Service providers or clients may be "tipped off" when clients manifest side effects associated with a specific drug (e.g., catatonic motor activity associated with haloperidol). Double-blind conditions may be

---

[1]Although a similar phenomenon occurs in psychological treatments, it should be pointed out that the effects are not entirely analogous to a placebo reaction. Kazdin (1980) states,

> A placebo in medicine is known in advance, because of its pharmacological properties (e.g., salt or sugar in a tablet), not to produce the effects that the patient experiences. In psychological treatment, one usually does not know in advance that the properties of the nonspecific treatment group are inert. For example, in many nonspecific treatment control groups, clients merely chat about topics tangetially related to their problem. Is this inert in the same sense that placebo medications might be? Chatting may be a relief for the client from his or her despair and, based upon sound psychological principles (yet to be enumerated) and psychological mechanisms (yet to be discovered), be a very active medical procedure. (p. 150)

similarly breeched when medication results in marked behavior changes that are obvious to participants in the study.

Table 1 provides a listing of various designs available for evaluation of drug effects. Each design is identified and classified according to its type (experimental or quasiexperimental) and its ability to accomodate blind assessments. Designs 1 through 15 are within-series designs commonly used in the clinical literature (see, e.g., Liberman, Davis, Moon, & Moore, 1973; Turner, Hersen, & Alford, 1974).

Recent applications have shown that this list may be extended. In Figure 2, for example, Bernard *et al.* (1981) employed a changing-criterion design (Design 16) to examine the efficacy of a procedure to reduce caffeine dependency. Similar strategies may be used to evaluate the withdrawal of a variety of psychoactive medications, with double-blind conditions possible through the gradual reduction of dosages to placebo levels. Designs 17 through 19 are com-

Table 1. Single-Case Experimental Drug Strategies[a]

Number	Design	Type	Blind possible
1	A-A_1	Quasi-experimental	None
2	A-B	Quasi-experimental	None
3	A_1-B	Quasi-experimental	Single or double
4	A-A_1-A	Experimental	None
5	A-B-A	Experimental	None
6	A_1-B-A_1	Experimental	Single or double
7	A_1-A-A_1	Experimental	Single or double
8	B-A-B	Experimental	None
9	B-A_1-B	Experimental	Single or double
10	A-A_1-A-A_1	Experimental	Single or double
11	A-B-A-B	Experimental	None
12	A_1-B-A_1-B	Experimental	Single or double
13	A-A_1-B-A_1-B	Experimental	Single or double
14	A-A_1-A-A_1-BA_1-B	Experimental	Single or double
15	A_1-B-A_1-C-A_1-C	Experimental	Single or double
16	Changing criterion A_1-B	Experimental	Single or double
17	MBD across subjects A-A_1-B	Experimental	Single or double
18	MBD across subjects A_1-B B-A_1 A_1-B B-A_1	Experimental	Signle or double
19	MBD across subjects Counter balanced for sequence effects	Experimental	Single or double

[a]Adapted from Hersen and Barlow (1976).

bined series, specifically multiple-baseline across subjects designs. These designs have the attractive feature of not requiring withdrawal of the medication to establish validity and permit investigation of sequence effects by counterbalancing placebo and drug phases across subjects (Design 19).

The application of time-series designs to assess the effect of medication and psychological intervention is illustrated in a study by Williamson, Calpin, DiLorenzo, Garris, and Petti (1981). In the treatment of a 9-year-old's overactive behavior, these researchers evaluated the therapeutic effects of Dexedrine (B); Dexedrine plus instructions (B, C); Dexedrine plus instructions and guided practice (B, C, D); and Dexedrine plus instructions, guided practice, and feedback and reinforcement (B, C, D, E). Observational measures were taken for on-task classroom behavior and appropriate lunchroom behavior. Activity levels were measured with an electronic device over the course of the A-B-A-B-BC-BCD-BCDE series. The results presented in Figure 5 suggest that Dexedrine was moderately successful in effecting desired change across all three measures. However, with introduction of each component of the treat-

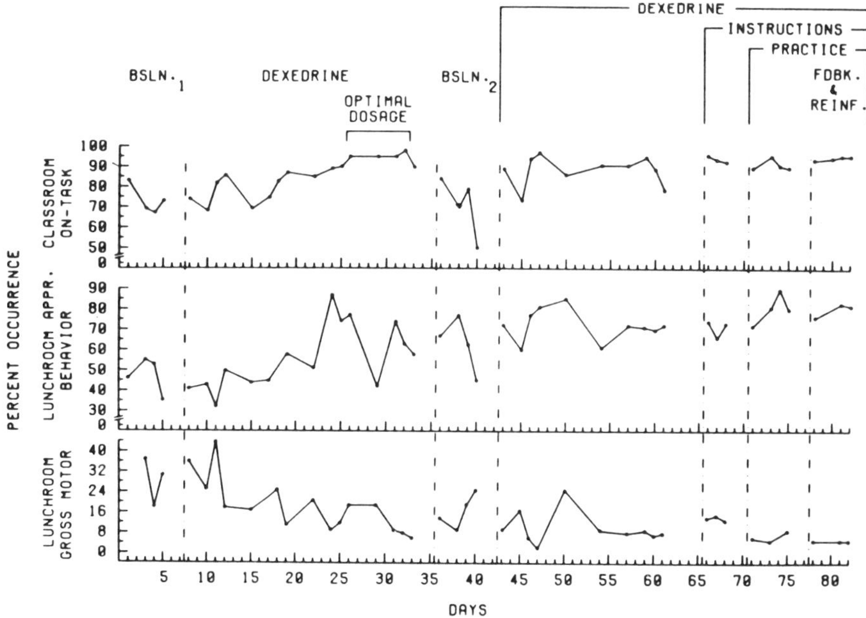

Figure 5. Behavioral observation data for classroom on-task behavior, lunchroom-appropriate behavior, and lunchroom gross motor behavior during all phases of the experiment (baseline phases are abbreviated as BSLN). (From "Treating Hyperactivity with Dexedrine and Activity Feedback" by D. A. Williamson, J. P. Calpin, T. M. DiLorenzo, R. P. Garris, & T. A. Petti, *Behavior Modification*, 1981, *5*, 399–416. Reproduced by permission.)

ment package, performance improves in some cases and is maintained in others. A note of caution, however, should be made with respect to the efficacy of the psychoactive medication. Since the medication phase (B) was not contrasted with a placebo phase (A_1), it is unclear whether effects were due entirely to the drug or whether client expectancies influenced the outcome.

Designs for Assessing Generalization

Design issues discussed thus far have been primarily concerned with demonstrating the causal relationships between the independent and dependent variables. Once this relationship is established, however, the clinical researcher will frequently be concerned with the nature and extent to which treatment effects generalize. Many authors have lauded the importance of assessing (Baer, Wolf, & Risely, 1968; O'Leary & Drabman, 1971) and actually programming generalization (Kazdin, 1976; Stokes & Baer, (1977). Until recently, however, there has been a conspicuous lack of methodology for achieving this end (Stokes & Baer, 1977).

In the present context, there are two types of generalization of concern: (1) generalization of treatment effects across situations (stimulus generalization) and (2) generalization of treatment effects to other client measures (response generalization).[2] When changes in the dependent variable occur in situations other than the treatment condition, stimulus generalization is said to have occurred. This may be the result of the subject's failure to discriminate changes in stimulus conditions (e.g., a different therapist providing the same therapy) or to the development of conditioned reinforcers capable of maintaining behavior after the original reinforcer is withdrawn. Response generalization, on the other hand, refers to changes in nontreated measures that can be attributed to the therapeutic intervention. For example, a phobic client receiving social reinforcement for successively longer contacts with the feared stimulus may also experience concomitant reductions in heart rate and blood pressure even though treatment contingencies were applied only to contact behavior.

Although the investigator may be interested in both stimulus and response generalization, the time-series design used places some restrictions on which type of generalization may be assessed (Kendall, 1981). The problem is that data patterns required to establish treatment effects are usually the converse of those needed to confirm generalization. Consider an A-B-A-B design in

---

[2]*Generalization* has also referred to the transfer of treatment effects to other situations (called stimulus generalization) and to other behaviors (called response generalization) in behavior therapy research (e.g., Stokes & Baer, 1977).

which strong treatment effects occur during the first B phase. If treatment is subsequently withdrawn (second A phase) and the data remain unaffected, is experimental control lacking or has stimulus generalization occurred? This dilemma makes assessing stimulus generalization problematic with within-series simple and complex phase-change designs. By contrast, however, evaluating response generalization with these types of designs is nonproblematic. That is, data collected on other client measures may suggest that effects have generalized without jeopardizing the experiment's internal validity. As indicated in Table 2, other time-series designs may also find assessment of certain types of generalization problematic and others nonproblematic.

Evaluating response generalization with the ATD and STD may also be achieved by concurrent assessment of nontarget behaviors. But since experimental control depends on changes in treatment conditions, looking at stimulus generalization again becomes troublesome. With the combined series designs, the MBD across situations or settings demonstrates control via behavior change that parallels the staggered introduction of treatment under different stimulus situations, making assessment of stimulus generalization difficult and response generalization possible when concurrent measures are available. In the case of the MBD across behaviors, the situation is reversed. Assessing generalization across different stimulus conditions is nonproblematic. However, because internal validity is linked to comparisons across behaviors, evaluating response generalization is typically not a viable option.

Several time-series design variations have been developed specifically to evaluate generalization of treatment effects. Among these are the (1) MBD-plus-generalization-phases design (Kendall, 1981), (2) sequential-withdrawal design (e.g., Rusch, Connis, & Sowers, 1979), (3) partial-withdrawal design

Table 2. Problematic and Nonproblematic Generalizations in Single-Subject Strategies

| | Generalization | |
Single-subject strategies	Problematic	Nonproblematic
Reversal	Treatment control or stimulus generalization	Testing for response generalization
Mutliple-baseline across situations	Treatment control or stimulus generalization	Testing for response generalization
Multiple-baseline across behaviors	Treatment control or response generalization	Testing for stimulus generalization
Multielement	Treatment control or stimulus generalization	Testing for response generalization

(e.g., Vogelsberg & Rusch, 1979), and (4) partial sequential-withdrawal design (Rusch & Kazdin, 1981).

Kendall (1981) offers a design strategy that is useful in assessing stimulus generalization. The design layout (depicted in Figure 6) involves obtaining concurrent measurements on multiple behaviors (three in this case) and staggering intervention in conventional MBD fashion. Following demonstration of treatment effects for each behavior, a series of generalization tests or phases are instituted. During these phases, stimulus conditions may be systematically varied in order to evaluate the generalization of treatment effects to the new stimulus situations. For example, holding the intervention constant, G_1 may represent treatment provided by a different therapist. G_2 may be treatment provided in a different setting (e.g., home vs. school), and G_3 may involve intervention using different materials (e.g., a token system using money vs. written check marks). The hypothetical data patterns appearing in Figure 6 represent some possible outcomes. For Behavior I, generalization is apparent across all three situations; for Behavior II, generalization appears nonexistent; and for Behavior III, effects were maintained in all but the G_2 phase. Data patterns in the latter two series may then lead the investigator to develop additional intervention strategies targeted at situations in which effects did not generalize.

Each of the withdrawal designs for assessing generalization are used with multicomponent interventions. As n components are individually withdrawn, data patterns are examined for the maintenance of treatment effects. That is, did therapeutic gains generalize to n-1, n-2, . . . , n-n conditions? In the sequential-withdrawal design, single components of a treatment package are withdrawn one at a time until all components have been discontinued. Effects of each n-i condition are evaluated in separate and consecutive experimental phases. This strategy has been used in conjunction with within-series withdrawal-type designs (e.g., O'Brien, Bugle, & Azrin, 1972) as well as combined-series multiple-baseline designs (e.g., Sowers, Rusch, Connis, & Cummings, 1980). A variant of this procedure, the partial-withdrawal design, involves withdrawing any of n treatment components from one series in a multiple-base-

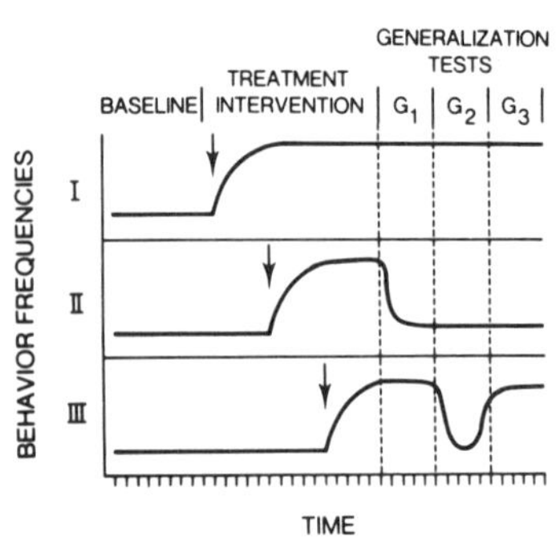

Figure 6. Hypothetical data patterns during postintervention generalization tests using a multiple baseline design. (From "Assessing Generalization and the Single-Subject Strategies" by P. C. Kendall, *Behavior Modification*, 1981, *5*, 307–319. Reproduced by permission.)

line design (see, e.g., Vogelsberg & Rusch, 1979). If performance is maintained in the treatment withdrawal series, additional components may be withdrawn from that series or like components withdrawn from other series. At any point in which performance deteriorates, intervention efforts would then focus on programming generalization to the specific n-i condition (Rusch & Kazdin, 1981). This strategy provides an advance look at how other series (i.e., behaviors, settings, or subjects) may respond to the withdrawal of particular components and avoids the loss of all treatment gains that may occur with complete withdrawal of intervention.

The partial-sequential withdrawal design represents a combination of the sequential and partial withdrawal designs (Rusch & Kazdin, 1981). In this procedure, all or part of a multicomponent treatment is withdrawn from one of the series in a multiple-baseline design. If the removal of treatment results in decreased performance, a sequential withdrawal of treatment components is instituted in the remaining series. As Rusch and Kazdin (1981) point out:

> Combining the partial and sequential withdrawal designs allows for the orderly withdrawal of the various components of the treatment in an effort to decrease the probability that subjects will discriminate the absence or presence of contingencies. [With this procedure] investigators can predict, with increasing probability, the extent to which they are controlling the treatment environment as the progression of withdrawals is extended to other behaviors, subjects, or settings. (p. 136)

Hypothetical applications of the partial–sequential withdrawal design appear in Figure 7. In the first example (Figure 7a), the introduction of praise and prompts (B) is staggered across two subjects. For Subject 1, the complete withdrawal of treatment produces a rapid loss of treatment gains. In order to avoid a similar loss in Subject 2, prompts only are withdrawn (C phase), followed by the withdrawal of prompts and praise (A phase) when effects were shown to maintain. The application of this procedure to evaluate three-component interventions with two and three subjects are presented in Figure 7b and 7c, respectively. In both examples, when the complete withdrawal of treatment (prompts, praise, and tokens) results in performance losses for Subject 1, the treatment package is reinstated and sequential withdrawal of tokens (D phase) and praise (A phase). In Figure 7c, Subject 3 benefits from the knowledge gain in subjects and does not suffer from treatment losses at all. Finally, Figure 7d illustrates the use of this strategy with a within-series design using two subjects. Following an A-B-A-B sequence for both subjects, treatment components are withdrawn one at a time for Subject 1. Like the other cases, when Subject 2 experiences performance losses with the removal of two components, both components are reinstated and sequentially withdrawn.

The purpose here has been to introduce the reader to various design options for assessing stimulus and response generalization in psychotherapy research. More detailed discussions of these procedures along with their rela-

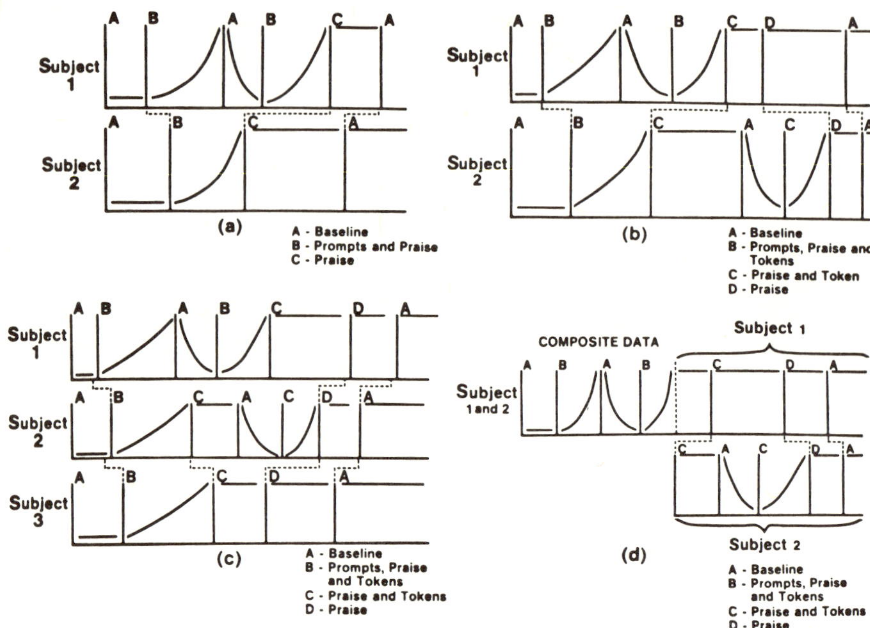

Figure 7. Within-series withdrawal designs used for assessing maintenance of treatment effects. The figure represents the withdrawal of a two-component treatment across two subjects (a), a three-component treatment across two subjects (b), a three-component treatment across three subjects (c), and a three-component treatment across two subjects within an A-B-A-B sequence (d). (From "Toward a Methodology of Withdrawal Designs for Assessment of Response Maintenance" by F. R. Rusch & A. E. Kazdin, *Journal of Applied Behavior Analysis*, 1981, *14*, 131–141. Reproduced by permission.)

tive advantages and disadvantages are available in Kendall (1981) and Rusch and Kazdin (1981).

TIME-SERIES RESEARCH: CONCEPTUAL AND METHODOLOGICAL ISSUES

In recent years the number of conceptual and methodological issues involved in time-series research has expanded greatly. Various developments in social science research have expanded, and many of these allow new conceptual and methodological issues to be addressed in time-series work. In this section of the chapter, we provide an overview of validity issues in time-series research with a focus on statistical conclusion validity, internal validity, construct validity, and external validity. Thereafter, we discuss the role of time-series research in the context of other methodological issues and its role in clinical practice.

Validity Issues

A major purpose of psychotherapy research is to reveal relationships between variables that may otherwise go undetected. Specifically, in experimental research the investigator is usually concerned with the effect that the independent variable (treatment) may have on the dependent variables (client outcome measures). However, the nature of research dictates that conclusions drawn as a result of investigation be considered tentative. Numerous extraneous variables may influence judgments about whether variables are related, whether cause-and-effect relations exist, whether the construct under study is the therapeutic variable, and the degree to which findings may be generalized. The extent to which the influence of these extraneous variables can be eliminated is the extent to which "true" or valid inferences may be drawn. Yet, all potential influences can never be controlled, and *absolute validity* is never achieved (Cook & Campbell, 1979; Kiesler, 1981). The task of the researcher is to minimize the threats to validity in order to increase confidence in the conclusions drawn.

Several discussions of the threats to validity have appeared in the applied research literature (Hersen & Barlow, 1976; Kazdin, 1980; Kratochwill, 1978) with most authors referring to the original work of Campbell and Stanley (1963) as a template for their treatment of internal and external validity. Campbell and Stanley (1963) noted that in order for valid inferences to be made about the causal relationship between the independent and dependent variables, the experiment must be internally valid. That is, internal validity is the degree to which the research eliminates alternative explanations for the results. External validity, on the other hand, refers to the extent to which the researcher's findings can be generalized to different subjects, settings, and experimenters. This conceptualization of validity issues has been extended by Cook and Campbell (1979). Internal validity is subdivided into statistical conclusion validity *and* internal validity, and external validity has been broadened to include construct validity *and* external validity. For more detailed treatment of these issues, the reader is referred to Cook and Campbell (1979).

Statistical Conclusion Validity

In order to determine if a treatment causes change in the dependent variable, the researcher must estimate whether and to what extent the treatment and outcomes covary or go together. Application of various data analytic methods in a time-series experiment complements the design to help draw conclusions from the data and therefore constitutes an important statistical conclusion validity issue (Kratochwill, 1979). However, use of various analytic methods to assess covariation poses threats to valid inference making. Cook and Campbell (1979) refer to these threats under the rubric of statistical conclusion validity and list the following threats: low statistical power, violated assumptions of statistical tests, fishing and the error-rate problem, the reliability of measures, the reliability of treatment implementation, random irrelevancies in the experimental setting, and random heterogeneity of respondents. Each of these threats must be considered in any experimental research and are completely relevant in time-series research in psychotherapy. In this section we will focus on some of the analytic issues that have received the most attention in time-series research. (For a more complete review of statistical conclusion validity threats, the reader is referred to Cook & Campbell, 1979, and Conger, Chapter 9.)

In time-series research in psychotherapy, a major source of controversy has been the use of visual analysis or inferential statistical techniques. As traditionally conceived, visual analysis consists of plotting the data from the experiment and representing them graphic form across various phases of the study. The researcher(s) then makes a judgment of the effect or effect pattern in interpretation of outcome. Statistical analysis consists of the application of an inferential statistical test with the decision rule based on probability estimates. However, both methods involve statistical qualities (e.g., stability, variability, overlap, number of scores, autocorrelation, etc.). Yet the distinguishing characteristic of the statistical test is the use of inferential probability statements.

Visual Analysis

Graphic analysis procedures have served an important function in time-series psychotherapy research. To begin with, these procedures have been easy to learn and apply. In time-series research, they have provided researchers with direct and continuing contact with the data and may prompt new ways of conceptualizing relations among variables. Second, particularly in applied behavior analysis, visual analysis has been a preferred method for detection of "large" experimental effects. It is argued that graphic inspection acts as a filter to eliminate small effects, and hence their use promotes large effects (cf. Baer, 1977; Michael, 1974; Parsonson & Baer, 1978). In contrast, statistical analysis is said to promote the study of a wide range of variables but detection of relatively small, nondurable, less replicable, less generalizable, and less meaning-ful effects.[3]

Despite the positive aspects of visual analysis, these procedures should be employed within the context of several considerations. First, there are a number of technical considerations that form the basis of the correct application of this method. Graphical data should be presented in a manner that promotes accurate interpretation. Usually, a variety of technical aids are employed to assist in visual analysis from the data plot (see Kazdin, 1976; Parsonson & Baer, 1978, for reviews). For example, such procedures as standard deviation bands and regression lines have been fitted to the data, although these methods have not received widespread support.

A second issue relates to the actual patterns of data stemming from base-line stability, variability, score overlap, number of data points, trends, changes in level, analysis across phases, and autocorrelation. Most of these can be handled through visual analysis, but special concerns have been raised when the data are autocorrelated (i.e., those series in which the present value of the data are to a degree predictable from past values). For example, Jones, Vaught, and Weinrott (1977) demonstrated that various patterns of data (e.g., stabilities, variabilities, averages), whether obtained visually or statistically, may be biased by serial dependency. When serial dependency is present, data may not be as interpretable as if the estimates were obtained from independent scores. Subsequently, Jones, Weinrott, and Vaught (1978) demonstrated that comparisons between visual and time-series analysis showed that serial dependency in scores is likely to reduce agreement between two data analysis methods. Thus, visual and statistical analysis will likely disagree most often when the data are highly autocorrelated.

A third consideration relates to the criterion applied to measures of outcome. In applied behavioral research, experimental and therapeutic criteria have been recommended (Kazdin, 1977; Wolf, 1978). The experimental criterion involves comparison of the dependent variable before and after introduction of the treatment and has typically been established through visual analysis. Visual analysis is said to be apropriate through use of a clinical or therapeutic criterion, making the use of a statistical analysis potentially redundant. The clinical criterion is usually established through social validation in

---

[3]This argument has been conceptualized within the form of conventional error probabilities (Baer, 1977; Parsonson & Baer, 1978). It is suggested that applied behavioral designs are interpreted with very low Type I error probabilities (concluding that a change in the dependent variable has occurred when in fact it has not) and corresponding with probabilities of Type II error (concluding that no change has occurred when in fact it has). This contrasts with conventional group designs, which are usually interpreted with moderately low Type I error probabilities and with lower Type II error probabilities relative to behavioral designs.

which behavior of the client is compared to peers who do not have problems and by soliciting evaluation of the target client's performance in the environment. Thus, outcomes are viewed as clinically important if the treatment brings the client's performance within the range of socially acceptable levels as judged normal by the client's peers or if the client's performance is judged by others as reflecting a qualitative improvement on global ratings (Kazdin, 1977a).

Example

A study reported by Matson (1981), in which three moderately mentally retarded females ranging in age from 8 to 10 years were treated for long-standing fears, provides an example of how social validity has been used to establish a therapeutic criterion. To establish a criterion for successful performance on the dependent measures (i.e., approaching and talking to strange adults as well as child ratings of overall fear), the children were matched on age, sex, and level of mental retardation with children having "normal" amounts of fear. Specifically, the teacher and teacher's aides were asked to rank the peers of the same set as the participant from their class in order of socially appropriate fear level. The behavior was rated based on the degree of fear displayed toward significant adult strangers in the child's life who were identified as "safe" (i.e., persons who would not harm the child). Ratings from both the teacher and the aide were summed. The child with the lowest total score was determined to be the most normal on fear of strangers. The child chosen from the class of each client (same age) was the only person matched with the classmate on clinical performance. Thus, each child so identified was run through the same assessment as those children displaying the clinical fears and was used as a clinical criterion for the success of the treatment program. Results of the study for each subject are reported in Figure 8. It can be observed that treatment produced outcomes that put the subjects within a normal range based on the responses of the matched normal child (represented by the dotted lines on each graph).

Considerations

The use of visual analysis in time-series research has a strong tradition and will likely continue in behaviorally oriented psychotherapy. Even when social validation procedures are employed, there are potential problems with visual analysis. As Kazdin (1977a) has noted, normative standards may be perceived as inappropriate standards against which to evaluate change; normative groups may be difficult to assess for certain types of problems; individuals conducting the subjective evaluation may not establish consistent criteria

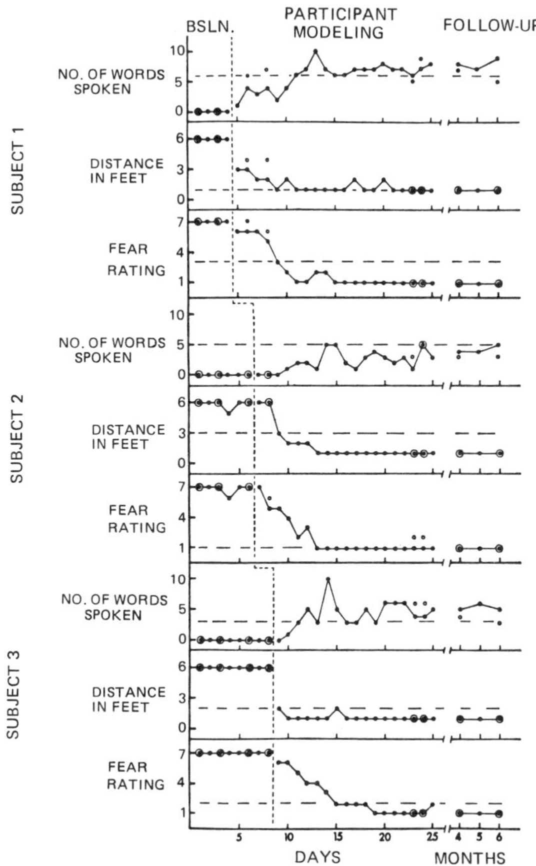

Figure 8. Levels of fear during baseline, treatment, and follow-up sessions. The open circles represent a measure of generalization at the children's homes. Dependent measures are represented for all three children in a multiple-baseline design. The dotted horizontal lines represent score of nonfearful children. (From "Assessment and Treatment of Clinical Fears in Mentally Retarded Children" by J. L. Matson, *Journal of Applied Behavior Analysis,* 1981, *14,* 287–294. Copyright 1981 by the Society for the Experimental Analysis of Behavior, Inc. Reproduced by permission.)

or may not view changes in line with the therapist; and/or scales employed may lack reliability and validity.

Aside from this issue, social validation still may not provide a *reliable* decision rule, since it will typically vary across studies (Kratochwill, 1979). This concern is present regardless of whether or not the focus of the investigation is on the same target problem.

Finally, visual analysis may not be a particularly good procedure for detection of subtle or weak variables. Such variables may be important, but typically they do not meet therapeutic criteria. For example, such variables could be important when added to treatment already of known benefit for maintenance or even enhancing effectiveness (Kazdin, 1976).

Statistical Analysis

Choice of an appropriate statistical test for a time-series experiment has become a major issue in the psychotherapy literature. Many of the issues raised in statistical analysis of time-series experiments have been discussed in several books (e.g., Kratochwill, 1978; McCleary, Hay, Medinger, & McDowell, 1980), and chapters (e.g., Kazdin, 1976; Wallace & Elder, 1980), and therefore only some major issues will be discussed in this chapter. A rationale for the use of statistical tests can be based on at least three considerations, including the characteristics of data, the role of a formal criterion in data analysis, and detection of small effects.

Data Characteristics

As noted in the previous section, correlated data may cause difficulties in data evaluation through visual methods. Statistical tests provide one of the best means to deal with this issue. However, it must be stressed that an appropriate statistical test must also be employed to deal with this problem (see below). In addition, certain characteristics of the data, such as variability and trend, may cause interpretive problems. Statistical tests generally provide a systematic means to deal with these issues. For example, time-series analysis provides the researcher with information about certain patterns in the data, such as seasonal components. Many such features are difficult to detect with the eye, and so researchers might be led astray. Also, as Kazdin (1976) has noted, statistical evaluation of the data provides the researcher with the possibility of analyzing continuous shifts across phases where no change in trend is apparent.

Formal Criterion

A major issue in the use of statistical analysis is that it provides researchers with a formal and reliable criterion on which to base conclusions. Jones *et al.* (1978) argued that researchers should question visual inference, since analysis based on these methods generally yields low reliability. On the other hand, given that an appropriate inferential test is established, statistical analysis provides a formal and reliable decision role on which the researcher can base conclusions in the study (see Glass *et al.,* 1975; Gottman & Glass, 1978; Jones *et al.,* 1977, 1978; Levin, Marascuilo, & Hubert, 1978).

Detection of Small Effects

As noted in the previous section of the chapter, researchers in some areas of psychotherapy (such as applied behavior analysis) have argued that large effects are to be supported in research. However, it is also possible that small effects might be important at some stage in the development of certain psychotherapeutic techniques. In such cases, statistical analysis may act as a filter to screen out variables that produce reliable but small effects (Kazdin, 1976). Study of variables that produce small effects could be supported on several grounds. To begin with, variables that independently produce small effects may produce large effects when combined with other variables. In addition, variables that initially produce small effects might demonstrate large effects when replicated with different types of therapists, subjects, and disorders. Generally, a case should be made for not dismissing studies that produce small effects until a fairly good data base is established in a particular area of psychotherapeutic application. Generally the statistical test is designed to help answer questions. The first is: Was the effect due to chance? A second is: Will it stand up to replication? (Elashoff & Thoresen, 1978). In summary, many issues in the use of statistical tests depend upon the area being investigated and the knowledge base developed therein.

Considerations

Several considerations have been raised in the use of inferential statistical tests in time-series research. First of all, consider the validity of a particular statistical test, which in itself is a statistical conclusion validity concern (Cook & Campbell, 1979). A rather large body of literature now suggests that adoption of conventional parametric tests such as t tests, analyses of variance, and multiple regression are generally not appropriate for most experiments of the time-series type (see Gottman & Glass, 1978; Kratochwill, 1978; Levin *et al.,* 1978; Thoresen & Elashoff, 1974). There are at least two major problems in employing traditional or classical parametric procedures. First, adoption of a parametric procedure such as analysis of variance would suggest that the researcher can make inferences about unobserved behavior samples of the subject or subjects under consideration. Yet since a nonrandomly selected time frame is selected for the experiment, an argument cannot be made for this type of generalization. Second, a major assumption in the use of traditional parametric procedures is that estimates of "error" in the data are independent (that is, what is left of an observation after it has been deviated by the model's parameters). Typically, the autocorrelated data that emerge from a time-series type of experiment render traditional tests inappropriate. Specifically, statistical tests such as those based on the F and t distributions will be inappropriate, because Type 1 error is seriously inflated with autocorrelated data. The data of Jones and his associates, discussed earlier, suggest that most time-series have autocorrelated data; therefore traditional procedures will generally be inappropriate. However, if the data are not correlated, a case might be made for the traditional statistical procedures; nevertheless, the burden is on the researcher to determine this (Levin *et al.*, 1978).

Aside from the validity of a particular test, there is also the issue of choosing the right type of test for answering a particular problem involved. Generally, in experiments of the time-series type, two general classes or domains of alternatives have emerged, including nonparametric methods (e.g., Edgington, 1980, 1982; Kazdin, 1976; Levin *et al.*, 1978) and time-series analysis (e.g., Glass *et al.*, 1975; Gottman & Glass, 1978; McCleary, Hay, Meidinger, & McDowell, 1980; McDowell, McCleary, Meidinger, & Hay, 1980). In the class of statistical tests called randomization tests, random sequencing is used as the basis for the procedure. A number of authors have described applications of these tests; there are now a number of examples of these procedures in the literature suggesting that they have applications to a number of time-series designs. Generally, such procedures are appropriate for within-, between-, and combined-series designs (discussed in the earlier section of this chapter). A major limitation of randomization tests is that they typically focus on a single descriptive summary measure, such as the mean, and the researcher is not always in a position to test changes in level or variability in the data. Nevertheless, these procedures seem useful, particularly during an exploratory phase of data analysis when a researcher may not have firm expectations about the results (Elashoff & Thoresen, 1978).

Another class of analysis procedures comprises time-series methods. These procedures are adopted in a typical time-series experiment to control the problem of autocorrelation by removing it through the specification of an appropriate model. After the researcher has established an appropriate model, the "residual" data series is viewed as a sequence of uncorrelated observations. Some time-series programs test level and slope, while others focus on other features of the data, such as change in level, variability, and seasonality. Time-series analysis requires the use of computer program for correct model identification and testing of the data. We have found the PACK program to be a very useful one. Applications of this program are relatively straightforward and can be used to accompany the review of time-series analysis provided by McCleary and Hay (1980).

Internal Validity

As noted above, the purpose of statistical testing is to draw valid inferences about the *covariance* between treatment and outcome. Valid conclusions regarding the causal relationship between these variables is the object of the experimental design. The central task is to decide whether A causes B. However, numerous extraneous variables may affect the measured outcome of B. Consider the situation where A is the treatment systematic desensitization and B is client self-report of fear. The researcher wishes to determine whether A→ B. It is quite possible that the therapist's experience with systematic desensitization (C) contributes significantly to the effectiveness of the treatment. Therefore, increases in C have a positive effect on A, which in turn affects B. This relationship may be characterized as C + A→B, which is much different than A→B. In order to find out how A affects B, the effects of C must be accounted for. Internal validity, then, is the degree of certainty that manipulation of the independent variable is responsible for observed changes in the dependent variable.

In the preceding example, therapist experience, if left uncontrolled, would constitute a threat to internal validity. Numerous threats to internal validity have been identified, along with suggestions for minimizing their influence (see Table 3; Cook & Campbell, 1979; Kazdin, 1980; Kratochwill, 1978). It should be noted that more than one threat to internal validity can affect a given study. Moreover, multiple threats that bias outcome in the same direction may serve to compound invalidity. Similarly, internal validity threats of equal magnitude operating in opposition to one another may tend to cancel their biasing effects (Cook & Campbell, 1979). Because the magnitude of the bias will generally vary from study to study, the researcher is presented with few options beyond controlling as many factors as possible if valid inferences are to be made.

Construct Validity

A major part of any research effort is the identification and definition of the variables under study. Upon defining the independent and dependent variables, the researcher wants to generalize from the operations or measures used to theoretical constructs. In order for such generalizations to be valid, there must be a good fit between the defined operations and the referent construct. As such, construct validity refers to the degree to which operations and constructs overlap. Threats to construct validity arise when (1) extraneous variables change inadvertently along with the treatment or (2) definitions inadequately represent the theoretical construct.

Table 3. Threats to Internal Validity[a]

History	The occurrence of events extraneous to the experimental treatment that may affect the dependent measure.
Maturation	Physical and/or psychological changes occurring within subjects that may affect the dependent measure over time. Becomes a threat to internal validity when such changes are not the focus of research.
Testing	Changes in the dependent variable due to the process of measuring subject performance. May result from subjects having taken a pretest because of the reactivity of the measurement process.
Instrumentation	Changes in the dependent measure due to the use of inconsistent measurement procedures over the course of evaluation. Instrumentation may occur when data collectors alter their method of recording performance as a result of experience, observer bias or drift, or the malfunction of mechanical recording devices. May also result from tests having unequal intervals, leading to so-called ceiling and floor effects.
Statistical regression	If subjects are assigned to groups or treatment conditions on the basis of unreliable pretest or baseline measures, high scores will tend to decrease their performance over subsequent measurement occasions while the performance of low scores will increase. Regression always occurs toward the population mean of a group; thus, scores in the midrange will likely be unaffected.
Selection	When groups are formed by arbitrary rather than random methods, their differential performance may be due to preexisting differences between groups rather than actual treatment effects.
Mortality	The withdrawal of some subject observed at the pretest or baseline period before the final assessment may result in unequal groups. Observed effects may be attributed to differences in subject characteristics or their response to treatment rather than the effects of the independent variable.
Interactions with selection	The interaction of history, maturation, and/or instrumentation threats with selection resulting in spurious treatment effects. History $\times$ maturation may occur when subjects or groups experience different historical factors which influence performance. Maturation $\times$ selection results when subjects or groups mature at different rates to increase the disparity between groups over time. Selection–instrumentation occurs when performance is scored differently for different groups due to observer factors or tests whose intervals are unequal.
Ambiguity about the direction of causal influence	For many correlational studies in which the temporal ordering of variables is not certain, it is unclear whether A causes B or B causes A. Measures collected at different points in time provide information about the temporal priority not available in correlational studies that are cross-sectional.
Diffusion of imitation of treatments	When subjects in the experimental and control groups are free to communicate with each other, it is possible that subjects may

Table 3. (continued)

	exchange information about the procedures of conditions of their particular group. The validity of the experiment is, therefore, threatened because the groups are no longer independent.
Compensatory equalization of treatments	When experimental treatments provide subjects with desirable services, administrators may find it unacceptable to "deprive" the no-treatment control group of these benefits and insist that comparable or compensatory services be provided. The intended contrast is thus nullified and causal statements about the independent variable are rendered invalid.
Compensatory rivalry by respondents receiving less desirable treatments	When subjects are aware of their group status (i.e., experiment or control), those not receiving treatment may compete with their experimental counterparts. Observed effects may be the result of this rivalry rather than the independent variable.
Resentful demoralization of respondents receiving less desirable treatments	Control subjects aware that they are receiving less desirable treatment may respond by lowering their standard of performance. Between-group differences following treatment could not be attributed to the effects of intervention.
Resentful demoralization of respondents receiving less desirable treatments	Control subjects aware that they are receiving less desirable treatment may respond by lowering their standard of performance. In such cases, between-group differences following treatment cannot be attributed to the effects of intervention.

[a]Adapted from Cook and Campbell (1979).

The following examples should help clarify the problem. This first condition is commonly known as "confounding," and it poses serious restrictions on making valid inferences that Construct A causes changes in Construct B. Consider the researcher who wishes to evaluate the impact of the independent variable client-centered therapy on the dependent variable adult depression. Client-centered therapy comprises several operations that, when well defined, represent that construct. However, a construct that often accompanies any form of therapy but is seldom defined is that of therapist attention. Because therapist attention alone has been shown to help lift adult depression (Kazdin & Wilson, 1978), confusion arises as to what extent each construct, client-centered therapy and/or therapist attention, is responsible for observed changes in adult depression. Using group designs, the inclusion of a control group of subjects receiving therapist attention only is a common method for teasing out the therapeutic impact of each construct. The objective may be achieved with time-series designs by alternating between treatment and therapist attention within a single series, between multiple series, or both.

The second condition leading to threats to construct validity involves definitions that inadequately represent the referent construct. Obtaining consensus on the meaning of a given construct may approach the impossible. Yet, some definitions clearly underrepresent or overrepresent their respective constructs. For example, defining social withdrawal in terms of length of time spent alone leaves unaddressed many other aspects of the construct used throughout the therapy literature.

Issues of construct validity also arise in drawing conclusions about the effectiveness of a specific treatment. As noted above, operations and constructs should be closely linked to avoid ambiguous conclusions. When therapeutic constructs are involved, operations should also reflect the optimal treatment strength and integrity (Yeaton & Sechrest, 1981). The strength of a treatment refers to the amount and purity of those factors that contribute to change. For example, social reinforcement may be considered a strong treatment when it comprises highly positive verbal comments in reference to the behavior to be reinforced, delivered immediately and contingent upon the target behavior, accompanied by positive facial expressions and physical contact, and without interference from competing factors. Treatment integrity, on the other hand, is the degree to which a specified treatment is delivered as prescribed (Sechrest & Redner, 1979). Deviations from the treatment plan, regardless of its effectiveness, make it unclear what construct is responsible for the therapeutic effects. Similarly, treatments of insufficient strength to achieve the desired results may underrepresent the treatment construct and lead to invalid conclusions about its effectiveness. This problem can be attenuated to some extent by monitoring the administration of the independent variable.

External Validity

In addition to drawing valid inferences between the independent and dependent variables, the researcher is usually interested in generalizing beyond the results. External validity concerns to the degree to which the findings of a given study can be generalized. Traditionally, time-series experiments and particularly those based on a single case have not been known for establishing or increasing the strength of external validity in a particular research area. For example, Kiesler (1971) noted that "ideographic study has little place in the confirmatory aspects of scientific activity which looks for laws applying to individuals generally" (p. 66). More recently, Kazdin and Wilson (1978) noted that "the results of single-case demonstrations provide no hint of the generality of the findings to other cases" (p. 169). Indeed, a major limitation of the single-case time-series experiment is that the researcher does not know if a particular therapeutic technique would be equally effective when applied to clients with a similar problem ("client generality"), or that different therapists employing the same technique would achieve the same results ("therapist generality"), or if the technique would work in a different setting ("setting generality") (cf. Hersen & Barlow, 1976). Despite these considerations, some definite conceptual advances have occurred in establishing the strength of external validity and the issues surrounding this construct.

Two dimensions of external validity are important in psychotherapy research. Of particular concern to the researcher is the extent to which the results generalize (1) to the population of subjects, disorders, settings, experimenters/therapists, and times targeted within the study and (2) *across* types of subjects, disorders, settings, experimenters or therapists, and times within and outside the study. Cook and Campbell (1979) distinguish between the two types:

> The former is crucial for ascertaining whether any research goals that specified populations have been met, and the latter is crucial for ascertaining which different populations (or subpopulations) have been affected by a treatment, i.e., for assessing how far one can generalize. (p. 71)

Like Cook and Campbell (1979), we would place greater emphasis on generalizing *across* populations. To begin with, most psychotherapy researchers will usually be interested in generalizing *across* populations rather than *to* populations. For example, many psychotherapeutic researchers working with agoraphobics would typically want to conclude that a particular treatment (exposure) had a particular effect with a sample in a study, independent of how well the population of agoraphobics can be specified. Second, the generalization *to* a population requires the drawing of random samples. Yet, this procedure is rare in most psychotherapy research, so that generalization *to* a population is usually not possible. Third, most time-series experiments involve small *n*'s or even one case, making the random sampling procedure impossible.

Yet in time-series experiments, several conceptual tactics might be employed to increase the strength of external validity. First of all, several tactics outlined by Cook and Campbell (1979) for increasing external validity in group designs might be employed in time-series experiments where large samples are used. For example, once a target population has been specified, the researcher can construct a sampling framework and select instances so that the sample is representative of the population (one hopes, randomly chosen and randomly assigned to groups). Also, the researcher may be able to define target classes of clients, settings, and times to ensure a wide range of instances and represent them in the design. This tactic does not require random sampling and is usually more feasible. Another tactic is to define the kinds of clients, settings, or times to which the researcher wants to generalize and then select one or more instances of each dimension in the design.

Each of these alternatives will be difficult to implement in the typical time-series experiment because most studies will not involve between-group designs and will usually be based on a small number of subjects. As another alternative, the researcher has a class of *replication* alternatives that can be implemented. This alternative is available to the field in establishing the external validity of research and is not limited to one researcher's efforts. That is, replication of experiments can be carried out in multiple settings by multiple researchers.

As traditionally conceived in single-case time-series research, replication involves three alternatives: direct replication, systematic replication, and clinical replication (Hersen, 1982; Hersen & Barlow, 1976; Sidman, 1960). In the direct-replication strategy, the researcher evaluates a single treatment administered by the same researcher in the same setting on a specific problem over more than one client. This replication strategy requires that (1) therapists and settings remain constant across replications, (2) the behavior problem be homogeneous across clients, (3) subject characteristics (e.g., age, sex, etc.) be as similar as possible, (4) the treatment be uniform (until failure to replicate occurs), and (5) one successful experiment and three replications be conducted. There are several considerations with this type of replication strategy. First of all, when a failure to replicate occurs (i.e., mixed replication in the series), there are no clear guidelines for seeking out the reasons why. The researcher must usually run additional studies and sort out the failure to replicate on certain dimensions (e.g., variations in setting, disorder, etc.). Second, a formal criterion for determining an experimental effect must be used so that a *reliable* replication can be established (Kratochwill, 1979). We would recommend that a formal statistical criterion be employed. Finally, direct replication is aimed at answering questions about generality across clients but does not address generality across therapists or settings.

Systematic replication can help address generality across therapists and settings in that findings from direct replications are replicated across these dimensions. Also, systematic replication series provide further information on the generality of findings across clients, since new subjects are included in the research. Hersen and Barlow (1976) consider systematic replication a search for exceptions. Generally, systematic replication will be a long process. Indeed, questions might be raised over when such a series is really finished (Kazdin & Wilson, 1978). Likely, such replication will be time-consuming and expensive, occurring over a period of many years.

A final class of replication alternatives, called clinical replication, refers to the administration of a treatment package (with two or more distinct treatment procedures) by the same researcher(s). Such a replication would be repeated within a specific setting on clients presenting similar combinations of multiple behavior problems. Such a strategy represents an example of treatment "technique building" and usually extends over many years.

Role of Time-Series Methods in Psychotherapy Research

Contributions to Research

Time-series methodology has been used in many settings for many years and has assisted psychotherapy researchers in developing an important knowledge base in the field. A question often arises in the psychotherapy field over what type of research methodology is best for advancing knowledge. Yet we believe that this question cannot be answered in the abstract. Actually, a number of different methodological strategies are useful for generating knowledge for psychotherapy. Attempts to address this issue must consider the knowledge base in a particular area, type of disorder, availability of subjects and therapists, and settings, among other features.

Several perspectives on clinical research have been helpful in elucidating the role of time-series methods in psychotherapy research (e.g., Agras & Berkowitz, 1980; Kazdin, 1981; Ross, 1981). One perspective offered by Agras and Berkowitz (1980) places research activities within a sequence for contributions to *clinical research*.[4] They present a "progressive model of clinical research" (see Figure 9) in which it is suggested that it is the accumulation of studies in a particular area that shapes the future course of research. The model begins with the development of a novel intervention beginning with short-term outcome studies. Single-case time-series strategies and analogue-population experimental strategies are said to be useful for defining which components of therapy are functional or for testing the effect of theoretically derived procedures.

This model has several important implications within the context of time-series research and its contributions to psychotherapy. First of all, the model suggests that many different types of research make contributions to clinical research. Thus, it is the interrelationships among the different approaches that create the knowledge base. Sometimes, too much emphasis has been placed on certain types of research, such as comparative outcome studies (e.g., Luborsky, Singer, & Luborsky, 1975; Smith & Glass, 1977). This has been particularly evident in the rise of meta-analytic studies to evaluate psychotherapy. We will not review the criticisms of this approach (see Kazdin & Wilson, 1978; Rachman & Wilson, 1980) but would like to emphasize that little or no emphasis has been placed on time-series investigation in these approaches.

It must also be emphasized that in certain areas of behavior disturbance it may be impossible to use a variety of alternative research strategies. For example, with certain types of rare disorders, case studies and single-case time-series designs may be the research strategy of choice. At this time, it would be

---

[4]Agras and Berkowitz (1980) note that the research strategies in their model extend beyond clinical research as usually conceived.

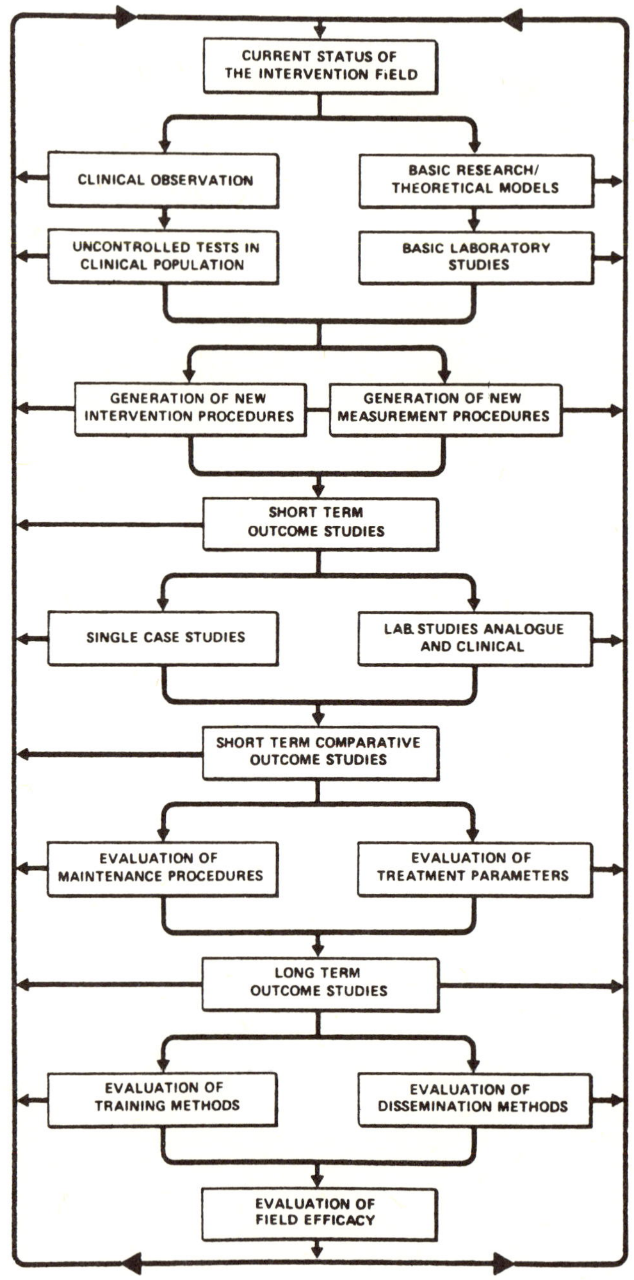

Figure 9. A progressive model of clinical research. (From "Clinical Research in Behavior Therapy: Halfway There?" By W. S. Agras & R. Berkowitz, *Behavior Therapy*, 1980, *11*, 472–487. Copyright 1980 by Associates for Advancement of Behavior Therapy. Reproduced by permission.)

desirable for the field to develop a strategy for integrating results from studies of the time-series type that extend beyond the more conventional literature review. This could probably by analogous to the meta-analysis strategy as long as researchers were mindful of the pitfalls of this approach.

Another major consideration in the contribution of time-series designs to research is that the results have direct implications for practice. A major issue here is whether or not the treatment generated in research has application in actual practice (Kazdin, 1981; Ross, 1981). A common perspective in the past is that research can be conceptualized as analogue and nonanalogue (i.e., clinical). Yet as Kazdin (1980) has argued (see also Kazdin, Chapter 7), conceptualizing research as analogue or nonanalogue has obscured important differences between all research and the clinical setting. Therapy conducted in a research context can be characterized by a number of dimensions (e.g., instructions, assessment, etc.) that limit generalizations to actual clinical practice. In this regard, even time-series research strategies will be limited in testing therapeutic efficacy.

As an alternative to traditional research procedures, Ross (1981) advocated an approach that combines detailed study of the individual case(s) when group comparison studies are conducted. In this *tracer method,* extensive information is gathered on each case after group-comparisons have been made. Thereafter, a *deviant-case analysis* is conducted (Ross, 1981).

> To do this, one first investigates a research question with one of the traditional experimental designs that employs groups of subjects and requires data analysis by appropriate statistical methods. Immediately after the data have been analyzed, while the subjects can still be identified and contacted, the results are examined for deviant cases, that is, for subjects whose performance did not "fit" the tested hypothesis. Depending on how many of these deviant cases are found, either all, or a sample, or the most extreme of them are then called back to the laboratory where they are examined by whatever case-study method is appropriate. Given time and resources, one can even go one step further and subject individuals whose performance was close to the mean of the group to similar individual scrutiny. By doing this, one can seek an answer to the question in what ways the deviant cases differed from the group. I suspect that this approach would generate new and interesting hypotheses that could then once more be tested in a well-controlled group design, thus leading to an ever greater refinement in our knowledge. (pp. 325–326)

While this approach holds promise for advancing knowledge of psychotherapeutic efficacy, it has limitations as well. To begin with, and as suggested above, not all disorders can be evaluated in group designs. Perhaps more problematic is the relevance and rigor characteristic of the approach. The actual constraints imposed on any therapy through research will limit generalizations that can be made (Kazdin, 1981).

As a further alternative, Kazdin (1981) proposed that *clinical replication* be used to obtain information on treatment efficacy. Clinical replication refers to a procedure in which treatment is introduced into an actual clinical setting (e.g., school, hospital, community) without the usual controls imposed in research (e.g., design, therapist training, treatment monitoring). For the strategy to be effective, the issues of internal validity would already have to be established. Clinical replication would then be focused on reestablishing the generalizability or external validity of a particular treatment. Time-series research strategies, along with other designs, would help establish the internal validity or treatment efficacy of a particular technique. Various methodological dimensions could then be loosened as the technique is applied in practice. However, it should be emphasized that clinical replication is itself subject to empirical evaluation.

Contributions to Practice

Various conceptual features of time-series designs have been extended beyond the methodology as conceived in this chapter. This is most clearly observed in the recent literature in which it is argued that single case time-series designs will promote research among practitioners in the psychotherapy field, perhaps better than traditional research approaches. Particularly in the behavior therapy field, a number of authors have suggested that single-subject time-series strategies offer an empirical method for practice (e.g., Barlow, 1980; Browning & Stover, 1971; Hayes, 1981; Hersen & Barlow, 1976; Jayarantne & Levy, 1980). The efficacy of this perspective has been discussed and debated in special issues of both the *Journal of Consulting and Clinical Psychology* (Barlow, 1981) and *Behavioral Assessment* (Nelson, 1981).

The practitioner's involvement in research has a number of benefits for the psychotherapy field (see Hayes & Nelson, 1981, for a brief review). Perhaps the major benefit is that practicing clinicians could perform research that is clinically relevant to certain behavior or personality disorders. As noted previously, much research in the psychotherapy field has been faulted for not focusing on clinical problems. Thus, research is said to be too analogue in nature to be of benefit to clinical practice. By having practicing clinicians involved in the research process, the research–practice gap may be reduced. Presumably, clinicians could produce research that, at some stage, is more relevant to future practice.

We would certainly support this position but would also argue that the single-subject time-series research strategy may be no more useful for making practitioners into researchers than traditional methods (see Agras, Kazdin, & Wilson, 1979; Franks & Wilson, 1979; Kazdin & Wilson, 1978; Wilson, 1981). Several conceptual and methodological features of time-series designs may actually militate against their application in practice as usually conceived

(Kratochwill & Piersel, 1983). First of all, single-subject time-series designs are usually recommended for practice because they share the single-subject similarity with practice. Yet, as portrayed in this chapter, not all time-series designs involve one subject. Many designs approach the complexity of traditional large-n–between-group designs based on the number of subjects participating. Moreover, the clinician may not always be concerned with one subject; rather, clinical treatment may be focused on a group or even a system.

Second, application of time-series designs for research requires that credible assessment strategies be employed. The requirements of assessment for research may differ considerably from what is possible in regular practice. For example, research measurement usually requires repeated direct assessment of the target problem. Yet direct measures on cognitive, behavioral, and physiological domains may prove very difficult.

Third, in order to evaluate treatment, various design and analysis structures are necessary. In some cases design (e.g., alternating treatments) and analysis (time-series analysis) may be difficult for the practicing clinician to apply, given the usual demands of practice (e.g., time and cost-efficiency).

Fourth, when treatment evaluation is done as *research,* special ethical concerns are raised (Kazdin, 1980). For example, research requires manipulation of variables, some of which may be undesirable for the client. Withholding information for research purposes may prove problematic in practice. Also, issues of invasion of privacy and confidentiality raise special concerns in research that are not always problematic in regular practice. Thus, practitioners hoping to integrate research in their practice face a number of ethical issues that make the research pursuit complex.

Finally, there are a host of variables associated with research generally and time-series research specifically that must be resolved for the research/ practice gap to be closed. Time, cost, and professional research competencies must be dealt with effectively for the issue to be resolved. Based on past examinations of practice, research as conceived here has not been a high priority among practitioners.

Resolution of these issues will not come easy, but some directions exist. One option is to involve more researchers in the direction of clinical research. For example, Wilson (1981) has noted that well-trained scientist–practitioners could participate on interdisciplinary clincal research teams. Another possibility is to have practitioners field test treatments developed and created by researchers (Barlow, 1981) or vice versa (Wickramasekera, 1981). Time-series designs could be useful in this endeavor, as well as other research strategies. The issue of loosening controls on research to more effectively generalize to practice must still be addressed (Kazdin, 1981).

CONCLUSIONS

In this chapter we have provided an overview of time-series research methodology in psychotherapy. These strategies were put in historical perspective and contrasted to the group and case study research procedures. Time-series research designs have made valuable contributions to the field of psychotherapy. Such strategies have provided an important alternative to large-*n*–between-group designs and have provided an option for conducting credible research with a single subject. Moreover, they have provided a strategy to improve on research traditionally conducted in psychotherapy investigation and called case study methods. These methods have also emphasized repeated measurement of client variables over the course of therapy and have thus facilitated the development of many new measures and assessment procedures in the field. Finally, they have provided a potential working model for practitioner evaluation of therapeutic work.

In this chapter we have provided an overview of methodological characteristics of time-series research. These types of research strategies are characterized by repeated measurement, allow specification of conditions in research, integrate replication into the research strategy, and promote a high degree of design flexibility in the research endeavor. Several different types of time-series designs were described, including within-series designs with simple and complex phases changes, between-series designs including alternating and simultaneous treatment designs, and combined series including multiple-baseline designs. Some special issues in various designs were discussed, including the length of phases and evaluation of psychoactive medication.

Like other research strategies in psychotherapy, time-series designs require researchers to consider the conceptual methodological issues that have been raised over their use. These include internal validity, construct validity, statistical conclusion validity, and external validity. Issues in each of these areas are described and discussed in the context of recent progress in the field. It is our opinion that time-series research has made valuable and unique contributions to the field of psychotherapy. It is also important to realize that time-series designs have a unique role to play in the field and will continue to do so in the future. Even more rapid progress will be made in conceptual and methodological features of these designs in the future.

REFERENCES

Agras, W. S., & Berkowitz, R. Clinical research in behavior therapy: Halfway there? *Behavior Therapy,* 1980, *11,* 472–487.

Agras, W. S., Kazdin, A. E., & Wilson, G. T. *Behavior therapy: Toward an applied clinical science.* San Francisco: W. H. Freeman, 1979.

Baer, D. M. "Perhaps it would be better not to known everything." *Journal of Applied Behavior Analysis,* 1977, *10,* 167–172.

Baer, D. M., Wolf, M. M., & Risley, T. R. Some current dimensions of applied behavior analysis. *Journal of Applied Behavior Analysis,* 1968, *1,* 91–97.

Barlow, D. H. Behavior Therapy: The next decade. *Behavior Therapy,* 1980, *11,* 315–328.

Barlow, D. H. On the relation of clinical research to clinical practice: Current issues, new directions. *Journal of Consulting and Clinical Psychology,* 1981, *49,* 147–155.

Barlow, D. H., & Hayes, S. C. Alternating treatments design: One strategy for comparing the effects of two treatments in a single subject. *Journal of Applied Behavior Analysis,* 1979, *12,* 199–210.

Barlow, D. H., & Wolfe, B. E. Behavioral approaches to anxiety disorders: A report on the NIMH-SUNY, Albany, Research Conference. *Journal of Consulting and Clinical Psychology,* 1981, *49,* 448–454.

Bergin, A. E. Some implications of psychotherapy research for therapeutic practice. *Journal of Abnormal psychology,* 1966, *71,* 235–246.

Bergin, A. E., & Strupp, H. H. *Changing frontiers in the science of psychotherapy.* Chicago: Aldine-Atherton, 1972.

Bernard, M. E., Dennehy, S., & Keefauver, L. W. Behavioral treatment of excessive coffee and tea drinking: A case study and partial replication. *Behavior Therapy,* 1981, *12,* 543–548.

Bijou, S. W., Peterson, R. F., Harris, F. R., Allen, K. E., & Johnston, M. S. Methodology for experimental studies of young children in natural settings. *Psychological Record,* 1969, *19,* 177–210.

Bolgar, H. The case study method. In. B. B. Wolman (Ed.), *Handbook of clinical psychology.* New York: McGraw-Hill, 1965.

Browning, R. M., & Stover, D. O. *Behavior modification in child treatment.* Chicago: Aldine/Atherton, 1971.

Campbell, D. T., & Stanley, J. C. Experimental and quasi-experimental designs for research on teaching. In N. L. Gage (Ed.), *Handbook of research on teaching.* Chicago: Rand McNally, 1963.

Chassan, J. B. *Research design in clinical psychology and psychiatry* (2nd ed.). New York: Irvington, 1979.

Cook, T. D., & Campbell, D. T. (Eds.). *Quasi-experimentation: Design and analysis issues for field settings.* Chicago: Rand McNally, 1979.

Dukes, W. F. N = 1. *Psychological Bulletin,* 1965, *64,* 49–74.

Edington, E. S. Random assignment and statistical tests for one-subject experiments. *Behavioral Assessment,* 1980, *2,* 19–28.

Edgington, E. W. Nonparametric tests for single subject multiple schedule experiments. *Behavioral Assessment,* 1982, *4,* 83–91.

Elashoff, J. D., & Thoresen, C. E. Choosing a statistical method for analysis of an intensive experiment. In T. R. Kratochwill (Ed.), *Single-subject research: Strategies for evaluating change.* New York: Academic Press, 1978.

Eysenck, H. J. The effects of psychotherapy: An evaluation. *Journal of Consulting and Clinical Pyschology,* 1952, *16,* 319–324.

Franks, C. M., & Wilson, G. T. (Eds.), *Annual review of behavior therapy: Theory and practice* (Vol. VII). New York: Brunner/Mazel, 1979.

Glass, G. V., Wilson, V. L., & Gottman, J. M. *Design and analysis of time-series experiments.* Boulder: University of Colorado Press, 1975.

Gottman, J. M., & Glass, G. V. Analysis of interrupted time-series experiments. In T. R. Kratochwill (Ed.), *Single subject research: Strategies for evaluating change.* New York: Academic Press, 1978.

Hall, R. V., & Fox, R. G. Changing criterion designs: An alternative applied behavior analysis procedure. In B. C. Etzel, J. M. LeBlanc, & D. M. Baer (Eds.), *New developments in behavioral research: Theory, method, and application.* Hillsdale, N.J.: Erlbaum, 1977.

Hartmann, D. P., & Hall, R. V. A discussion of the changing criterion design. *Journal of Applied Behavior Analysis,* 1976, *9,* 527–532.

Hayes, S. C. Single case experimental design and empirical clinical practice. *Journal of Consulting and Clinical Psychology,* 1981, *49,* 193–211.

Hayes, S. C. & Nelson, R. O. Clinically relevant research: Requirements, problems, and solutions. *Behavioral Assessment,* 1981, *3,* 209–215.

Heermann, E. F., & Braskamp, L. E. (Eds.). *Readings in statistics for the behavioral sciences.* Englewood Cliffs, N.J.: Prentice-Hall, 1970.

Hersen, M. Single case experimental designs. In A. S. Bellack, M. Hersen, & A. E. Kazdin (Eds.), *International handbook of behavior modification and therapy.* New York: Plenum Press, 1982.

Hersen, M., & Barlow, D. H. *Single-case experimental designs: Strategies for studying behavior change.* New York: Pergamon Press, 1976.

Jayaratne, S., & Levy, R. L. *Empirical clinical practice.* New York: Columbia University Press, 1980.

Johnston, J. *Econometric methods,* (2nd ed.). New York: McGraw-Hill, 1972.

Jones, R. R., Vaught, R. S., & Weinrott, M. Time-series analysis in operant research. *Journal of Applied Behavior Analysis,* 1977, *10,* 151–166.

Jones, R. R., Weinrott, M. R., & Vaught, R. S. Effects of serial dependency on the agreement between visual and statistical inference. *Journal of Applied Behavior Analysis,* 1978, *11,* 151–166.

Kazdin, A. E. Statistical analysis for single-case experimental designs. In M. Hersen & D. Barlow (Eds.). *Single case experimental designs: Strategies for studying behavior change.* New York: Pergamon Press, 1976.

Kazdin, A. E. Assessing the clinical or applied significance of behavior change through social validation. *Behavior Modification,* 1977, *1,* 427–452. (a)

Kazdin, A. E. Methodology of applied behavior analysis. In T. Brigham & A. C. Catania (Eds.), *Handbook of applied behavior research: Social and instructional processes.* New York: Irvington/Halstead, 1977. (b)

Kazdin, A. E. *History of behavior modification.* Baltimore: University Park Press, 1978.

Kazdin, A. E. *Research design in clinical psychology.* New York: Harper & Row, 1980.

Kazdin, A. E. Drawing valid inferences from case studies. *Journal of Consulting and Clinical Psychology,* 1981, *49,* 183–192.

Kazdin, A. E. *Methodology of psychotherapy outcome research: Recent developments and remaining limitations.* Paper presented as master lecture at The American Psychological Association, Los Angeles, August 1981.

Kazdin, A. E. *Single-case research designs: Methods for clinical and applied setting.* New York: Oxford University Press, 1982.

Kazdin, A. E., & Hartmann, D. P. The simultaneous-treatment design. *Behavior Therapy,* 1978, *8,* 682–693.

Kazdin, A. E., & Kopel, S. A. On resolving ambiguities in the multiple-baseline design: Problems and recommendations. *Behavior Therapy,* 1975, *6,* 601–608.

Kazdin, A. E., & Wilson, G. T. *Evaluation of behavior therapy: Issues, evidence, and research strategies.* Cambridge, Mass.: Ballinger, 1978.

Kendall, P. C. Assessing generalization and the single-subject strategies. *Behavior Modification,* 1981, *5,* 307–319.

Kiesler, D. J. Experimental designs in psychotherapy research. In A. E. Bergan & S. L. Garfield (Eds.), *Handbook of psychology and behavior change.* New York: Wiley, 1971.

Kiesler, D. J. Empirical clinical psychology: Myth or reality? *Journal of Consulting and Clinical Psychology,* 1981, *49,* 212–215.

Kolko, D. J., Dorsett, P. G., & Milson, M. A. A total-assessment approach to the evaluation of social skills training: The effectiveness of an anger control program for adolescent psychiatric patients. *Behavioral Assessment,* 1981, *3,* 383–402.

Kratochwill, T. R. (Ed.). *Single-subject research: Strategies for evaluating change.* New York: Academic Press, 1978.

Kratochwill, T. R. Intensive research: A review of methodological issues in clinical, school, and counseling psychology. In D. C. Berliner (Ed.), *Review of research in education.* Itasca, Ill.: Peacock, 1979.

Kratochwill, T. R., & Levin, J. R. On the applicability of various data analysis procedures to the simultaneous and alternating treatment designs in behavior therapy research. *Behavioral Assessment,* 1980, *2,* 253–360.

Kratochwill, T. R., & Piersel, W. C. Time-series research: Contributions to empirical clinical practice. *Behavioral Assessment,* 1983, *5,* 165–176.

Kratochwill, T. R., Schnaps, A. P., & Bissell, M. S. Research design in school psychology. In J. R. Bergan (Ed.), *School psychology in contemporary society.* Columbus, Ohio: Merrill, in press.

Lazarus, A. A., & Davidson, G. Clinical innovation in research and practice. In A. E. Bergan & S. L. Garfield (Eds.), *Handbook of psychotherapy and behavior change.* New York: Wiley, 1971.

Leitenberg, H. The use of single-case methodology in psychotherapy research. *Journal of Abnormal Psychology,* 1973, *82,* 87–101.

Levin, J. R., Marascuilo, L. A., & Hubert, L. J. N = nonparametric randomization tests. In T. R. Kratochwill (Ed.), *Single-subject research: Strategies for evaluating change.* New York: Academic Press, 1978.

Liberman, R. P., Davis, J., Moon, W., & Moore, Jr. Research design for analyzing drug–environment–behavior interactions. *Journal of Nervous and Mental Disease.* 1973, *156,* 432–439.

Luborsky, L., Singer, G., & Luborsky, L. Comparative studies of psychotherapies. Is it true that everyone has won and that all must have prizes? *Archives of General Psychiatry,* 1975, *32,* 995–1008.

Marks, I. M. Toward an empirical clinical science: Behavioral psychotherapy in the 1980s. *Behavior Therapy,* 1982, *13,* 63–81.

Matson, J. L. Assessment and treatment of clinical fears in mentally retarded children. *Journal of Applied Behavior Analyses,* 1981, *14,* 287–294.

McCleary, R., Hay, R. A. Jr., Meidinger, E. E., & McDowell, D. *Applied time-series analysis for the social sciences.* Beverly Hills, Calif.: Sage, 1980.

McDowell, D., McCleary, R., Meidinger, E. E., & Hay, R. S., Jr. *Interrupted time-series analysis.* Beverly Hills, Calif.: Sage 1980.

Meehl, P. E. Theoretical risks and tabular asterisks: Sir Karl, Sir Ronald, and the slow progress of soft psychology. *Journal of Consulting and Clinical Psychology,* 1978, *46,* 806–835.

Michael, J. Statistical inference for individual organism research: Mixed blessing or curse? *Journal of Applied Analysis,* 1974, *7,* 647–653.

Mischel, W. *Personality and assessment.* New York: Wiley, 1968.

Morris, R. J., & Kratochwill, T. R. *Assessment and treatment of children's fears and phobias.* New York: Pergamon, 1983.

Myers, J. L. *Fundamentals of experimental design.* Boston: Allyn & Bacon, 1979.

Nelson, R. O. Realistic dependent measures for clinical use. *Journal of Consulting and Clinical Psychology,* 1981, *49,* 168–182.

O'Brien, F., Bugle, E., & Azrin, N. H. Training and maintaining a retarded child's proper eating. *Journal of Applied Behavior Analysis.* 1972, *5,* 449–465.

O'Leary, K. D., & Drabman, R. S. Token reinforcement programs in the classroom: A review. *Psychological Bulletin,* 1971, *75,* 289–296.

Ollendick, T. H., Shapiro, E. S., & Barrett, R. P. Reducing stereotypic behaviors: An analysis of treatment procedures utilizing an alternating treatments design. *Behavior Therapy,* 1981, *12,* 570–577.

Parsonson, B. D., & Baer, D. M. The analysis and presentation of graphic data. In T. R. Kratochwill (Ed.), *Single-subject research: Strategies for evaluating change.* New York: Academic Press, 1978.

Paul, G. L. Behavior modification research: Design and tactics. In C. M. Franks (Ed.), *Behavior therapy: Appraisal and status.* New York: McGraw-Hill, 1969.

Rachman, S. J., & Wilson, G. T. *The effects of psychological therapy* (2nd ed.). Oxford: Pergamon Press, 1980.

Robinson, P. W., & Foster, D. F. *Experimental psychology: A small-n approach.* New York: Harper & Row, 1979.

Ross, A. O. Of rigor and relevance. *Professional Psychology,* 1981, *12,* 318–327.

Rusch, F. R., & Kazdin, A. E. Toward a methodology of withdrawal designs for the assessment of response maintenance. *Journal of Applied Behavior Analysis,* 1981, *14,* 131–140.

Rusch, F. R., Connis, R. T., & Sowers, J. The modification and maintenance of time spent attending to task using social reinforcement, token reinforcement and response cost in a restaurant setting. *Journal of Special Education Technology,* 1979, *2,* 18–26.

Sechrest, L., & Redner, R. Strength and integrity of treatments in evaluation studies. In *Evaluation reports.* Washington, D.C.: National Criminal Justice Reference Service, 1979.

Shapiro, M. B., & Ravenette, P. T. A preliminary experiment of paranoid delusions. *Journal of Mental Science,* 1959, *105,* 295–312.

Shontz, F. C. *Research methods in personality.* New York: Appleton Century Crofts, 1965.

Sidman, M. *Tactics of scientific research.* New York: Basic Books, 1960.

Skinner, B. F. *The behavior of organism.* New York: Appleton Century Crofts, 1938.

Skinner, B. F. *Science and human behavior.* New York: Macmillan, 1953.

Smith, M. L. & Glass, G. V. Meta analysis of psychotherapy outcome studies. *American Psychologist,* 1977, *32,* 134–137.

Sowers, J., Rusch, F. R., Connis, R. T., & Cummings, L. T. Teaching mentally retarded adults to time-manage in a vocational setting. *Journal of Applied Behavior Analysis,* 1980, *13,* 119–128.

Stokes, T. F. & Baer, D. M. An implicit technology of generalization. *Journal of Applied Behavior Analysis,* 1977, *10,* 349–367.

Strupp, H. H. Clinical research, practice, and the crises of confidence. *Journal of Consulting and Clinical Psychology,* 1981, *49,* 216–219.

Thoresen, C. E., & Elashoff, J. D. An analysis-of-variance model for intrasubject replication design: Some additional comments. *Journal of Applied Behavior Analysis,* 1974, *7,* 639–641.

Turner, S. M., Hersen, M., & Alford, H. Case histories and shorter communications. *Behaviour Research and Therapy,* 1974, *12,* 259–260.

Vogelsberg, T., & Rusch, F. R. Training three severely handicapped young adults—Walk, look and cross uncontrolled intersections. *AAESPH Review,* 1979, *4,* 264–273.

Wallace, C. J., & Elder, J. P. Statistics to evaluate measurement accuracy and treatment effects in single subject research designs. In M. Hersen, R. M. Eisler, & P. M. Miller (Eds.), *Progress in behavior modification* (Vol. 10). New York: Academic Press, 1980.

Watson, N., Caddy, G. R., Johnson, J. H., & Rimm, D. C. Standards in the education of professional psychologists: The resolutions of the conference at Virginia Beach. *American Psychologist,* 1981, *36,* 514–519.

Wells, K. C., Conners, C. K, Imber, L., & Delamater, J. Use of single-subject methodology in clinical decision-making with a hyperactive child on the psychiatric inpatient unit. *Behavioral Assessment,* 1981, *3,* 359–369.

Wickramasekera, I. A. Clinical research in a behavioral medicine private practice. *Behavioral Assessment,* 1981, *3,* 265–271.

Wilson, G. T. Some thoughts about clinical research. *Behavioral Assessment,* 1981, *3,* 217–225.

Williamson, D. A., Calpin, J. P., DiLorenso, T. M., Garris, R. P., & Petti, T. A. Treating hyperactivity with dexedrine and activity feedback. *Behavior Modification,* 1981, *5,* 399–416.

Wolf, M. M. Social validity: The case for subjective measurement or how applied behavior analysis is finding its heart. *Journal of Applied Behavior Analysis,* 1978, *11,* 203–214.

Yates, A. A. Research methods in behavior modification: A comparative evaluation. In M. Hersen & R. M. Miller (Eds.), *Progress in behavior modification* (Vol. 2). New York: Academic Press, 1976.

Yeaton, W. H., & Sechrest, L. Critical dimensions in the choice and maintenance of successful treatments: Strength, integrity, and effectivness. *Journal of Consulting and Clinical Psychology,* 1981, *49,* 156–167.

7

Therapy Analogues and Clinical Trials in Psychotherapy Research

ALAN E. KAZDIN

INTRODUCTION

The evaluation of psychotherapy continues to be a major topic in clinical psychology and psychiatry.[1] Research is frequently designed to determine the efficacy of particular techniques, the relative effectiveness of alternative treatments, the combination of treatments that maximize change, and a variety of related questions (see Kazdin, 1980). Identification of effective psychotherapy techniques is a high priority for several reasons. First, estimates of the number of persons in the population at large who experience psychiatric dysfunction or problems of living and who might benefit from treatment have been as high as 15% to 25% (e.g., President's Commission on Mental Health, 1978). Thus, the social, economic, and personal repercussions that mental health problems present are extensive (Kiesler, 1980). Although psychotherapy might not be expected to alleviate all mental health problems, certainly the availability of effective treatments would be of interest to a large body of consumers who might benefit directly.

---

[1]In the present chapter, the term *psychotherapy* is used to delineate psychosocial interventions in general and encompasses a variety of other terms such as *insight-oriented psychotherapy, behavior therapy, cognitive therapy,* and others that reflect particular conceptual views about the putatively critical features of treatment.

ALAN E. KAZDIN ● Department of Psychiatry, Western Psychiatric Institute and Clinic, University of Pittsburgh School of Medicine, Pittsburgh, Pennsylvania 15213. Completion of this chapter was supported by a Research Scientist Development Award (K02 MH00353) from the National Institute of Mental Health.

Second, professional interest in the multiple questions of therapy outcome is high (Agras, Kazdin, & Wilson, 1979; Garfield, 1981; Smith, Glass, & Miller, 1980; VandenBos, 1980). Attempts to unravel the underlying bases of clinical disorders and the mechanisms of change and to evaluate the effects and relative efficacy of alternative treatments continue. The need to develop the scientific basis of psychotherapy is widely recognized.

Third, the most recent impetus for the scrutiny of therapy outcome in the United States comes from congressional interest in the cost of psychological services (Marshall, 1980). Because reimbursement for psychotherapy services is included in national health care policy proposals, Congress has a keen interest in evaluating whether demonstrably effective treatments exist and whether the benefits warrant the costs (Kiesler, 1980; Parloff, 1979).

The need for clear answers to questions about treatment effects is great. Nevertheless, the answers are not likely to come quickly; perhaps simple answers are not likely to come at all. Indeed, the means of seeking answers to basic questions are still a matter of active debate. Fundamental issues surrounding the appropriate outcome measures (Kazdin & Wilson, 1978; Strupp & Hadley, 1977), research designs (Barlow, 1980; Strupp & Hadley, 1979), and methods of evaluating existing studies (Rachman & Wilson, 1980; Smith & Glass, 1977) continue to be debated. In short, several facets of the methodology of treatment outcome research continue to evolve (Malan, 1973; VandenBos & Pino, 1980).

A major concern within clinical psychology and psychiatry is the utility of outcome research in providing information relevant to clinical practice. Much of the outcome research has been conducted under highly controlled laboratory conditions with subjects (college students) whose severity of dysfunction typically only faintly approximates that of patients usually seen in treatment. Investigations of this sort have frequently been referred to as *analogue research* because they tend only to resemble in varying degrees the clinical situation to which the results might be generalized (Bernstein & Paul, 1971; Borkovec & O'Brien, 1976; Kazdin, 1978). General agreement exists that findings obtained from analogue studies need to be replicated in *clinical trials*—that is, research in clinical settings administered under conditions that reflect or closely resemble those normally associated with treatment delivery (Parloff, 1979).

The present chapter examines analogue research and clinical trials and their benefits and limitations in developing effective treatment techniques for clinical use. The chapter focuses on the differences and similarities of these alternative research strategies. In addition, the hiatus between analogue research and clinical trials and clinical trials and clinical practice is also discussed.

THERAPY ANALOGUES

Analogue research is designed to evaluate treatment or some processes related to treatment under conditions that depart from the usual conditions in which therapy is administered. When defined broadly, a variety of different types of research can be viewed as therapy analogues. For example, laboratory research with infrahuman subjects designed to test basic processes related to the development or amelioration of maladaptive behavior falls squarely within the domain of analogue research (see Adams & Hughes, 1976).

A prime example is research on "experimentally-induced neuroses" (Pavlov, 1927). Experimental neuroses refer to laboratory-induced emotional states that include heightened irritability, avoidance, withdrawal, and multiple physiological disturbances (Russell, 1950). The reactions can be induced in a variety of ways (e.g., presentation of a difficult discrimination task; repetitive mild shock) and have been demonstrated among several different species. There are obvious differences between experimentally-induced neuroses in infrahumans and human neurotic reactions seen in clinical settings (Hunt, 1964). Nevertheless, investigation of emotional reactions in the laboratory has contributed to the understanding of human neuroses and their treatment. Laboratory procedures designed to eliminate induced emotional reactions in infrahumans have provided the conceptual and procedural underpinnings of many current treatments for human neuroses (e.g., systematic desensitization, flooding, modeling; Masserman, 1943; Wolpe, 1958).

Laboratory research with humans has also been designed to provide an analogue of various processes of psychotherapy (Heller, 1971). A major example is the work on verbal conditioning (see Krasner, 1955, 1965). In the typical verbal conditioning experiment, an experimenter provides reinforcing consequences (verbal approval, nods) for specific statements that the subject makes in ordinary conversation or for words (e.g., personal pronouns) that the subject selects during a sentence-construction task. The reactions of the experimenter influence the types of verbalization on the part of the subject. The purpose of the research has been to duplicate, under laboratory conditions, selected aspects of the interaction that may occur between a therapist and patient. Evidence that experimenters can influence verbalizations on the part of subjects and that these changes have impact on measures of personality (Kanfer & Phillips, 1970) may reflect fundamental processes that go on in ordinary psychotherapy (e.g., Truax, 1966).

Laboratory analogues of psychotherapy have permitted investigation of variables that would be difficult to explore experimentally in the context of clinical situations. A case in point is the influence of the client's behavior on the reactions of the therapist. In laboratory situations, the influence of the subject on the behavior of an interviewer can be examined in an interview or quasi-therapy situation. The subject (actually a confederate working for the investigator) can be trained to engage in certain types of behavior to influence the interviewer (who unknowingly is the actual subject; e.g., Heller, Myers, & Kline, 1963). The demands and obligations of treatment in clinical situations would not readily permit manipulation of client behavior to evaluate therapist reactions.

Several paradigms of analogue research can be identified with both infrahuman and human laboratory research. Indeed, insofar as psychotherapy is concerned with factors that influence affect, perception, cognition, personality, and behavior, virtually all areas of psychological research might be viewed as analogues of therapy. In fact, the overriding purpose of analogue research is not merely or even primarily to examine treatments that may be of use in clinical settings. Rather, analogue research seeks to examine basic behavioral processes (e.g., underlying bases of fear, the interrelations among alternative response modalities; Bandura, 1978, 1979; Borkovec & Rachman, 1979). Basic research may provide insights about the mechanisms of affective, cognitive, behavioral, and psychophysiological changes. In the long run, uncovering the basic mechanisms of change may make the greatest contribution to clinical treatment.

Although a variety of research paradigms can be viewed as analogue research, the term has been used in recent years to refer primarily to outcome investigations where treatments are evaluated under conditions that depart from those evident in clinical settings. A major concern of such research is that the findings may not generalize or apply to patients seen in clinical settings.

General Characteristics

Investigations of treatment outcome under conditions that depart from the usual clinical settings are referred to as analogue research because they differ from clinical research in a number of ways. First, the *target problem* that is the focus of treatment may differ greatly from what is seen in clinical settings. For example, in behavior therapy, extensive research has been conducted on the treatment of mild or subphobic levels of fear and avoidance. Typically, college students serve as subjects and are identified as fearful based on their responses to a questionnaire or to a behavioral test where they must approach a feared object (e.g., harmless snake). The type of fear that is studied and its severity may depart from the fears usually seen in a clinical population.

The circumscribed fears studied in college student populations may respond differently to treatment than more diffuse fears and fears embedded in multiple problems that persons evince in clinical settings. Even within college student populations, some sources of fear or anxiety (e.g., social situations) may more closely resemble reactions usually seen in clinical patients than other such sources (e.g., small animals; Borkovec & O'Brien, 1976). The severity of fear is also quite relevant in evaluating the results of analogue research, because alleviation of relatively mild problems would be expected to be much easier than alleviation of more severe problems. Of course, the extent to which fears of college students resemble clinical problems is a matter of degree. Studies that use stringent criteria to identify subjects with intense fear (e.g., Lang, 1968) may be more likely to produce results that can be generalized to clinical disorders than studies that apply more lenient criteria (e.g., Robinson & Suinn, 1969).

Second, *characteristics of the population* that receive treatment contribute to the extent to which an investigation is an analogue of the clinical situation. The phrase *characteristics of the population* refers to a variety of subject and demographic variables apart from the target problem. Analogue studies frequently utilize college students because they represent a captive subject pool and are available in sufficient numbers to meet the requirements of multiple-group designs. College students may differ from the usual clinical population in such characteristics as level of education, age, marital status, socioeconomic class, and, of course, occupation. Differences in these subject or demographic variable may be relevent to the efficacy of treatment. Indeed, population differences may contribute to the types of problems that college students and adults no longer in college bring to treatment. Thus, the generality of findings from an investigation with college students may depend upon the similarity of such students to clinical populations.

Third, the *manner of recruiting* persons for treatment studies often distinguishes analogue studies. In analogue research, college student subjects are recruited through classes and given course credit or money for participation in research. Students may be interested in learning about different types of treatment or in the small incentives that are occassionally offered. Clinical patients typically seek treatment to resolve a particular problem or have others (e.g., parents) who seek treatment for them. The provision of incentives for participation in treatment makes research an analogue of the clinic situation.

Fourth, the *therapists* who provide treatment may make an investigation an analogue of clinical treatment. In a clinical situation, experienced therapists usually provide treatment. In analogue research, therapists are often graduate or undergraduate students. Students and professional therapists may differ on a host of factors such as age, experience, and credibility as a provider of treatment. Insofar as treatment effects depend upon characteristics of the therapist or the therapist–patient relationship, the use of students as therapists may be an important dimension that distinguishes analogue research.

Fifth, the *selection of treatment* may contribute to the analogue nature of treatment research. Patients who seek treatment often exert choice in terms of whom they see and where they go. Also, if the client is dissatisfied, he or she can go elsewhere for treatment. In analogue research, persons often agree to participate in a project and, having agreed, are assigned randomly to one of several conditions. The manner in which they select treatment may be an important difference between analogue research and clinical treatment, because the opportunity to select treatment is related to the treatment outcome (Devine & Fernald, 1973; Gordon, 1976).

Sixth, the subject's *set or expectations* about treatment may be an important difference between therapy analogue research and clinical treatment. Clients come to therapy expecting to receive an effective treatment, a set likely to be augmented by the professional stature of the therapist and treatment facility and by the convincing description of treatment provided by the therapist (see Frank, 1973). In analogue research, subjects may have a very different set of expectations from persons who seek treatment in clinical settings. In analogue research, subjects are often informed that they are participating in an experiment and that treatment or amelioration of a problem may be ancillary.

Seventh, the *setting* where treatment is conducted may also distinguish analogue research from clinical work. Many analogue studies are conducted in academic psychology departments, although most clinical treatment is carried out in clinics and hospitals (Parloff, 1979). The setting may influence the expectations that subjects have about what will transpire. Indeed, the importance of the setting has been suggested in assessment research showing that subjects evince more severely problematic behavior when they believe their responses are being measured in a clinic rather than a laboratory setting (Bernstein, 1973; Bernstein & Nietzel, 1973, 1974). Analogue research is usually conducted in settings that depart from the usual clinic facilties, a factor that may influence the effects that treatments have on performance.

Finally, the *variations of treatment* explored in analogue research often differ from those used in actual treatment. In analogue research, treatment techniques may be altered so that they are controlled or implemented in a standardized fashion. For example, the number of treatment sessions may be held constant, the type of situations presented to the persons may be standardized rather than individualized across subjects, or equipment (such as tapes or slides) might be introduced to ensure that the treatment administration does not systematically vary across conditions, therapists, or subjects within a condition. For example, analogue research on desensitization has occasionally used slides of feared scenes rather than presenting these stimuli verbally for the clients to imagine (Brown, 1973; Wilson, 1973). The use of slides is a major procedural variation that could influence the effects of treatment. The extent to which results might generalize to more typical variations of treatment is a question often raised in analogue studies.

The above characteristics are not the only ones that are used as the basis for referring to treatment studies as analogue research. For any particular study, not all of the characteristics may depart from the conditions ordinarily evident in clinical settings. Also, for any given characteristic, the study may vary in degree in terms of its departure from the clinical settings. Consequently, analogue studies include a variety of types of investigations that differ markedly in their departure from the conditions of clinical work. Among the characteristics discussed, the target problem and population constitute the major concern over therapy analogues (Bernstein & Paul, 1971; Cooper, Furst, & Bridger, 1969; Levis, 1970). However, other characteristics may also be critical to the generality of results from analogue to clinical settings.

Utility of Analogue Research

There are many arguments for investigation of treatment under well-controlled laboratory conditions. Treatment research in clinical settings is beset with several practical and ethical obstacles (Foa & Steketee, 1981). To begin with, in most treatment studies in clinical settings, it may be difficult to procure a sufficient number of subjects with the same or similar problems to meet the requirements of a multiple-group experiment. A sufficient number is needed within each group to provide a powerful (sensitive) test of alternative variations of the treatment. A homogeneous group of subjects is needed so that subjects can be assessed with the same set of measures and that variability among subjects on the measures is relatively small.

When a sufficient number of clients with a particular problem can be found, there are usually constraints in clinical settings in assigning them to specific treatment or control conditions. For example, in inpatient treatment studies, administrative demands often dictate that more severely disturbed patients be assigned to certain conditions (Gripp & Magaro, 1971). Similarly, in outpatient studies, those who evince serious dysfunction are sometimes placed into treatment rather than assigned to control (e.g., waiting-list) conditions (Rogers & Dymond, 1954). Random assignment of subjects to conditions is much more feasible in the context of analogues of the clinical situation than in the clincial situation itself.

In clinical settings, it may be difficult to control the influence of competing factors that can affect the results of the investigation. For example, in outpatient therapy, individuals may seek specific experiences to help improve their plight (e.g., encounter groups, counseling from a physician, self-prescribed medication). The likelihood of seeking alternative experiences might be increased by assignment to a treatment that is not very credible or to a waiting-list control group (Pande & Gart, 1968). In analogue research, subjects are usually not participating because of the need to overcome an immediate crisis. Hence, they may be less likely to seek alternative experiences that could interfere with evaluation of the treatment to which they are assigned.

Apart from the difficulty of selecting clients and assigning them randomly to conditions in clinical settings, it is often difficult to obtain therapists who agree to engage in treatment research and meet the demands that research typically prescribes. The incentives for practicing clinicians to participate in experiments may be minimal. Also, most clinicians are interested in conducting treatment in their own way rather than following a treatment manual or standardized procedure that may be confining. On the other hand, in analogue studies, students usually serve as therapists and, by virtue of their status and level of training, are quite receptive to the opportunities provided by a treatment research project and to the procedures and concomitant supervision attendant upon that participation.

Ethical issues often make research in clinical settings difficult to conduct. Many questions about the effects of therapy require control groups that withhold specific, possibly crucial components of treatment, that delay treatment, or that even withhold treatment altogether. Researchers often wish to determine the necessary and sufficient conditions to produce change. The primary means of accomplishing this goal is to provide variations of treatment with selected components deleted or altered. It may be that clients in some of the conditions are less likely to improve, but this is not known in advance. For clients who seek treatment, assignment to conditions that do not maximize the chances of improvement may be difficult to defend ethically. Delaying or withholding treatment illustrate the ethical issues even more sharply. Presumably, clients coming to treatment wish improvement as soon as possible. It is difficult to justify assignment for research purposes to delayed-treatment or no-treatment control conditions.

In light of the practical and ethical issues noted above, the range of questions that can easily be addressed about treatment in clinical settings is relatively restricted. Analogue studies are especially useful because they allow the investigator to control the conditions of experimentation to a much greater extent than in clinical investigations. The control permits the investigator to minimize the several sources of variance that might obscure the effects of treatment. Clients who receive treatment can be selected because of their homogeneity on the type and severity of the problem as well as on subject and demographic variables. Also, delivery of treatment can readily be controlled; different features of treatment delivery—such as the precise materials that are presented, the number and duration of sessions, and therapist training—can be held constant, all of which can further minimize sources of variability. Minimization of variability among subjects and therapists increases the power of the test of treatments. Investigations in nonclinical settings permit the use of control groups that might otherwise be unavailable in clinical settings.

The advantages of laboratory, as opposed to clinical, research result from the different priorities of the settings. In the laboratory, volunteer subjects, rather than patients who have sought treatment, participate in treatment. Hence, greater priority can be given to the demands of the experiment. Many of the demands, such as standardization of treatment or use of various control groups, can be readily superimposed. In contrast, in clinical settings with clients who have sought and pay for treatment, the priority is to provide individually-tailored treatment. High priority is on the clinical care of the individual patient. Sacrifices required by experimentation are usually secondary. Thus, multiple treatments may be applied simultaneously and duration of treatments for similar patients may vary as a function of judgment of what is best for the patient.

Generality of the Results

A major concern of analogue research is that the results may not generalize to a clinical situation (Cooper *et al.*, 1969; Marks, 1978). Yet, merely because the study departs from conditions of the clinical situation does not necessarily mean that the results will not be generalizable from one setting to another. Little research has directly evaluated the generality of particular findings obtained in laboratory-based research to the clinical situation. The concern that findings will not generalize stems from the seemingly large discrepancies in the conditions of analogue research and clinical practice. For example, in analogue research, subjects often present relatively circumscribed problems that might be expected to respond to even relatively weak forms of treatment. In clinical work, clients often present multifaceted problems; a circumscribed treatment focus would probably ignore many of the concomitant problems that patients present and would have minimal impact on their dysfunction. Related analogue research often shows marked changes with only a few treatment sessions. The complexity of the problems clients often bring to clinical settings makes a small number of sessions inadequate even to sort out the problems, leaving aside their treatment. Despite the obvious differences between analogue and clinic conditions, the generality of results from the former to the latter has not been adequately tested.

As noted earlier, analogue studies can vary in a variety of dimensions (e.g., target problem, population). On each dimension, a study can vary in the degree to which it resembles the clinical situation to which the investigator may wish to generalize. Thus, analogue studies are not necessarily a distinct category of research. Rather, a study can be evaluated in terms of its standing on various dimensions to determine the extent to which the study approaches the clinical situation. Dimensions discussed earlier included the target problem, the population, the manner in which clients are recruited, the therapists, client selection and set, the setting of treatment, and variation of treatment. The generality of results from an analogue study to the clinical situation may depend on which of these dimensions the study varies from the clinical situation and to what degree.

Some dimensions may be more relevant to generality of the results than others. For example, whether the subjects are college students or older adults (population dimension) may be less relevant than the target problem that is the focus of treatment. When the target problem consists of a clinically debilitating dysfunction (e.g., agoraphobia, sexual dysfunction, obsessive-compulsive rituals), whether the subjects are college students seen in a university clinic may be relatively unimportant. The difficulty in evaluating the generality of findings is that analogue studies usually differ from the clinical situation on several dimensions simultaneously and in different degrees.

Severity of the target behavior is usually seized upon as the most relevant dimension that distinguishes analogue research, since such research typically focuses on problems of subclinical severity. One might expect that treatments shown to alter mildly problematic behavior (e.g., fear in college students) would have little generality to more severe albeit possibly related problems (e.g., clinical phobias; see Emmelkamp, 1979). Indeed, less severe problems might change readily in response to such influences as experimental demand characteristics, nonspecific treatment factors, and placebos. For example, persons who experience relatively severe insomnia do not respond as readily to placebo treatment as do those who experience mild or moderate insomnia (Nicolis & Silvestri, 1967). Consequently, treatments shown to be effective with mild problems may not produce similar effects with more severe problems.

Not all characteristics of the clinical situation may make therapeutic changes more difficult to achieve than in analogue research. Indeed, in relation to laboratory-based conditions, some features of the clinical situation might even increase the likelihood that treatment produces change. For example, the expectancies for improvement on the part of patients, persuasive claims and a sincere commitment to these claims on the part of therapists, and the prestige and experience of the therapist may augment the effectiveness of treatment in clinical rather than in laboratory-based applications of treatment.

Overall, the generality of findings from analogue research to the clinical situation is not easily resolved without additional empirical evidence. Analogue research has been rejected by many on the grounds that it provides, by its very nature, a weak test of the relationship between treatment and therapeutic change in the clinical situation. Yet, the generality or lack of generality of the findings is an area in need of further research. Research needs to test the influence of departures from the clinical situation along various dimensions and the implications of such departures for generalizing to the clinical situation.

CLINICAL TRIALS

Although analogue studies can vary markedly in their degree of resemblance to clinical applications of treatment, they can generally be distinguished from research where the characteristics closely approximate if not actually reflect the conditions of clinical settings, (i.e., clinical trials). Clinical trials are important to discuss because of their special role in evaluating treatment techniques and because of their relation to analogue research.

General Characteristics

Clinical trials generally involve outcome investigations conducted in clinical settings. In relation to analogue research, characteristics of clinical trials are usually easily discerned. Instead of students or volunteers who may not actively seek treatment, patients in need of immediate care are included in clinical studies; instead of graduate students, professional therapists and clinicians provide treatment; instead of treatment of mild, subclinical, and circumscribed problems, relatively more severe or multifaceted clinical disorders are treated. In general, treatment in a clinical trial is tested under conditions where it would ordinarily be applied.

Actually, clinical trials are not qualitatively different from analogue research. On a variety of dimensions related to the resemblance of research to the clinical situation, analogue research and clinical trials represent endpoints. However, resemblance to the clinical situation is a matter of degree on each dimension. Table 1 illustrates several dimensions discussed earlier and different points on the continuum of degree of resemblance to the clinical situation. As shown in the table, an investigation with conditions listed under analogue research or clinical trials, respectively, depart considerably from or very closely resembles the usual conditions of clinical settings. Once the extreme points are identified, it is clear that research can in varying degrees fall somewhere between these extremes.

Much of contemporary research probably falls between the categories of clear analogue research or clinical trials. For example, in the extreme case, analogue studies have focused on college students with subclinical problems. In contrast, clinical trials have focused on patients who have come for treatment at a clinic under the usual conditions for obtaining therapy. A consider-

Table 1. Selected Dimensions along Which Investigations May Vary in Their Degree of Resemblance to the Clinical Situation

	Resemblance to the clinical situation		
Dimension	Relatively little resemblance (analogue research)	Moderate resemblance	Identity with or relatively high resemblance (clinical trials)
Target problem	Nonproblem behavior, laboratory task performance, mild problem at subclinical levels	Similar to that seen in the clinic but probably less severe or more circumscribed	Problem seen in the clinic, intense or disabling
Population	Infrahuman subjects, nonclinical group such as college students chosen primarily because of their accessibility	Volunteers screened for problem and interest in treatment	Clients in outpatient clinic
Manner of recruitment	Captive subjects who receive therapeutically related incentives (e.g., course credit) for participating	Persons recruited especially for available treatment	Clients who have sought treatment without solicitations from the clinic
Therapists	Nontherapists, nonprofessionals, students, automated presentation of major aspects of treatment (audio- or videotapes)	Therapists in training with some previous clinical experience	Professional therapists
Selection of treatment	Client assigned to treatment with no choice of specific therapist or condition	Client given choice over some alternatives in an experiment	Client chooses therapist and specific treatment
Client set	Expect an experimental arrangement with a nontreatment focus	Expect "experimental" treatment with unclear effects	Expect a veridical treatment and improvement
Setting of treatment	Laboratory; academic, psychology department	University clinic devised for treatment delivery with established clientele	Professional treatment facility with primary function of treatment delivery
Variation of treatment	Standardized, abbreviated, or narrowly focused version of treatment	Variation that permits some individualization and flexibility in content and/or duration	Treatment tailored to the individual or determined on the basis of the client's problems

able amount of contemporary research utilizes subjects who fall somewhere in between in terms of how they are recruited. Through the use of newspaper, television, and radio advertisements, researchers frequently solicit volunteers from the community setting who are interested in receivng treatment. The stringency of the screening criteria that are invoked further determines the extent to which the study will resemble conditions of clinical practice. If stringent criteria are invoked to identify persons with clear dysfunctions, then, of course, the severity of the problems that are studied may be very close to what is seen in clinical work. If lenient criteria or no criteria are invoked, the investigation may constitute only a trivial step above the usual analogue population.

Even if subjects are recruited because of the clinical severity of their problems, they may differ slightly from persons with similar problems who ordinarily seek treatment. Perhaps volunteers have not sought treatment previously because they did not perceive their dysfunction as sufficiently debilitating or because other aspects of their lives (e.g., work, marital adjustment, social network) were quite satisfactory. The advertisements for free treatment may have provided a sufficient impetus to try treatment that would otherwise have not been sought. A clinical population with the "same" presenting problem might differ from a volunteer population in a variety of background or contextual dimensions that make the need for treatment more pressing. The differences between volunteers solicited for treatment and clients with similar problems and the relevance of these differences for generality of the conclusions that are reached with volunteer populations remain to be addressed in research.

Consider, as a concrete example, variations of an investigation that is designed to identify effective treatments for obesity. An analogue variation of the study might include college students who are 5% to 10% overweight and who do not fall within the usual range of obesity. A clinical trial might be conducted at a treatment facility and include a more extreme group of persons who, say, are 20% or more overweight. A study falling between these two might seek volunteers from the community who are interested in losing weight. Suppose only those persons who meet stringent standards (e.g., 20% or more overweight) are included in the investigation. Thus, the presenting problem is of equal severity to that seen in clinic settings. The fact that in one case (clinical trials) subjects were not solicited to come to a treatment setting and in the other case they were may not be relevant in terms of generality of the results to the clinical situation. Treatments shown to be effective in reducing weight might not depend on how the subjects were recruited.

On the other hand, it is possible that volunteers recruited for research differ in important ways from persons who seek and perhaps repeatedly have sought treatment. Volunteers may be more (or less) responsive to an intervention because of fewer previous formal treatment trials, may have special motivation to adhere to treatment demands, may have less dysfunction in other areas of their lives, may have stronger support systems (e.g., spouses, children) who can sustain gains produced with treatment, and so on. The existence of such differences and their plausibility in influencing generality of the results to clinical samples are a matter of surmise. However, the general point is that even when volunteer and clinical samples share the same presenting problems, generality may still be an open question because of other differences between the samples.

Generality of Results

There is general agreement that clinical trials represent somewhat of a final achievement or endpoint in outcome research in terms of the evolution of evaluation strategies (Parloff, 1979). Positive leads from case studies, uncontrolled trials, and analogue studies can culminate in a clinical demonstration. Once a controlled clinical trial attests to the efficacy of treatment, the research process has attained a major accomplishment. Even though clinical trials test treatments under clinic conditions, generality of the results to clinical settings is still a relevant concern.

Clinical trials can vary markedly along a variety of dimensions and in their resemblance to the clinical situation where treatment is ordinarily practiced. In clinical trials, as in analogue research, many features of treatment delivery may be altered to permit evaluation of the intervention. The research exigencies may make the situation slightly different from clinical situations. Thus, in some clinical trials, the most severely disturbed or impaired patients may be excluded. For example, because available treatments with known efficacy exist, hypertensive or depressed patients whose dysfunctions are severe may be intentionally excluded from the project and placed under immediate care. Similarly, screening criteria of patients for clinical trials often identify patients who have relatively more circumscribed or well-delineated dysfunctions than those seen in clinical practice. With a heterogenous clinical sample, some exclusion criteria are likely to be invoked to make the sample more homogeneous for research purposes. In any case, characteristics of patients included in clinical trials are not always the same as those seen in routine treatment. Clinical trials are conducted under varying conditions that closely resemble clinical settings, but the conditions are *not* necessarily identical.

The temptation is to assume that the results of the clinical trial illustrate what the effects of treatment might be when the techniques are applied clinically outside the confines of research. Yet, clinical trials, to meet the demands of research, often introduce special features that depart from most clinical applications of treatment (Agras & Berkowitz, 1980; Emmelkamp, 1979; Kazdin, 1982). The degree of experimental control, the careful application of treatment, and monitoring of treatment administration are some of the features that characterize research rather than clinical applications of treatment. The differences in rigor and care in administering treatment may be quite relevant to the outcome. And whether the effects of treatment applied in clinical practice achieve the effects demonstrated in research is an open question.

An example may convey the issue of generality raised by clinical research. In an investigation with hospitalized psychiatric patients, Paul and Lentz (1977) compared the effects of social-learning treatment (token economy), milieu therapy, and routine hospital care. The results indicated that the social-learning treatment was consistently more effective than the other approaches on a large number of measures of symptoms, including interpersonal functioning and self-care skills within the hospital, discharge of patients into the community, and status of patients over a 1½-year follow-up. The demonstration would seem to provide strong evidence for the positive effects of a social-learning program after placement of chronically hospitalized patients in the community.

Several features of the study may have contributed to the program's success. First, training of hospital staff was extensive and included academic instruction, practice and rehearsal, and on-the-job supervision. Second, special research staff monitored all facets of the program including staff and patient changes. Third, and related, staff behavior was assessed daily and supervisory feedback was provided for staff execution of the program. Fourth, treatment manuals were developed to describe the program and were constantly updated to ensure that the program was conducted as intended.

The supervisory and monitoring systems represent excellent procedures for ensuring the integrity of treatment (i.e., that treatment is carried out as intended;) (Sechrest, West, Phillips, Redner, & Yeaton, 1979). The procedures that help ensure treatment integrity may have a significant bearing on the clinical outcome. It is an open question whether the results obtained by Paul and Lentz would be obtained in other hospital settings that adoped the social-learning program. The program might not achieve the remarkable results without adopting the elaborate staff training and monitoring procedures to sustain treatment integrity.

As another example, consider the study by Beck and his colleagues, who compared cognitive therapy and pharmacotherapy (imipramine) to treat unipolar depressed outpatients (Beck, Rush, Shaw, & Emery, 1979; Rush, Beck, Kovacs, & Hollon, 1977). Although both treatments significantly decreased depressive symptoms, cognitive therapy produced greater changes at 3-, 6-, and 12-month follow-up assessments (Kovacs, Rush, Beck, & Hollon, 1981; Rush *et al.,* 1977). A few features included in this clinical trial may have important implications for successful extension of the techniques to clinical work. Specifically, therapists (residents, pre- and post-doctoral clinical psychology trainees, psychiatrists) received relevant didactic and instructional experiences, completed at least one supervised case of cognitive therapy, and received weekly supervision over the course of the investigation. The importance of intensive training has been suggested by Beck *et al.* (1979), who maintain that from several months to 2 years may be essential to implement treatment effectively. Of course, the matter needs to be investigated empirically. However, the general point is that cognitive therapy may be quite effective in a clinical trial with depressed patients. Yet, the results reflect not only on the specific treatment but also on the manner in which it was implemented. Clinical trials often include special procedures (therapist training, monitoring of treatment implementation) that may maximize the benefits that treatments are likely to produce. The special procedures may be no less important than the content of treatment in determining the generality of results from research to practice. The degree of experimental control, specification of treatment, and monitoring of treatment administration that characterize research departs from circumstances of clinical practice. And this is how research should be. The purpose of research is to show what *can* happen under certain well-specified conditions. A separate issue, of obvious importance in clinical research, is whether in fact this is what *does* happen under similar circumstances outside the context of research (i.e., clinical practice). In short, clinical trials address many of the concerns posed by analogue studies about generality of results. However, clinical trials raise their own issues of generality. Special features of the treatment included as part of research evaluation may influence the generality of the results. In terms of generality of the results, the hiatus from clinical trials to clinical practice may be as great or even greater as that from analogue research to clinical trials.

Matching Research Strategies to the Questions

Analogue research and clinical trials are often pitted against each other in arguments about the more beneficial way in which questions about treatment and mechanisms of behavior change can be examined. The contributions of these different types of research need to be evaluated in the context of the different types of questions that can be posed about treatment. Analogue research emphasizes the application of rigorous experimental methods to treatment evaluation. Under conditions that are relatively free from the obligations, obstacles, and constraints of treatment settings, several conditions can be imposed. The control that the investigator can exert over the assignment of subjects to various treatment and control conditions permits evaluation of questions that would be difficult to examine in the context of clinical settings. Questions about the reasons why treatment produces change and the necessary and sufficient conditions for change are especially well suited to analogue conditions. These questions often require special control conditions that are difficult to implement in most treatment settings.

For example, to control for "nonspecific" treatment factors, an "attention placebo" group might be required in the design. Alternatively, a particular treatment variation that is unlikely to produce therapeutic change might be included. The groups might be included to address an important question about the treatment and its underlying basis. Use of such groups is much more feasible and ethically defensible in analogue situations. Similarly, no-treatment groups may be feasible in analogue studies because subjects have not sought treatment for a particular dysfunction, agree to a delay or no treatment, and are not likely to seek alternative treatments outside of the context of the study. Evaluating the parameters that contribute to treatment effects may also require special variations of treatment or control conditions where portions of treatment are altered or omitted. The requisite control conditions to ask these types of questions raise serious ethical issues for clinical situations.

Apart from ethical issues, analogue research has major methodological advantages in studying treatment questions. The experimental control that laboratory-based conditions afford increases the power of the investigation to detect treatment differences. By standardizing or holding constant many conditions of the experiment, treatment differences are more likely to be detected than in clinical situations, where sources of variation are more difficult to control.

Clinical trials are especially well suited to examination of the effectiveness of alternative techniques with clinical problems and populations. The complexities and priorities of clinical settings make difficult the evaluation of subtle questions about particular treatments and their conceptual bases. Yet, the critical question for a given technique is whether treatment produces change in a clinical population. The accumulation of analogue studies, however useful for other purposes, cannot address the clinical effectiveness of treatment. Hence, direct tests of treatment are essential to ensure that the findings obtained in analogue research are not academic. Clinical trials are also frequently utilized to compare the relative effectiveness of alternative treatments or combinations of treatment. Comparative outcome research is frequently conducted in clinical trials. The research provides a direct test of two or more treatments in the situation where such information is critical. From a methodological standpoint, comparative research is quite suitable to clinical trials because it often does not necessarily require the use of other control groups (e.g., no-treatment, waiting list) that are frequently employed in analogue studies. The main question is the relative efficacy of two or more treatments; all participants in the clinical trial can receive an active treatment.

As a general statement, the main difference between analogue research and clinical trials pertains to their orientation with respect to questions about treatment. Analogue research tends to be *technique-oriented* in the sense that it is especially suitable for addressing the multiple questions about treatment and variables that contribute to its efficacy. Clinical trials tend to be *problem-oriented* in the sense that they are useful for addressing what is needed to alter the clinical problems that patients bring to treatment. Analogue research is essential for examining variations of treatment that are likely to maximize therapeutic change. The difficulty of conducting clinical research makes the completion of several studies on alternative variations of treatment prohibitive (Agras & Berkowitz, 1980). Analogue studies provide the opportunity to examine questions about how techniques can be implemented and about nuances of treatment that would be difficult to study in a longer-term and more elaborate clinical trial. The accumulated knowledge about outcome obtained in analogue research can be tested directly in clinical trials.

The difficulty of contemporary research is that clinical trials of a given technique or comparison of alternative techniques are often conducted with little antecedent laboratory-based research. Of course, the development and evaluation of clinically effective techniques need not necessarily be preceded by analogue research. For example, effective treatments for phobic and obsessive-compulsive disorders such as *in vivo* exposure have been developed primarily in clinical trials (Emmelkamp, 1981). The difficulty in comparative research, however, is that the absence of evidence about constituent techniques makes unknown the optimal variations of treatment. Thus, variations of treatment examined in a clinical trial may not represent the most effective versions of the treatment. Conclusions reached may be premature and quite different from what would be reached if different variations were applied.

CONCLUSION

Analogue studies and clinical trials are not qualitatively different types of research. In extreme cases, they can be readily distinguished because they may depart from each other on several different dimensions, discussed previously. However, on separate dimensions, studies can vary in their degree of resemblance to the clinical situation. On any particular dimension and across multiple dimensions, the distinction between analogue and clinical research may become blurred.

Certainly the major dimensions that are used to distinguish analogue research and clinical trials are the subject population and severity of the target problem, dimensions that usually go together. Analogue studies typically focus on college students and relatively mild problems; clinical trials focus on a more heterogeneous group with more severe problems. Yet, there are a variety of intermediate points on these and other dimensions along which analogue and clinical research might vary.

Analogue research and clinical trials were discussed as differentially suited to different questions about treatment. It would be misleading to imply that whether an analogue study or clinical trial is selected is determined purely or perhaps even primarily by the question of interest to the investigator. Whether analogue or clinical research is conducted often has to do with the setting in which the investigator is employed (Emmelkamp, 1981). Many investigators are affiliated with academic psychology or counseling departments where university students can be recruited in large numbers for relatively short-term treatment projects. Consequently, investigations are conducted with persons whose dysfunctions do not represent the usual clinical populations. Also, therapy may be administered by other students who are in training and may be limited in duration by academic terms rather than by clinical necessities. The constraints of the setting in terms of access to clinical populations and demands for publication may also help promote short-term studies of treatment (Agras *et al.,* 1979).

Other investigators are affiliated with hospitals and clinics and have direct access to clinical populations. These investigators are more likely to conduct clinical trials. Studies tend to take longer to complete and are occasionally plagued by problems that methodology teaches one to try to avoid (e.g., ensuring that only one treatment is provided to patients within a particular group). The setting usually fosters interest in applied questions for which answers are urgently needed. Thus, although analogue studies and clinical trials are suited to different types of questions, the setting in which one is employed partially limits or dictates the types of questions that are more readily addressed.

Both analogue investigations and clinical trials, as discussed in the present chapter, constitute strategies of experimental *research.* A major issue in contemporary clinical psychology is the relationship of research to practice. Despite the obvious resemblance of the conditions of clinical trials to clinical practice, important differences remain. The requirements of research necessarily lead to departures from clinical work. Research in general tends to examine one or two variables at one time, so that the influence of these variables can be evaluated. In clinical practice, clinicians intentionally vary multiple factors simultaneously for purposes of maximizing therapeutic change.

For example, in practice, clinicians consider each patient individually and provide multiple treatments that seem to be needed for the diverse problems that may be presented. During the course of treatment, different techniques may be started or stopped, goals of treatment may be altered, and other changes may be made that seem to be in the best interest of the patient at the time (Fishman, 1981). Also, because the goal is to ameliorate the patient's problems, treatment may be continued and altered until improvements have been achieved. In clinical research, open-ended, free-flowing, and nonstandardized applications of treatment would preclude conclusions to be drawn about the factors that accounted for change. Consequently, there is a necessary difference in research and practice.

The differences in priorities and *modi operandi* have led to frequent discussions of the hiatus between clinical research and practice and to suggestions for narrowing the gap. Several suggestions have focused on utilizing clinical practice as a place for research, so that the conditions of clinical work are experimentally evaluated. The suggestions include using single-case experimental designs with patients who are seen in clinical practice (Hayes, 1981; Hersen & Barlow, 1976); collecting information to rule out specific threats to internal validity so that valid inferences can be drawn in uncontrolled case studies (Kazdin, 1981); and collecting systematic information across large numbers of individual cases which, when accumulated over time and across settings, will permit analyses of treatments and variables that contribute to their efficacy (Barlow, 1980).

Another solution is to make better use of clinical practice as a source of variables for investigation than is currently the case (Lazarus & Davison, 1971; Woolfolk & Lazarus, 1979). In practice, clinicians may combine techniques in different ways or reach decisions about what treatments are needed for particular sorts of patients. The technique applications and decisions are based primarily on judgment rather than clear evidence. Yet, judgment cannot be criticized given the absence of research addressing the types of problems presented to the clinician.

The gap between research and practice might be narrowed by surveying therapists or by observing how therapy is actually conducted (e.g., Klein, Dittman, Parloff, & Gill, 1969). The results should then be followed up with clinical research that investigates as best as possible the effects of how treatment is conducted in practice. For example, the decisions that clinicians make in combining treatment or identifying what treatments are best for certain sorts of patients can be obtained from soliciting responses of clinicians to various real or hypothetical cases. Alternatively, information obtained directly from case files may suggest how treatment decisions are made. The results can be extended into experimental research, where the empirical bases for such decision making can be tested.

The evaluation of variables identified from clinical practice in the context of clinical trials still does not reproduce the conditions of practice. Clinical trials by necessity will still need to meet the demands of research, so that conclusions can be drawn about a circumscribed set of variables. However, the particular variables that are selected will be maximally relevant to clinical practice.

The present chapter has discussed analogue research, its characteristics, and differences from the conditions of clinical applications of treatment. Clinical trials were also discussed because they attempt to evaluate treatment in nonanalogue conditions, where techniques ultimately need to be tested. A critical issue is not whether a particular investigation is classified as an analogue study or clinical trial but rather how the study departs from the clinical situation and whether the dimension on which it departs and the degree of departure are related to the generality of the results. At the present time, the dimensions that influence generality of findings from either analogue research or clinical trials to the clinical situation are not well understood.

Apart from their differences, analogue investigations and clinical trials are similar insofar as they are both *research* strategies. Hence, they both depart from the clinical situation in important ways. Consequently, a significant issue is the relationship between findings obtained in research and their application to practice. The frequently lamented gap between research and practice indicts both analogue and clinical research. Consequently, research strategies need to be developed that attempt to examine conditions of clinical practice. The present chapter suggested the scrutiny of clinical practice to identify variables for further investigation in clinical trials.

REFERENCES

Adams, H. E., & Hughes, H. H. Animal analogues of behavioral treatment procedures: A critical evaluation. In M. Hersen, R. M. Eisler, & P. M. Miller (Eds.), *Progress in behavior modification* (Vol. 3). New York: Academic Press, 1976.

Agras, W. S., & Berkowitz, R. Clinical research in behavior therapy: Halfway there? *Behavior Therapy,* 1980, *11,* 472–487.

Agras, W. S., Kazdin, A. E., & Wilson, G. T. *Behavior therapy: Toward an applied clinical science.* San Francisco: Freeman, 1979.

Bandura, A. On paradigms and recycled ideologies. *Cognitive Therapy and Research,* 1978, *2,* 79–103.

Bandura, A. On ecumenism in research perspectives. *Cognitive Therapy and Research,* 1979, *3,* 245–248.

Barlow, D. H. Behavior therapy: The next decade. *Behavior Therapy,* 1980, *11,* 315–328.

Beck, A. T., Rush, A. J., Shaw, B. F., & Emery, G. *Cognitive therapy of depression.* New York: Guilford, 1979.

Bernstein, D. A. Behavioral fear assessment: Anxiety or artifact? In H. Adams & I. P. Unikel (Eds.), *Issues and trends in behavior therapy.* Springfield, Ill.: Thomas, 1973.

Bernstein, D. A., & Nietzel, M. T. Procedural variation in behavioral avoidance tests. *Journal of Consulting and Clinical Psychology,* 1973, *41,* 165–174.

Bernstein, D. A., & Nietzel, M. T. Behavioral avoidance tests: The effects of demand characteristics and repeated measures of two types of subjects. *Behavior Therapy,* 1974, *5,* 183–192.

Bernstein, D. A., & Paul, G. L. Some comments on therapy analogue research with small animal "phobias." *Journal of Behavior Therapy and Experimental Psychiatry,* 1971, *2,* 225–237.

Borkovec, T. D., & O'Brien, G. T. Methodological and target behavior issues in analogue therapy outcome research. In M. Hersen, R. M. Eisler, & P. M. Miller (Eds.), *Progress in behavior modification,* (Vol. 3). New York: Academic Press, 1976.

Borkovec, T., & Rachman, S. The utility of analogue research. *Behaviour Research and Therapy,* 1979, *17,* 253–261.

Brown, H. A. Role of expectancy manipulation in systematic desensitization. *Journal of Consulting and Clinical Psychology,* 1973, *41,* 405–411.

Cooper, A., Furst, J. B., & Bridger, W. H. A brief commentary on the usefulness of studying fears of snakes. *Journal of Abnormal Psychology,* 1969, *74,* 413–414.

Devine, D. A., & Fernald, P. S. Outcome effects on receiving a preferred, randomly assigned, or nonpreferred therapy. *Journal of Consulting and Clinical Psychology,* 1973, *41,* 104–107.

Emmelkamp, P. M. G. The behavioral study of clinical phobias. In M. Hersen, R. M. Eisler, & P. M. Miller (Eds.), *Progress in behavior modification,* (Vol. 8). New York: Academic Press, 1979.

Emmelkamp, P. M. G. The current and future status of clinical research. *Behavioral Assessment,* 1981, *3,* 249–253.

Fishman, S. T. Narrowing the generalization gap in clinical research. *Behavioral Assessment,* 1981, *3,* 243–248.

Foa, E. B., & Steketee, G. S. The interplay between scientific gain and ethical concerns in outcome research. *Behavioral Assessment,* 1981, *3,* 255–264.

Frank, J. D. *Persuasion and healing: A comparative study of psychotherapy* (2nd ed.). Baltimore: Johns Hopkins University Press, 1973.

Garfield, S. L. Psychotherapy: A 40-year appraisal. *American Psychologist,* 1981, *36,* 174–183.

Gordon, R. M. Effects of volunteering and responsibility on the perceived value and effectiveness of a clinical treatment. *Journal of Consulting and Clinical Psychology,* 1976, *44,* 799–801.

Gripp, R. F., & Magaro, P. A. A token economy program evaluation with untreated control ward comparisons. *Behaviour Research and Therapy,* 1971, *9,* 137–149.

Hayes, S. C. Single case experimental design and empirical clinical practice. *Journal of Consulting and Clinical Psychology,* 1981, *49,* 193–211.

Heller, K. Laboratory interview research as an analogue to treatment. In A. E. Bergin & S. L. Garfield (Eds.), *Handbook of psychotherapy and behavior change: An empirical analysis.* New York: Wiley, 1971.

Heller, K., Myers, R. A., & Kline, L. V. Interviewer behavior as a function of standardized client roles. *Journal of Consulting Psychology,* 1963, *27,* 117–122.

Hersen, M., & Barlow, D. H. *Single-case experimental designs: Strategies for studying behavior change.* New York: Pergamon, 1976.

Hunt, H. F. Problems in the interpretation of "experimental neurosis." *Psychological Reports,* 1964, *15,* 27–35.

Kanfer, F. H., & Phillips, J. S. *Learning foundations of behavior therapy.* New York: Wiley, 1970.

Kazdin, A. E. Evaluating the generality of findings in analogue therapy research. *Journal of Consulting and Clinical Psychology,* 1978, *46,* 673–686.

Kazdin, A. E. *Research design in clinical psychology.* New York: Harper & Row, 1980.

Kazdin, A. E. Drawing valid inferences from case studies. *Journal of Consulting and Clinical Psychology,* 1981, *49,* 183–192.

Kazdin, A. E. Methodology of psychotherapy outcome research: Recent developments and remaining limitations. In J. H. Harvey & M. M. Parks (Eds.), *Psychotherapy research and behavior change.* Washington, D.C.: American Psychological Association, 1982.

Kazdin, A. E., & Wilson, G. T. Criteria for evaluating psychotherapy. *Archives of General Psychiatry,* 1978, *35,* 407–416.

Kiesler, C. A. Mental health policy as a field of inquiry for psychology. *American Psychologist,* 1980, *35,* 1066–1080.

Klein, M. H., Dittmann, A. T., Parloff, M. B., & Gill, M. M. Behavior therapy: Observations and reflections. *Journal on Consulting and Clinical Psychology,* 1969, *33,* 259–266.

Kovacs, M., Rush, A. J., Beck, A. T., & Hollon, S. D. Depressed outpatients treated with cognitive therapy or pharmacotherapy. *Archives of General Psychiatry,* 1981, *38,* 33–39.

Krasner, L. The use of generalized reinforcers in psychotherapy research. *Psychological Reports,* 1955, *1,* 19–25.

Krasner, L. Verbal conditioning and psychotherapy. In L. Krasner & L. P. Ullmann (Eds.), *Research in behavior modification.* New York: Holt, 1965.

Lang, P. J. Fear reduction and fear behavior: Problems in treating construct. In J. M. Shlien (Ed.), *Research in psychotherapy* (Vol. 3). Washington, D.C.: American Psychological Association, 1968.

Lazarus, A. A., & Davison, G. C. Clinical innovation in research and practice. In A. E. Bergin & S. L. Garfield (Eds.), *Handbook of psychotherapy and behavior change: An empirical analysis.* New York: Wiley, 1971.

Levis, D. J. The case for performing research on nonpatient populations with fears of small animals: A reply to Cooper, Furst, and Bridger. *Journal of Abnormal Psychology,* 1970, *76,* 36–38.

Malan, D. H. The outcome problem in psychotherapy research: A historical review. *Archives of General Psychiatry,* 1973, *29,* 719–729.

Marks, I. Behavioral psychotherapy of adult neuroses. In S. L. Garfield & A. E. Bergin (Eds.), *Handbook of psychotherapy and behavior change: An empirical basis* (2nd ed.). New York: Wiley, 1978.

Marshall, E. Psychotherapy works, but for whom? *Science,* 1980, *207,* 506–508.

Masserman, J. H. *Behavior and neurosis.* Chicago: University of Chicago Press, 1943.

Nicolis, F. B., & Silvestri, L. C. Hypnotic activity of placebo in relation to severity of insomnia: A quantitative evaluation. *Clinical Pharmacology and Therapeutics,* 1967, *8,* 841–848.

Pande, S. K., & Gart, J. J. A method of quantify reciprocal influence between therapist and patient in psychotherapy. In J. M. Shlien (Ed.), *Research in psychotherapy* (Vol. 3). Washington, D.C.: American Psychological Association, 1968.

Parloff, M. B. Can psychotherapy research guide the policymaker? A little knowledge may be a dangerous thing. *American Psychologist,* 1979, *34,* 296–306.

Paul, G. L., & Lentz, R. J. *Psychosocial treatment of chronic mental patients: Milieu versus social-learning programs.* Cambridge, Mass.: Harvard University Press, 1977.

Pavlov, I. P. *Conditioned reflexes: An investigation of the physiological activities of the cerebral cortex.* London: Oxford University Press, 1927.

President's Commission on Mental Health. *Report to the President* (Vol. 1). Washington, D.C.: U.S. Government Printing Office, 1978.

Rachman, S. J., & Wilson, G. T. *The effects of psychological therapy* (2nd ed.). Oxford: Pergamon, 1980.

Robinson, C., & Suinn, R. M. Group desensitization of a phobia in massed sessions. *Behaviour Research and Therapy,* 1969, *7,* 319–321.

Rogers, C., & Dymond, R. *Psychotherapy and personality change.* Chicago: University of Chicago Press, 1954.

Rush, A. J., Beck, A. T., Kovacs, M., & Hollon, S. Comparative efficacy of cognitive therapy and pharmacotherapy in the treatment of depressed outpatients. *Cognitive Therapy and Research,* 1977, *1,* 17–38.

Russell, R. W. The comparative study of "conflict" and "experimental neurosis." *British Journal of Psychology,* 1950, *41,* 95–108.

Sechrest, L., West, S. G., Phillips, M. A., Redner, R., & Yeaton, W. Some neglected problems in evaluation research: Strength and integrity of treatments. In L. Sechrest, S. G. West, M. A. Phillips, R. Redner & W. Yeaton (Eds.), *Evaluation studies: Review annual* (Vol. 4). Beverly Hills: Sage, 1979.

Smith, M. L., & Glass, G. V. Meta-analysis of psychotherapy outcome studies. *American Psychologist,* 1977, *32,* 752–760.

Smith, M. L., Glass, G. V., & Miller, T. I. *The benefits of psychotherapy.* Baltimore: Johns Hopkins University Press, 1980.

Strupp, H. H., & Hadley, S. W. A tripartite model of mental health and therapeutic outcomes. *American Psychologist,* 1977, *32,* 187–196.

Strupp, H. H., & Hadley, S. W. Specific vs. nonspecific factors in psychotherapy. *Archives of General Psychiatry,* 1979, *36,* 1125–1137.

Truax, C. B. Reinforcement and non-reinforcement in Rogerian psychotherapy. *Journal of Abnormal Psychology,* 1966, *71,* 1–9.

VandenBos, G. R. (Ed.). *Psychotherapy: Practice, research, policy.* Beverly Hills, Calif.: Sage, 1980.

VandenBos, G. R., & Pino, C. D. Research on the outcome of psychotherapy. In G. R. VandenBos (Ed.), *Psychotherapy: Practice, research, policy.* Beverly Hills, Calif.: Sage, 1980.

Wilson, G. T. Effects of false feedback on avoidance behavior: "Cognitive" desensitization revisited. *Journal of Personality and Social Psychology,* 1973, *28,* 115–122.

Wolpe, J. *Psychotherapy by reciprocal inhibition.* Stanford, Calif.: Stanford University Press, 1958.

Woolfolk, R. L., & Lazarus, A. A. Between laboratory and clinic: Paving the two-way street. *Cognitive Therapy and Research,* 1979, *3,* 239–244.

8

Comparative Outcome Research

RICHARD G. HEIMBERG and ROBERT E. BECKER

INTRODUCTION

After one has asked whether psychotherapy is effective, a next logical step is to ask whether one approach to psychotherapy works better than another (Gottman & Markman, 1978; Rachman & Wilson, 1980). Unfortunately, it is neither easy, logical, nor productive to phrase the question in such global terms. As a result, the area of comparative outcome research has been a controversial one and the quality of our findings has not justified our efforts.

For our purposes, comparative outcome research (COR) will be defined as that body of research which evaluates the relative efficacy of two or more (psycho)therapeutic techniques or compares one such technique to the therapeutic techniques of other disciplines. Included in our definition are studies comparing (1) two distinct techniques derived from similar theoretical backgrounds, such as a study comparing the behavior therapy techniques of systematic desensitization and assertion training; (2) two therapeutic techniques representative of different "schools" of therapy (Cross, Sheehan, & Khan, 1982; Sloane, Staples, Cristol, Yorkston, & Whipple, 1975); and (3) a psychotherapeutic approach to a nonpsychological therapeutic approach, such as pharmacotherapy (Klerman, DiMascio, Weissman, Prusoff, & Paykel, 1974; Rush, Beck, Kovacs, & Hollon, 1977). The reader is referred to the excellent and comprehensive review of studies in the first category by Kazdin and Wilson (1978b). We will concentrate the majority of our efforts on the second and third categories.

RICHARD G. HEIMBERG • Department of Psychology, State University of New York at Albany, Albany, New York 12222.   ROBERT E. BECKER • Department of Psychiatry, Albany Medical College, 47 New Scotland Avenue, Albany, New York 12208.

The volume of COR studies is large. Anyone who attempts to review the literature comprehensively will face an overwhelming task indeed. We, therefore, have adopted a selective approach. Several of the major comparative studies will be reviewed. With these studies as background, the myriad conceptual and methodological issues confronting the COR investigator will be detailed and confronted; solutions will be recommended whenever possible.

Before proceeding to our review of the major studies, a number of issues should be addressed. First of all, the form of the experimental question posed by COR studies is critical. Simply to ask whether Method A is better than Method B is to ask a question that may be misleading or uninformative (Bergin & Lambert, 1978; Edwards & Cronbach, 1952; Kiesler, 1966; Paul, 1967). To do so is to set up a battle between the faithful of various schools of therapy; it turns the supposedly systematic process of research into an emotional win–lose situation. Paul's (1967, p. 111) eloquent and oft-quoted outcome statement provides a more realistic approach. When posing an outcome question, we should ask, *"What* specific treatment, by *whom,* is most effective for *this* individual with *that* specific problem, and under *which* set of circumstances?"* Edwards and Cronbach, some 30 years ago, provided an alternative statement of the problem:

> Is Method A better than Method B? If the question is stated so badly, it is inappropriate for research. To obtain an answerable problem, the criterion must be carefully defined, the methods must be specified, and the range of persons and conditions to be considered must be identified. (Edwards & Cronbach, 1952, p. 52)

Even if an appropriate question can be specified, there is concern about the utility of COR. While DiLoreto (1971) asserts that treatment–control comparisons are of little value and that all outcome studies should compare two active treatments, other authors adopt a more conservative stance. Kazdin and Wilson (1978b) and Rachman and Wilson (1980) criticize traditional comparative research for its poorly formulated experimental questions, heavy reliance on large-*n* between-group designs, and general methodological inadequacy. It is lamented that the evaluation strategies employed in these studies have been "ill-suited for identifying the mechanisms of therapeutic change or comparing the relative merits of alternative treatment methods" (Kazdin & Wilson, 1978b, p. 47). Both sets of authors contend that before COR can be usefully conducted, powerful evidence of the efficacy of each treatment and the appropriate applications of each must be available. Rarely in the course of our review of the COR literature have these conditions been met.

Although Kazdin, Rachman, and Wilson are suspicious of traditional COR, these outstanding scientists are aware that COR studies will continue to be conducted, that they may receive a great deal of professional and public scrutiny, and that they may be disproportionately relied upon in various funding and public policy decisions. Hollon and Beck (1978), in their review of psychotherapy–pharmacotherapy comparisons, raise an additional frightening notion—that conclusions based on inadequately designed comparative studies may be more likely to be accepted than the conclusions of single-treatment studies! The greater number of statistical comparisons typical of COR studies increases the probability that significant findings will occur on a chance basis. Also, a number of design problems (to be outlined in a later section of this chapter) may impede the demonstration of treatment efficacy for one modality but not others. In this situation, joint occurrence of significant and nonsignificant results may spuriously increase the confidence with which an overall pattern of findings is accepted. Finally, COR studies may be more likely to be accepted for publication when subtle confounds produce controversial results. The above arguments underscore the need for an enlightened, informed approach to the conduct of comparative studies and a respectful appreciation of the many pitfalls that one may encounter along the way. Such is the purpose of this chapter.

RESEARCH REVIEW

Progress over the Last Several Years

In this section, we will review and critique five of the better known comparative treatment outcome studies (DiLoreto, 1971; Klerman *et al.,* 1974; Paul, 1966; Rush *et al.,* 1977; Sloane *et al.,* 1975). These studies were selected from the literally hundreds of available reports because of their notoriety and because their methodological sophistication makes their contributions valuable. Criticisms of these studies become all the more telling if the reader remains aware of the considerable accomplishments of these investigators and the high regard in which their experiments are held. Many other good studies have been bypassed due to space limitations, and still others will be cited below in connection with specific issues.

Paul (1966)

Gordon Paul's seminal work represents the first well-controlled comparison of a behavior therapy technique, systematic desensitization, to insight-oriented therapy. Subjects were 96 speech-anxious undergraduates selected from a pool of 710 students enrolled in a public speaking course. Assessment consisted of a variety of self-report measures of anxiety and related states (e.g., extraversion, emotionality) administered before treatment and at a 6-week follow-up as well as behavioral and physiological measures collected immediately before and after treatment. Subjects were randomly assigned to five 1-hour sessions of systematic desensitization, insight therapy, or a credible attention–placebo group or they received no treatment. All therapists were highly experienced in the conduct of client-centered or neo-Freudian therapy but had no prior experience in the conduct of systematic desensitization. Nevertheless, each therapist treated an equal number of subjects in the desensitization, insight, and attention–placebo conditions. In spite of this potential source of bias in favor of insight-oriented treatment, desensitization produced the most marked improvement. The desensitization treatment surpassed controls on all measures and insight therapy on measures of behavioral performance, anxiety, and overall improvement. On no measure did insight therapy produce more favorable outcomes than the attention–placebo treatment.

In an additional set of comparisons, reported in Paul and Shannon (1966), a nine-session group treatment combining systematic desensitization and discussion was compared to the treatments evaluated by Paul (1966; same data). "Combined group desensitization" produced results comparable to those of individual desensitization and superior to those of the insight or attention–placebo conditions.

Critique of Paul (1966), Paul and Shannon (1966). Paul's (1966) study was a landmark effort and a methodological inspiration to those who would carry out later COR studies. However, it has not been free of criticism. Several authors (e.g., Luborsky, Singer, & Luborsky, 1975; Smith & Glass, 1977) have cited the use of college-student subjects as a major drawback and discarded this study because it did not use "real patients." Three other criticisms, we believe, are more central. First, by restricting treatment to five sessions, insight therapy may not have been given a fair test. Insight therapy, by its very nature, sets loftier goals and may reasonably require a longer time to reach them. However, we and others (Kazdin & Wilson, 1978a,b; Pokorny & Klett, 1966; Shlien, 1964; Shlien, Mosak, & Dreikurs, 1962) believe this to be a more complex issue, and we will return to it in a later section of the paper. Second, while Paul did an excellent job in specifying the activities of systematic desensitization and attention–placebo therapists, we know little of the actual activities of the insight group. In a review of early behavioral-versus-insight studies, Roback (1971) concluded that none adequately defined the insight therapy condition. Since many different types of insight therapy exist, the conclusions to be drawn from Paul's study are extremely limited. Finally, Paul's selection of insight therapists, as noted above, posed a serious threat to the internal validity of the study that was avoided only because of the superior results for systematic desensitization. Had insight been most effective, therapist preference or greater expertise in that approach would have posed very reasonable competing hypotheses.

Three issues about the Paul and Shannon (1966) comparisons deserve comment. First, combined group–desensitization subjects received more treatment sessions (9) than subjects in the other conditions (5). Second, group subjects received a more broadly based treatment for anxiety than did subjects in individual treatment. While individual subjects were treated specifically for speech-related anxiety, group subjects also received treatment for test anxiety. Third, group subjects were treated after all other subjects had finished their treatment and had completed one extra assessment. These issues clearly prohibit straightforward interpretation of results.

DiLoreto (1971)

DiLoreto (1971) conducted a controlled comparison of systematic desensitization, Albert Ellis's rational-emotive therapy, Carl Rogers's client-centered therapy, attention-placebo, and no-contact control conditions for the treatment of interpersonal anxiety. This study is one of the few that attempted to evaluate the relative effectiveness of the treatments for subjects known to differ on specific characteristics, in this case introversion versus extraversion. Based on Eysenck's (1961) notions of conditionability, DiLoreto predicted that introverts would respond most favorably to systematic desensitization while extraverts would fare best with client-centered treatment. Subjects were 42 male and 58 female college students who reported high levels of interpersonal anxiety and a strong desire for treatment. A battery of self-report and behavioral measures was administered to subjects pretreatment, posttreatment and at a 3-month follow-up. Several of these measures were developed specifically for this study and are of questionable validity. Subjects were randomly assigned to treatment groups of five subjects each, with equal numbers of introverts and extraverts receiving each treatment. Each group (except the no-contact group) received approximately 11 hours of treatment with graduate-student therapists who had experience and commitment to the treatments they administered. Results provided only partial support for DiLoreto's (1971) hypotheses. All treatments produced greater changes in anxiety and reports of extratherapy behavior change than controls. Rational-emotive and client-centered therapies were most effective with introverted clients, while systematic desensitization was equally effective with both introverts and extraverts.

Critique of DiLoreto (1971). This impressive study is often cited as demonstrating the superiority of behavior therapy. However, Bergin and Lambert (1978) cite the use of college-student subjects and relatively inexperienced therapists to discount this claim. While we find much to admire in the design and conduct of this study, there are several additional points of concern. Already noted is the use of unvalidated outcome measures. The attention-placebo group also strikes us as poorly conceived, focusing on general and academic aspects of campus life rather than focusing inert or "nonspecific" treatment ingredients on the same problem areas. In a positive vein, DiLoreto took a careful look at the behavior of each therapist in each condition as they interacted with introverts and extraverts. However, this careful analysis revealed a growing tendency towards nonadherence to protocol as treatment progressed. This tendency was systematically related to therapists' amount of experience and personal psychotherapy. Thus, as the study progressed, treatments were no longer evaluated at full strength. We suspect that rational-emotive therapy and client-centered therapy would have suffered more than systematic desensitization in this regard, since they are inherently more complex. Our viewpoint is supported by the critiques of Ellis (1971), Boy (1971), and Goldstein and Wolpe (1971), who were each concerned about the method of presentation of their own therapy techniques.

Sloane *et al.* (1975)

The Sloane *et al.* study is a detailed, comprehensive evaluation of the effects of behavior therapy and psychoanalytically oriented psychotherapy and is generally recognized as the best COR study conducted to date (Bergin & Lambert, 1978; Kazdin & Wilson, 1978b; Rachman & Wilson, 1980). Ninety psychiatric outpatients, of whom approximately two-thirds were diagnosed as neurotic and one-third as personality disorders, were matched for sex and problem severity and randomly assigned to either behavior therapy, psychotherapy, or a waiting-list control group. Assessments—conducted pretreatment, after 4 months of treatment, and at a 1-year follow-up—included a series of personality inventories, structured interviews by an independent assessor, an analysis of target symptoms, and ratings by patients, therapists, informants, and the assessor. In contrast to many other comparative studies, this one was staffed by therapists who were leaders in their fields (behavior therapy: Wolpe, Lazarus, Serber; psychotherapy: Urban, Vispo, Freed) and who would reasonably be expected to provide their respective treatments at maximal strength. A detailed list of definitions of each therapy was drawn up, but therapists were given a fair degree of latitude for practice as usual. Tape recordings of sessions were utilized to assess compliance with protocol and to conduct a series of process analyses of each treatment procedure. Posttreatment assessment demonstrated greater improvement in treated patients than in controls, with some edge to behavior therapy in social functioning and ratings of adjustment. Follow-up analyses again revealed behavior therapy and psychotherapy to be superior to controls, but little difference between these treatments was evident. In summary, Sloane *et al.* (1975) assert that behavior therapy is at least as effective as psychotherapy and may actually surpass psychotherapy in the treatment of *severe* neuroses and personality disorders. However, Bergin and Lambert (1978) question the validity of the latter conclusion.

Critique of Sloane *et al.* (1975). Sloane *et al.* have conducted an exquisitely designed study with a number of outstanding features. Kazdin and Wilson (1978b) list the following good points: (1) use of experienced therapists, (2) a large *n* of highly motivated clients, (3) random assignment, (4) inclusion of a no-treatment control group, (5) a relatively long follow-up period, and (6) minimal subject loss due to attrition. Bergin and Lambert (1978) add: (1) matching of subjects on several important dimensions, (2) use of a clinical population of subjects, (3) a serious effort to assess the extent to which therapeutic activities were actually representative of the different therapies, (4) administration of treatments in equal amounts and in duration sufficient to assess the effects of each, and (5) administration of a wide variety of outcome measures of relevance to the goals of each therapy. We believe that a serious psychoanalytic therapist might question Bergin and Lambert's fourth point, and we question their fifth. Personality inventories, interviews, and clinical ratings do not achieve the operational specificity desired by most behavior therapists. Several other difficulties will be noted throughout our review, but three problems require mention. First, the independent assessor, a psychoanalytically oriented psychiatrist, was not blind to patient assignment. Second, patients were selected on the basis of their appropriateness for psychotherapy; thereby introducing a potential bias against behavior therapy. Finally, follow-up comparisons are difficult to interpret since many patients (15 behavior therapy, 9 psychotherapy, and 22 waiting-list subjects) received treatment of some sort between the posttreatment and follow-up assessments.

Klerman *et al.* (1974)

Klerman *et al.* (1974) investigated the efficacy of the tricyclic antidepressant amitriptyline and psychotherapy for the prevention of relapse in women who had suffered from neurotic depression. Effects of these treatments on social adjustment were reported in a second paper (Weissman, Klerman, Paykel, Prusoff, & Hansen, 1974). Two-hundred seventy-eight depressed women were given an open trial of 100 to 200 mg/day of amitriptyline for 4 to 6 weeks. One-hundred fifty women who had experienced at least a 50% reduction in symptoms (as measured by the Raskin Depression Scale) were then assigned to one of six cells in the design of this maintenance study. Subjects received either amitriptyline, placebo, or no pill *and* either weekly psychotherapy with an experienced social worker or minimal contact. The maintenance period lasted 8 months or until a patient relapsed (returned to original symptom levels). Subjects receiving amitriptyline showed similar relapse rates whether they received psychotherapy (12.5%) or not (12%). Placebo treatment also produced similar rates across psychotherapy (28%) and minimal contact (30.8%) conditions. When no medication or placebo was administered, 36% of minimal-contact patients relapsed. However, psychotherapy and no pill reduced this rate to 16.7%. Klerman *et al.* (1974) interpret these data as indicating the superiority of amitriptyline, presumably because amitriptyline did show meaningful effects independent of psychotherapy. However, as noted by Hollon and Beck (1978), the low relapse rates for psychotherapy alone do not support this conclusion.

Patients who did not relapse were further evaluated by Weissman *et al.* (1974). Patients receiving psychotherapy showed positive effects on social adjustment, while amitriptyline had little effect. Thus, there is some evidence to support the combined use of amitriptyline and psychotherapy, since positive effects on relapse and social adjustment may be expected.

Critique of Klerman *et al.* (1974), Weissman *et al.* (1974). This study is reasonably designed in most respects. However, two specific factors demand caution in interpreting results. The content of psychotherapy is poorly specified, so it is difficult to know its precise nature. Also, patient selection criteria seriously bias the study in favor of amitriptyline. All patients were selected from the pool of positive responders in the open drug trial. However, no selection criteria involving response to psychotherapy were employed. The conclusions of these studies may therefore be generalized only to previously depressed women with a history of positive response to amitriptyline.

Rush *et al.* (1977)

Rush *et al.* compared Aaron Beck's cognitive therapy to the tricyclic antidepressant imipramine for the treatment of unipolar depression in outpatients. Forty-one patients were randomly assigned to cognitive therapy $(n = 19)$ or imipramine $(n = 22)$ and were assessed with the Beck Depression Inventory, the Hamilton Rating Scales for Depression and Anxiety, and the Raskin Depression Scale before treatment, after treatment, and at 3-month and 6-month follow-ups. Patients with a history of poor drug response were excluded. Treatment lasted an average of 12 weeks; cognitive therapy patients received a maximum of 20 1-hour sessions while drug patients received a maximum of twelve 20-minute sessions. Treatment was administered by psychiatric residents with relatively little experience in cognitive therapy but substantial experience with drug treatment. Both treatments led to a reduction in depressive symptomatology, but cognitive therapy patients were more improved according to self-reports and clinical ratings. Rated as markedly improved were 79% of cognitive therapy patients as opposed to 23% of drug patients. Significantly higher rates of attrition and reentry to treatment occurred among drug patients during follow-up. Differences in self-reported depression were still demonstrated at the 3-month follow-up regardless of whether the analysis included completers or completers plus dropouts, but only the latter analysis was significant at 6 months. A 1-year follow-up of completers (Kovacs, Rush, Beck, & Hollon, 1981) revealed that both treatment groups remained generally symptom-free. Self-reported depression was still significantly less for the cognitive therapy group, but no other differences remained.

Critique of Rush *et al.* (1977). The Rush *et al.* study has received much scrutiny because of the current popularity of cognitive therapy and the favorable showing of this version of psychotherapy in comparison to pharmacotherapy. In the midst of favorable reviews (e.g., Bergin & Lambert, 1978; Kazdin & Wilson, 1978b), a number of concerns have arisen. Since this study was conducted by the cognitive therapy group, experimenter bias in favor of cognitive therapy remains a competing hypothesis. However, this is by no means a criticism unique to this study and simply awaits independent replication for its resolution. Also of concern is the use of assessors who were not blind to treatment assignment, the absence of control groups that might further delineate the significance of reported changes, and the greater amount of treatment time devoted to cognitive therapy. The use of therapists inexperienced in cognitive therapy and the exclusion of patients with a history of poor drug response are not problematic, but only because of the superiority of cognitive therapy.

Finally, in a recent paper Becker and Schuckit (1978) ask whether the type of medication, dosage level, and duration of the medication trial provide an optimal regimen for comparison to cognitive therapy. They assert that lithium carbonate may be a better treatment for the several patients who reported chronic or recurrent depressions and that the dosage levels of imipramine were too low and maintained for too brief a period of time. In rebuttal, Rush, Hollon, Beck, and Kovacs (1978) state that the level of drug response obtained in their study was similar to that typically obtained in the literature. Furthermore, no differences in response were found between chronically and acutely depressed patients. In their defense, it should be noted that lithium treatment of unipolar depression and large-dose administration of tricyclics were relatively uncommon when the study was initiated.

Current State of the Field

Review of representative studies may provide a glimpse into the problems encountered by specific investigators and the validity of the results they report. On the basis of the studies reviewed above, we may conclude that COR is a difficult undertaking even for the most outstanding investigators, but it is not possible to summarize the findings of COR research in general without a much grander effort. In this section we will look at two strategies for accomplishing that lofty goal and also at the controversy stirred by each of them.

Luborsky *et al.* (1975): A Trip to the Ballot Box

One potential method of evaluating the effectiveness of different kinds of psychotherapy is to look at each COR study as a vote for one kind of therapy over another. The Rush *et al.* (1977) study might be considered a vote for cognitive therapy, the Paul (1966) study a vote for systematic desensitization, and the Sloane *et al.* (1975) study a "tie" between behavior therapy and psychoanalytically oriented psychotherapy. A tally of positive votes for each therapy can then provide an index of the relative effectiveness of that therapy. This is the approach of Luborsky *et al.* (1975).[1]

Luborsky *et al.* conducted a variety of comparisons, including psychotherapy versus no treatment, psychotherapy versus behavior therapy, and psychotherapy versus pharmacotherapy. As a preliminary step, all 113 studies to be included were "graded" for methodological adequacy according to a list of criteria generated by Fiske, Hunt, Luborsky, Orne, Parloff, Reiser, and Tuma (1970), including:

1. Random or matched assignment of patients to groups.
2. Use of "real patients." Studies using student volunteers were excluded.
3. Sufficiently experienced therapists.
4. Equally competent therapists for each treatment.
5. Treatments equally valued by patients.
6. Outcome measures congruent with goals of treatment.
7. Outcome evaluated by independent measures (beyond patient–therapist ratings).
8. Information about concurrent treatments obtained and equated across groups.
9. Therapist activity evaluated for adherence to protocol.
10. Treatments given in equal and reasonable amounts.
11. Adequate sample size.

---

[1] A similar analysis of the relative effectiveness of behavior therapy and cognitive behavior modification has been reported by Ledwidge (1978). Most of the same comments and criticisms applied to Luborsky *et al.* (1975) are also relevant to Ledwidge. Additional discussion of that paper may be found in Locke (1979), Meichenbaum (1979), Mahoney and Kazdin (1979), and rebuttals by Ledwidge (1979a, b).

Grades assigned to particular studies served as an index of the certainty with which certain findings might be accepted. On the basis of this "voting method," Luborsky *et al.* assert that "Everyone has won and all must have prizes." When this well-known phrase from *Alice in Wonderland* is applied to COR, it reflects their conclusions that most comparative studies have found nonsignificant differences in the proportions of patients improved at the end of therapy and that a high percentage of patients who go through any kind of therapy benefit from the experience. Specifically excluded from this sweeping statement were comparisons of combined treatments versus single treatments (combined treatments were superior) and pharmacotherapy versus psychotherapy (pharmacotherapy was superior).

Critique of Luborsky *et al.* (1975). This study has been resoundingly criticized from a number of vantage points. The most thorough review is provided by Rachman and Wilson (1980). We are in unanimous agreement with their critique and find little additional to add. The following criticisms are drawn from that outstanding review.

1. It is difficult to understand why Luborsky *et al.* (1975) chose to exclude certain studies and include others. They have set out to focus on "real patients," presumably those who might be encountered in community mental health centers or psychiatric clinics. Yet the addictions, sexual dysfunctions, psychoses, and childhood disorders were excluded. More predictably, studies employing student volunteers were excluded. It has been argued, however, that if one can assess the limits of generalizability of these analogue studies, they may usefully contribute to the determination of treatment efficacy (Borkovec & Rachman, 1979; Kazdin, Chapter 7, this volume). Luborsky *et al.* (1975) did include studies of the effects of treatment on ethnocentrism scores (Pearl, 1955) and on the soiling behavior of chronic psychiatric patients (Tucker, 1956). What is the relevance of these studies to the "typical patient"? Why do they contribute more than the studies by Paul (1966) and DiLoreto (1971), which were excluded?

2. Luborsky *et al.* (1975) accept the therapy uniformity myth (Kiesler, 1966). Within each broad class of psychotherapy, behavior therapy, and pharmacotherapy, all treatments are assumed to be equivalent and equally representative of all others. However, there is much heterogeneity among therapeutic techniques, and there is increasing evidence in the field of behavior therapy that all techniques are not equally effective for the treatment of specific problems (Agras, Kazdin, & Wilson, 1979; Kazdin & Wilson, 1978; Rachman & Wilson, 1980). Although it was not mentioned by Rachman and Wilson, the assumption of therapy uniformity may appear least tenable in the analysis of pharmacotherapy studies. It would dictate the equivalence of different medicines or treatments given in different doses for different lengths of time for

different disorders with different presumed pathogeneses. We could hardly recommend the physician who seriously adopted this stance.

Another variation of this same criticism is that Luborsky *et al.* (1975) ask the wrong questions. They simply make a series of comparisons on Method A versus Method B, with no regard for the concerns earlier raised by Paul (1967) and Edwards and Cronbach (1952).

3. Rachman and Wilson (1980) raise numerous concerns about the grading system. First, they analyze several of the studies included by Luborsky *et al.* (1975) and conclude that most studies are more poorly designed than the latter investigators acknowledged. In their review, the most common concern was the unsatisfactory measurement of treatment outcome. Typical problems were failure to keep assessors blind to treatment assignment and sole reliance on psychological tests of questionable validity for the measurement outcome. A second problem in grading is that the same study might receive different grades in different comparisons. The same study might receive a grade of "B" in the psychotherapy-versus-behavior therapy comparison but an "A" in some other analysis. While there may be valid reasons for this practice, none are delineated by Luborsky *et al.* (1975).

4. A final concern about the use of the "box-score" approach is that it assigns equal weight to all studies regardless of their quality. Two poorly designed studies with negative results outweigh one well-designed study with positive findings. Although Luborsky *et al.* (1975) intended the grading scheme to overcome this problem, the difficulties with that procedure render it inadequate to the task.

A series of similar arguments may be raised about the psychotherapy-versus-control comparisons of Luborsky *et al.* (1975). However, in keeping with the comparative thrust of this chapter, these will be bypassed. In closing this section, we are left with two serious questions: Would it not be best to base any comparison on the best information available rather than relying so heavily on poor studies? Even if poor studies are excluded, can the "trip to the ballot box" result in a meaningful "election," or does the method itself contribute to the high number of no-difference findings?

Smith and Glass (1977): Another Meta-View through the Looking Glass

Smith and Glass (1977) attempted to accomplish the same goal pursued by Luborsky *et al.* (1975) but approached it from a decidedly different direction. Almost 400 studies of counseling and psychotherapy were reviewed and statistically integrated in order to test the general effectiveness of treatment and the differential effectiveness of the various types of treatment. The primary measure of interest was *effect size,* defined as the *"mean difference between treated and control subjects divided by the standard deviation of the control group,"* that is, $ES = (\overline{X}_T - \overline{X}_C) / s_C$ (Smith & Glass, 1977, p. 753, original italics). The effect sizes of various treatments, based on the outcomes of specific measures in specific studies, were compared to answer the questions of interest. This technique has been labeled "meta-analysis" and contrasted to other types of data analyses by Glass (1976).

Eight-hundred thirty-three effect sizes were derived from the studies reviewed by Smith and Glass (1977) and served as the dependent variable in the meta-analyses. In addition to type of therapy (e.g., psychodynamic, rational-emotive, behavioral), 15 additional independent variables (mostly concerned with patients' demographic characteristics, therapist characteristics, and research design) were utilized but are of less concern here. The first finding reported by Smith and Glass (1977) concerned the general efficacy of psychotherapy. The average study showed an effect size of .68 for treated groups over control groups. Thus, the average treatment group produced .68 of a standard deviation improvement, and the average treated client fared better than 75% of controls. However, for present purposes, other findings reported by Smith and Glass (1977) may be of greater interest.

As a next step, average effect sizes were calculated for 10 different approaches to therapy. There was considerable variability in average effect size, with systematic desensitization "scoring" the highest (.9) and gestalt therapy the lowest (.26). Other scores of interest were rational-emotive therapy (.77), behavior modification (.76), implosion therapy (.64), client-centered therapy (.63), and psychodynamic therapy (.59).

These comparisons involved an extremely heterogeneous group of studies and left a number of important factors (e.g., duration, problem severity, type of outcome) uncontrolled. In order to cope with this problem, Smith and Glass (1977) used multidimensional scaling techniques to derive four broader classes of therapy. Average effect sizes were not reported for this analysis, however, and therapy classes were further aggregated into "superclasses." Gestalt therapy was dropped from the analysis because it contributed too few studies. Systematic desensitization, implosion, and behavior modification made up the "behavioral superclass," while the remaining techniques constituted the "nonbehavioral superclass." These aggregates achieved mean effect sizes of .80 and .60 respectively. Smith and Glass (1977) discount this difference because the behavioral studies measured outcome more quickly after temination and the behavioral group relied on *more subjective outcome measures* (a somewhat questionable conclusion). An analysis of approximately 50 studies that directly compared behavioral and nonbehavioral therapies revealed a difference in mean effect size of only .07. On this basis, Smith and Glass (1977) reached a conclusion similar to that of Luborsky *et al.* (1975), that all psychotherapy is effective and that there is little difference between the different types of therapy.

Critique of Smith and Glass (1977). The effort by Smith and Glass (1977) not only garnered results similar to those of Luborsky *et al.* (1975) but has been subjected to the same type of intense scrutiny. In addition, a series of rebuttals and rejoinders have appeared and the debate has been lively (Eysenck, 1978; Gallo, 1978; Glass, 1978; Glass & Smith, 1978; Mansfield & Busse, 1977; Presby, 1978). Additional commentaries have been provided by Kazdin and Wilson (1978b), Rachman and Wilson (1980), and Gottman and Markman (1978).

Several criticisms leveled at Luborsky *et al.* (1975) would seem to apply here as well. One first wonders how studies were selected for inclusion. The large number of studies would indicate comprehensiveness, but two points make one wonder otherwise. First, 144 studies were dissertations or fugitive documents not easily available to the rest of us. Second, a large percentage of studies reviewed by Kazdin and Wilson (1978b) are not included in the analyses. Rachman and Wilson (1980) cogently argue that such selective (or uneven) inclusion of studies might seriously bias results. For instance, the excellent work on treatment of anxiety by Marks, Gelder, and colleagues (reviewed in Rachman & Wilson, 1980) in the United Kingdom and a number of others around the world clearly points to the superiority of performance-based exposure (e.g., flooding) treatments over systematic desensitization. Exclusion of these studies certainly contributed to the reported superiority of desensitization over implosion.

One might similarly wonder how the 10 specific treatments were selected for special attention. Certainly research on other types of treatment is common enough for inclusion. Also, how can a specific technique designed for the treatment of a specific disorder (e.g., systematic desensitization) be directly compared to a therapy system designed to treat diverse clients with diverse concerns?

Smith and Glass (1977) may justifiably be accused of endorsing the uniformity myth (Kiesler, 1966) in several of its forms. Several types of psychodynamic therapy are treated equivalently, as are several types of behavior modification. All patients are considered similar regardless of severity or type of problem. Underachieving college students are considered together with schizophrenics and other seriously afflicted persons (Rachman & Wilson, 1980). Little reference is made to differences among therapists, although several therapist factors were included in their independent variable list.

As in Luborsky *et al.* (1975), studies of poor quality are considered equal to well-designed studies. This practice has been soundly criticized (Eysenck, 1978; Mansfield & Busse, 1977). The concern, of course, is that poor data will yield inaccurate or uninterpretable results. Glass (1976, 1978; Glass & Smith,

1978) maintains that poor studies yield roughly the same results as good ones and to eliminate these studies on methodological grounds would be to throw out much useful data. It is further claimed that quality of design accounted for only 1% of the variance in effect size in the meta-analyses and is therefore a minor nuisance (Glass, 1978). Eysenck (1978, p. 517, our italics) responds in his typical iconoclastic manner: *"Garbage in—garbage out!"* Our reading of Glass's papers (1976; Smith & Glass, 1977) lead us to follow Eysenck. In building his case for the equivalence of good and bad studies, Glass (1976) cites Astin and Ross's review of studies on the effects of glutamic acid on the IQ of mentally retarded children. The studies reviewed produced conflicting results. Early studies supported the beneficial effects of glutamic acid, while later ones did not. When the quality of results was cross-tabulated with the quality of experimental design, some interesting patterns emerged. "The seemingly 'positive' findings came from the poorly designed experiments; good experiments showed no effects" (Glass, 1976, p. 5).

Another area of concern is the development and treatment of the superclasses. Although the classes were based on a multidimensional scaling procedure, there are still large differences between the therapies in each superclass. Of particular concern is the grouping of rational-emotive therapy, an approach that makes heavy use of behavioral procedures, with the nonbehavioral superclass (Presby, 1978). We find the grouping decisions, in general, to be a questionable practice, since there were fairly large differences among therapies within the *same* superclass. Note Smith and Glass's (1977, p. 757) statement that "implosive therapy is demonstrably inferior to systematic desensitization." Although this assertion has been criticized on other grounds, it would seem to provide sufficient reason not to put these two therapies together.

Smith and Glass (1977) conclude that there is little difference between therapies. The behavioral class did achieve a greater effect size than the nonbehavioral class, but the meaning of this is unclear. Neither we nor Rachman and Wilson (1980) find their analyses of superclasses compelling, although all the criticisms mentioned above make it equally difficult to accept the possible differences among the 10 original therapy types. As a final note, Smith and Glass (1977) present hypothetical situations (composites of several independent variables) and use unstandardized regression weights to estimate the effect sizes of various therapies. For example, 20-year-old, intelligent phobic clients might have been treated by therapists with 2 years experience. If evaluated with highly subjective measures immediately after treatment, psychodynamic therapy would have achieved an effect size of .919, compared to 1.049 for systematic desensitization and 1.119 for behavior modification. For the treatment of a 30-year-old neurotic of average intelligence seen in individual therapy by a highly trained therapist, the numbers were .643, .516, and

.847, respectively. The latter example in particular points to differences in the (hypothetical) effectiveness of treatment. It is supported by a recent reanalysis of the Smith and Glass (1977) data by Gallo (1978), who found that being in therapy at all and type of therapy accounted for precisely the same amount of variance (10%).

In our rather extensive criticism of Smith and Glass (1977), we do not wish to imply that meta-analyses have no place in COR. On the contrary, we believe that meta-analysis has a very useful place. However, the quality of the studies employed and the diversity of treatments, patients, outcome measures, and so on, make it unproductive to answer "the big question." In line with our previous comments, no techniques can provide useful answers to useless questions.

A meta-analysis conducted by Blanchard, Andrasik, Ahles, Teders, and O'Keefe (1980) provides a more rational model for the aspiring meta-analyst. Blanchard *et al.* simply asked which of a few types of relaxation or biofeedback treatments are most effective for the treatment of tension and migraine headaches. Separate analyses were conducted for each type of headache, and acceptable dependent measures and exclusion criteria were carefully specified. Short- and long-term follow-ups were separately evaluated. For the treatment of migraines, temperature biofeedback (alone or combined with autogenic training) and relaxation training were found to be equally effective and superior to medication placebo. For tension headaches, frontal EMG biofeedback, relaxation training, and their combination were equally effective and more so than medication placebo, psychological placebo, or headache monitoring.

CURRENT ISSUES

With the exceptions of our critiques of Luborsky *et al.* (1975) and Smith and Glass (1977), we have steered clear of any attempt to evaluate the overall quality of the comparative literature or to specify in detail which treatments may work best for what patients with what problems. We leave this to others and refer the reader to Meltzoff and Kornreich (1970) and the more recent volumes by Kazdin and Wilson (1978b), Agras *et al.* (1979), and Rachman and Wilson (1980). In the following sections, we adopt a different strategy. Rather than offering one more review of the literature, we will attempt to provide a list of the several conceptual and methodological issues that should be addressed in the design and conduct of COR. We will rely heavily on the studies reviewed earlier in this paper. Examples of failure to consider these issues and resultant problems of interpretation will be provided and potentially productive strategies outlined whenever possible.

Major Issues

The Uniformity Myth

Although we have described the uniformity myth in some detail in previous sections, its ubiquitousness requires that we give it further attention. The uniformity myth is an assumption of equivalence that is naively applied to almost every aspect of COR studies. Kiesler (1966) describes these assumptions as applied to patients and therapists. All patients are treated as equivalent, often with blatant disregard for the differences in their presenting problems, personal histories or demographic characteristics (e.g., Luborsky *et al.,* 1975; Smith & Glass, 1977). Therapists are considered equivalent despite differences in their training, experience, theoretical orientation, or personality characteristics. Obviously, this practice impedes the search for effective treatments for specific individuals with specific problems (Paul, 1967). It also injects unnecessary variance into our research designs, which may keep us from seeing answers that do appear in nature. Thus, Kiesler (1966), Edwards and Cronbach (1952), and DiLoreto (1971) call for the inclusion of individual difference variables and factorial designs in COR.

The uniformity myth may be applied in any number of insidious ways and potentially affects *every* COR study. In addition to patients and therapists, treatments may fall victim as well. Any study that refers to "insight therapy," "psychotherapy," or "behavior therapy" as if these were meaningful terms exemplifies the uniformity myth (Kazdin & Wilson, 1978b; Roback, 1971). The forms that any of the treatments may take are infinitely varied and of potentially differential effectiveness.

One additional example of the uniformity myth that is seldom recognized but implicitly accepted is the homogeneity of laboratories or investigators. We refer here not to quality of research design but to more ephermeral issues. Consider the study by Rush *et al.* (1977) as an example. Its findings have been widely accepted despite the fact that it is the only study to find substantial advantage of a psychotherapy over a pharmacotherapy. Replications have been attempted, but only by those favorable to cognitive therapy. While the findings of this group may, in fact, stand the test of time, replication by different investigators, from different laboratories, in different parts of the country or world, and from divergent theoretical orientations seems essential (Weissman, 1979).

Issues Pertaining to the Administration of Treatment

Equation of Treatment Content. Replicability requires that any experimental operation be carefully specified. However, the valid conduct of COR imposes more severe demands. Although the specific procedures included in different treatments will obviously vary, the *content* of therapeutic sessions should be equated as closely as possible. Consider the problems of interpretation posed by Paul and Shannon's (1966) comparison of group desensitization, individual desensitization, and individual insight therapy. The group treatment's inclusion of a test anxiety hierarchy confounds treatment format and content. This error leads to any number of competing hypotheses even if other design confounds are ignored: group treatment was superior to insight treatments *because of* the extra content, group treatment was actually inferior to individual desensitization but pulled even because of extra content, and so on. A recent study by the senior author (Heimberg, Madsen, Montgomery, & McNabb, 1980) confronted this issue. Imaginal, rehearsal, and performance-based treatments for the reduction of heterosexual–social anxiety were compared. In order to equate treatment content as closely as possible, the imaginal hierarchy also served as the basis for the sequence of rehearsals and *in vivo* homework assignments. Content was thus removed as a competing hypothesis for treatment effects.

Equation of Treatment Parameters. Here we refer to issues such as the number of sessions, length of sessions, time between sessions, and so on. Unless such factors constitute the independent variables in an investigation, they should be held constant. Several studies have, unfortunately, violated this injunction. In a study by Gelder, Marks, Wolff, and Clarke (1967), systematic desensitization, individual psychotherapy, and group psychotherapy were compared for the treatment of phobic patients. The treatments were administered weekly but for varying periods of time. While desensitization and individual therapy sessions lasted an hour, group sessions lasted 90 minutes. More seriously, desensitization sessions covered a period of 9 months, while individual therapy sessions continued for 12 months and group sessions for 18 months! Similarly, Paul and Shannon's (1966) group-treatment subjects received 9 sessions while all other subjects received only 5.

Issues of amount of treatment bring up a conflict between experimental control and the requirements of different therapies. As noted in the discussion of Paul (1966), insight therapies may require a longer time period to produce similar effects than certain behavior therapies. A 5- or 10-session trial may not provide a reasonable opportunity for treatment effects to emerge, and the insight therapy may "artificially" produce poorer results. However, provision of differential treatment time simply confounds number of sessions with treatment and provides no solution. A better strategy might be to include number of sessions as an independent variable and measure outcome at the end of each treatment period and at repeated follow-ups. Whether number of sessions ultimately makes a difference is, of course, an empirical issue. In an investigation of methods of job interview preparation, we found little difference between a 4-session program and a 2-session program of social skills training (Heimberg,

Cunningham, Stanley, & Blankenberg, 1982). However, a COR study by Shlien (1964; Shlien *et al.*, 1962) sheds additional light. Time-limited (maximum of 20 sessions) Adlerian and client-centered therapies were compared with unlimited client-centered therapy (average of 37 sessions). All groups produced substantial gains, but the time-limited group achieved them much faster. The provision of time limits appeared to alter the process of therapy, so that improvement actually began at an earlier session.

Value, Credibility, and Acceptability of Treatment. Fiske *et al.* (1970) assert that treatments in comparative studies must be equally valued by patients. Value may be conceptualized in a number of ways, but two specific aspects of value have received recent attention—credibility and acceptability of treatment. *Credibility* refers to the extent to which clients believe the rationale on which a particular treatment is based and the extent to which they expect to benefit from the specific therapeutic procedures. It seems apparent that credibility and positive expectancy should be necessary preconditions for therapeutic change. In a study by Borkovec and Nau (1972), various therapeutic rationales received very different credibility ratings, and credibility has been shown to relate to therapeutic outcome (Nau, Caputo, & Borkovec, 1974). However, we are unaware of any study that assesses therapeutic outcome in groups of "believers" and "nonbelievers."

Whether differences in credibility present a problem depends on the experimental question under study (Kazdin & Wilson, 1978b). When a simple outcome question is asked, no real problem exists. However, if the mechanisms of change are being considered, differential credibility of treatment poses a serious competing hypothesis. It should be controlled or at least assessed. The importance of credibility has been well demonstrated in the study of systematic desensitization (Kazdin & Wilcoxon, 1976). Desensitization has been shown to be superior only to control conditions of lesser credibility. When compared to equally credible controls, no differences typically emerge.

Acceptability of treatment techniques refers to "judgments by laypersons, clients, and others of whether treatment procedures are appropriate, fair and reasonable for the problem or clients (Kazdin, 1981, p. 493)." Kazdin has conducted a series of studies on the acceptability of different techniques for the treatment of children's behavior problems, and his findings should be seriously considered. Acceptability is positively influenced by the severity of the problem behavior and the extent to which parents and children are mutually involved in the formulation of the treatment plan (Kazdin, 1980a,b). Acceptability is negatively affected by the presence of adverse side effects, but it is *unaffected by treatment efficacy* (Kazdin, 1981). Although this research needs to be extended to other techniques and client populations, the implications are clear. Acceptability may pose competing hypotheses for comparative treatment effects in that it may mediate effort, compliance, and attrition. As with credibility, acceptability needs to be assessed.

Adherence to Treatment Protocol. Stated simply, you must know what you've done! In order to assess the effectiveness of two specific treatments, one must know that two different treatments have been administered and that they have been administered correctly. In order to accomplish this goal, the content and procedures of each treatment must be carefully specified. However, treatments must be monitored repeatedly over the course of the treatment period, and measures must be taken to evaluate the correspondence between treatment-as-administered and treatment-as-intended. Sloane *et al.* (1975) obtained tape recordings of the fifth treatment session for each patient and derived a number of measures that indicated the differences between psychotherapy and behavior therapy. While this procedure is better than many others, its inadequacies are pointed out by the experiences of DiLoreto (1971). DiLoreto monitored session content several times over the course of therapy. While early sessions indicated adherence to protocol, later sessions showed an increasing gap between intended treatment procedures and actual therapist performance for one therapist (of only two) in each of the rational-emotive and client-centered therapy conditions. These assessments revealed a clear threat to the internal validity of the DiLoreto study. One wonders what the implementation of continuous training and feedback for therapists might have done to alleviate this problem.

Minimizing Technique Overlap. Therapeutic methods, even those derived from very different theoretical perspectives, may share the same techniques. If two such therapies are the objects of a COR study, technique overlap may pose a competing hypothesis for a "Method A = Method B > Controls" outcome. This concern arises from several different directions. First, treatments must be sufficiently distinct for a comparative study to be justifiably conducted and a spurious finding of no difference to be avoided. Second, it may be difficult to know whether a lack of difference is due to common techniques or to the similar effectiveness of nonoverlapping techniques (Kazdin & Wilson, 1978b). Of course, if efficacy is the only concern, these issues are unimportant.

Issues Pertaining to Therapists

Therapist Competence. Therapists should be highly trained and able to deliver their treatments at *maximal strength*. This statement follows from our value judgment that we should want to know our upper limits of effectiveness rather than what could be done with pedestrian effort. The amount of training required depends upon the complexity of the techniques under study and does not automatically translate into the possession of the Ph.D. or M.D. However, of the studies we have reviewed, only Sloane *et al.* (1975) have clearly met this criterion by soliciting highly reputed behavior therapists and psychoanalysts to participate in their study.

Therapist Bias. Experimenters are not the only ones who are invested in the outcomes of their studies. Therapists are typically devoted to particular techniques or schools of therapy. Paul's (1966) therapists were devotees of insight therapy, while Rush *et al.* (1977) predominantly utilized psychiatric residents who appeared to favor pharmacotherapy. Therapist bias is a factor that must be equated across conditions, so that therapists do not engage in self-fulfilling prophecies or communicate to patients in the various treatment conditions differential expectancies for success.

Should Therapists Be Crossed with or Nested within Treatment Conditions? The goal of therapist assignment to conditions should be to ensure that each treatment is delivered in exemplary fashion and that treatment effects are not threatened by nuisance factors like therapist bias or the subtle effects of therapist personality characteristics. Two primary strategies have been followed in the literature. Paul (1966) and Rush *et al.* (1977) crossed therapists with treatments; that is, each therapist saw equal numbers of patients in each condition. DiLoreto (1971), Sloane *et al.* (1975), and Klerman *et al.* (1974) nested therapists within conditions. A larger number of therapists were utilized, but each therapist treated patients in only one condition.

Crossing therapists with conditions controls a number of important factors (e.g., therapist sex, appearance, general interpersonal style), since the same person administers the several treatments. But it raises several other important issues. First of all, crossing may be impossible at some points and impractical at others. For instance, psychotherapy–pharmacotherapy studies may cross therapists with conditions only if all therapists are medically trained and legally competent to administer medications. Therefore, crossing would limit this literature to pharmacotherapy versus "psychotherapy as conducted by psychiatric residents," hardly a fair test. Second, crossing may result in a situation in which therapists are more skilled at the delivery of one technique than another or clearly prefer one technique to the other. Such appeared to be the case in the Paul (1966) and Rush *et al.* (1977) studies. However, *in these specific cases,* their conclusions are actually strengthened, since the nonpreferred treatment proved most effective.

DiLoreto (1971) poses an eloquent argument in favor of nesting therapists within conditions. He suggests that we start by assuming

> that the therapist's personality and his treatment techniques are integrally and inseparably linked. Then, by securing therapists who are committed to techniques which one wishes to compare, by having each administer the techniques they respectively deem effective, and by comparing what they say they do with what

> they actually do . . . one is in a much better position to assess treatment conditions
> as they are most often administered with little or no loss in scientific rigor. (pp.
> 16–17)

We are inclined to agree with DiLoreto on this issue. The problems inherent in using therapists as their own controls (crossing) may make it difficult to draw conclusions from these studies. Nesting circumvents several of these problems. However, if a nesting strategy is to be reasonably employed, two points must be heeded. Monitoring of therapy quality remains important, since one must know that each group of therapists administers their treatment equally well. Second, the larger the number of therapists employed, the safer a nesting strategy becomes. With small numbers, an aberrant therapist will seriously skew outcome and lead to a "therapist × treatment confound" (Kazdin & Wilson, 1978b); with larger numbers, this threat is minimized.

Issues Pertaining to Patients

"Real" Patients versus Volunteers. This issue has attracted much attention and produced an ongoing debate in the field. Luborsky *et al.* (1975) excluded studies using volunteers on the grounds that they provide little useful information. In contrast, Borkevec and Rachman (1979) assert that volunteer studies make a useful contribution as long as one specifies the limits of reasonable generalizability of their findings. We refer the reader to Kazdin (Chapter 7, this volume) for a well-balanced view on the issue.

Sample Composition. Sloane *et al.* (1975) and Cross *et al.* (1982) accepted patients into their studies if they were highly motivated, nonpsychotic, between the ages of 18 and 55, and appropriate for psychotherapy. Thus, their study samples consisted of patients of diverse backgrounds, personal characteristics, and presenting problems. Sloane *et al.* (1975) have been praised by Bergin and Suinn (1975) on precisely these grounds—that diversity of patients gives the treatments a stronger test. However, we must disagree. The selection of a heterogeneous subject sample impedes our ability to search for the best approach to the treatment of specific problems and forces us onto a less useful, global course. Unless sufficient numbers of patients are included for the investigators to examine the interactions between techniques and patient characteristics (as in DiLoreto, 1971), this strategy will provide us with little information. Careful specification of sample characteristics and replication of effects across diverse samples seems a safer course.

It should go without saying that, regardless of the characteristics of the total sample, there should be no preintervention differences among patients receiving the various treatments. Randomization or matching subjects on important characteristics and then randomizing within matched groups should handle potential problems easily. However, the initial equivalence of groups should not be assumed. In the previously mentioned study by Gelder *et al.* (1967), patients were poorly matched for length of illness. Patients receiving group psychotherapy had been ill an average of 12.2 years, while subjects in the desensitization and individual psychotherapy conditions had been ill for only 8.2 and 6.3 years respectively. The potential problems of interpretation posed by these differences requires little elaboration.

Differential Selection. Patients to be included in COR studies should be equally appropriate for all treatments under study. However, it is often the case that the procedure by which subjects are selected for inclusion may bias the sample in favor of one or the other treatments (Hollon & Beck, 1978). Three of the five COR studies we reviewed earlier may be faulted on this point. In their psychotherapy-versus-pharmacotherapy comparisons, Klerman *et al.* (1974) included only those patients who had a history of positive drug response and Rush *et al.* (1977) excluded any patients with a history of poor response. Previous psychotherapy failures were not excluded. Similarly, Sloane *et al.*'s (1975) patients were screened by a psychoanalytically oriented assessor on appropriateness for verbal psychotherapy. In the former cases, a bias toward drug treatment might reasonably be expected. As such, selection bias competes with pharmacologic effect as an explanation for the results of Klerman *et al.* (1974). Since Rush *et al.* (1977) found cognitive therapy superior to imipramine therapy, bias is not a serious competing hypothesis. In the case of Sloane *et al.* (1975), however, selection bias may have hidden any differences in favor of behavior therapy.

Differential Help-Seeking during Follow-up. The interpretation of comparative treatment effects becomes a dangerous exercise if the efforts of subjects to obtain additional treatment are not carefully monitored. The desire to seek extra treatment may be strongest during the follow-up period when no regular treatment is provided as part of the study. Control subjects may, of course, pose the greatest risk, and comparisons of each active treatment to controls may be invalidated. However, it remains possible that subjects receiving active treatments will desire additional treatment and that differences among conditions in rate of additional help-seeking may hide other differences that do exist between the groups. Consider the problems of interpretation posed for Sloane *et al.* (1975). Of 30 patients assigned to each condition, 15 receiving behavior therapy, 9 receiving psychotherapy, and 22 assigned to the waiting list received at least three sessions of additional treatment during follow-up. The differences among conditions in amount of extra treatment make the interpretation of follow-up data almost impossible.

Differential Attrition. Patients may drop out of a study for an infinite number of reasons. These may be broken down into treatment-related and non-treatment-related reasons (Lasky, 1962). Non-treatment-related reasons (e.g., subject is transferred by employer or drops out because the family baby-sitter quits) should be randomly distributed across conditions and should do little to bias outcome. The biggest threat posed by non-treatment-related attrition is a reduction in statistical power due to a reduced sample size or unequal cell frequencies. However, treatment-related attrition (i.e., attrition due to improvement, nonimprovement, deterioration, side effects, etc.) poses a serious threat to the validity of COR studies.

Dropout rate is, of course, a valid dependent measure and may reasonably be employed in the evaluation of treatments. Rush *et al.* (1977) found that attrition was significantly higher in pharmacotherapy than it was in psychotherapy. According to descriptions provided by Rush *et al.* (1977, p. 25), all dropouts did so for treatment-related reasons. The lone dropout from cognitive therapy did so because a crisis had been resolved. Pharmacotherapy patients dropped out because of side effects, a suicidal crisis, and failure to respond to treatment. Because of conservative steps taken by the investigators, attrition may not have biased results too seriously. However, the issue remains. If patients drop out because they fail to respond, remaining patients in that treatment may show an artifactually positive response. They should be more highly motivated and show greater improvement. Similarly, dropouts after improvement produce a negative bias against the treatment since remainers probably are not doing so well. If the reasons for attrition are unknown, its effects on the data cannot be specified. Consider the study by Zitrin, Klein, and Woerner (1976) on the treatment of phobia by psychotherapy, behavior therapy plus imipramine, and behavior therapy plus placebo. Dropout rates for the three conditions were 42%, 29%, and 10%, respectively. Without knowledge of the reasons for dropping out, the outcome data are uninterpretable. Zitrin *et al.* (1976) speculate that remainers were probably the most motivated, thus biasing results in favor of psychotherapy.

Lasky (1962) examined the results of a study by Overall, Hollister, Pokorny, Casey, and Katz (1962) for the effects of attrition. Although this is not strictly a "psychotherapy study," the findings are instructive. This study compared imipramine, isocarboxazid, and amorbarbital-dextroamphetamine for treating 210 newly admitted depressed males at a VA hospital. Subjects were evaluated at pretest and 3 weeks and 12 weeks later. Original data analysis based on the 108 completers showed no differences between treatments. However, when the 3-week data were reanalyzed using the 108 completers and 73 other patients who would later drop out, imipramine was superior. Apparently, many of the dropouts were patients who had improved with imipramine. Their loss hid the superiority of that treatment.

Lasky suggests a number of potential solutions to the dropout problem. First, one may replace dropouts with new patients or those on a waiting list. However, this procedure is risky because the replacement subject may be different from the dropout in subtle ways. In fact, the replacement must be different or this subject would drop out too! Thus, the final sample will be dissimilar to the original sample. This problem is compounded if attrition rates are different for two treatments in the same study. At the end of the study, the investigator may be measuring the effects of two different treatments on two different populations! A second solution is to assess each subject at the time he or she is dropped and use the scores from that assessment as the subject's score at each later measurement point. While this strategy assumes no additional change beyond the terminal assessment, it has much to offer.

Rush *et al.* (1977) made good use of the terminal assessment strategy. First, they compared pretreatment scores of dropouts and completers, and no differences emerged. Second, they performed outcome analyses for completers only and also for completers and dropouts combined. Thus, the direction of bias could be assessed—significant differences favoring cognitive therapy occurred most often in the combined analyses, revealing that dropouts biased results in favor of drug treatment. However, the careful handling of dropouts and comprehensive statistical analyses allows readers to draw their own conclusions.

Issues Pertaining to the Measurement of Outcome

The Criterion Problem. In comparing treatments derived from different theoretical orientations, selection of appropriate outcome measures may become a monumental task. While Fiske *et al.* (1970) suggest that all COR studies include measures from diverse theoretical orientations, we are pessimistic that such practices might ever develop.[2] The prototypical problem arises in studies comparing behavior therapy and insight therapy. Of course, it occurs because behaviorists require behavior change and eschew insight, while insight therapists require insight and eschew behavior change. The unfortunate solution adopted by some (e.g., Cross *et al.*, 1982; Sloane *et al.*, 1975) is to adopt measures that are not very satisfactory to either group. Both of these studies relied exclusively on rating scales or structured interviews. While some of the rating scales were quite specific, they cannot be equated with behavior samples (Kazdin & Wilson, 1978b). Likewise, neither strategy is likely to reveal changes in unconscious motivation or ego boundaries.

The criterion problem has vexed researchers for decades and we cannot presume to solve it here. A partial solution, however, may be to look at different types of outcome measures. Kazdin and Wilson (1978a) have specified a number of such measures:

---

[2]Shlien (1966) suggests that COR researchers rely on a series of measures that are unrelated to specific methods or theories. While this suggestion rings true, attempts to detail his notions fall short. For instance, Shlien recommends that pupil size might serve as a generally acceptable outcome measure. We propose that pupil size is a valid measure if and only if a functional relationship can be demonstrated between pupil size and the specific problem under study.

1. The extent to which treatments restore clients to an acceptable level of functioning.
2. The proportion of clients who achieve this level.
3. The breadth of change—that is, the assessment of improvement or deterioration in areas outside the therapeutic focus on the defined problem.
4. Durability of improvement over time.
5. The efficiency of treatment. How rapidly are gains achieved? It was mentioned above that limits on the number of sessions may artifactually handicap insight therapies in comparison with more structured therapies. According to Kazdin and Wilson's (1978a) notion of efficiency, the slower therapy is also relatively inefficient and should be evaluated in that context. If two treatments ultimately produce similar effects, the one that does so more quickly is to be preferred.
6. Financial cost.
7. Emotional cost—that is, how much stress does the client have to endure in order to achieve relief?
8. Cost-effectiveness. Are the changes achieved worth the effort invested? Kazdin and Wilson (1978a) criticize Sloane *et al.* (1975) on a cost-effectiveness basis. They question whether the same amount of change achieved by the behavior therapists or psychotherapists might also have been achieved by (cheaper) nonprofessional personnel.

The reader is referred to Kazdin and Wilson (1978a,b) for a fuller exposition of these alternative means of evaluating outcome.

Who Should Perform Outcome Assessments? Two examples clearly demonstrate the *wrong way* to perform outcome assessments (and several other problems as well). Ellis (1957) and Lazarus (1966) each compared three treatments. *Each served as the only therapist and each served as the only assessor.* In each case the author–therapist–assessor had a vested interest in the outcome and in each the favored treatment emerged superior. Obviously, rater bias cannot be ruled out as a serious competing hypothesis. However, problems of this sort can easily be avoided.

In order to avoid problems of rater bias, evaluations should be conducted by someone other than the experimenter or therapist. The assessor should be systematically deprived of knowledge about a patient's initial level of functioning and the treatment to which he or she is assigned. In comparative psychotherapy studies, this arrangement is not difficult to achieve as long as sufficient personnel are available. However, in psychotherapy–pharmacotherapy comparisons, side effects of the medications may provide dramatic clues about treatment assignment and thereby destroy well-intentioned attempts to keep assessors blind. This is a difficult issue with no clear solution. However, prudent choice of a pharmacologic placebo may help to reduce the scope of the problem.

Follow-up Assessments. Follow-up assessments are a necessary and important part of any psychotherapy research (Cross *et al.,* 1982; Garfield, 1980; Gottman & Markman, 1978; Kazdin & Wilson, 1978b). However, such assessments are often neglected. In our review of assertion training studies (Heimberg, Montgomery, Madsen, & Heimberg, 1977), follow-ups of longer than 3 months were rare. Gottman and Markman (1978) report similar findings in a survey of systematic desensitization research. However, in comparative outcome research, follow-ups should be a regular part of every study.

Cross *et al.* (1982) argue that multiple follow-ups should be included. Behavior may vary toward improvement or deterioration over the course of time, and "multiple follow-up assessment isolates the patterning or configuration of effects, thereby providing detailed information on the specific consequences of treatment" (p. 103). In COR studies, multiple follow-ups are imperative, since the changes produced by each treatment may follow different temporal patterns. Consider the example displayed in Figure 1, which was suggested by the work of Gottman and Markman (1978). Treatments A and B are compared at posttreatment and at follow-ups of 6 months and 1 year. At posttest, Treatment A produces significantly larger effects than Treatment B. Gradual improvement is noted in Treatment A until the first follow-up, but thereafter effects begin to deteriorate. On the other hand, Treatment B results in large gains between posttest and the first follow-up, and gains continue to increase until the end of the follow-up year. At each assessment point, a different conclusion is reached. At posttest, A > B; at 6 months, A = B; at 1 year, B > A. While this example may not represent the routinely expected

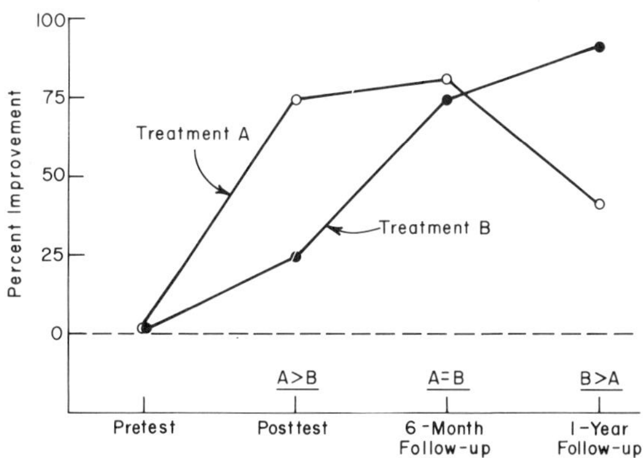

Figure 1. Hypothetical comparison of the effects of two treatments at posttest and two follow-up points.

outcome, it is not an impossibility. Treatment A could represent any number of behavioral treatments for habit disorders like smoking or overeating. Treatment B could represent any approach that required the buildup of sophisticated interpersonal skills (e.g., heterosexual–social skills training). In any case, only with extended follow-ups will we eventually know the fate of our treatment efforts.

Problems in Psychotherapy–Pharmacotherapy Comparisons

In general, our comments have been applicable to all COR studies regardless of the specific treatments being compared. However, several issues may have specific impact on studies attempting to compare a psychotherapy with a pharmacologic treatment. Space limitations prohibit a thorough analysis of these issues; the reader is referred to Hollon and Beck (1978) for a more detailed analysis. We restrict ourselves to one issue—how the design of most psychotherapy–pharmacotherapy comparisons may provide a poor representation of psychotherapy.

In a thought-provoking paper, Hollon and DeRubeis (1981) note that 78% of all psychotherapy–pharmacotherapy studies utilize 2 by 2 factorial designs. An active drug and a pharmacologic placebo are crossed with psychotherapy and its control. Thus, the condition that is intended to represent psychotherapy is not psychotherapy at all. It is more accurately described as psychotherapy plus placebo. Since these studies are typically conducted in double-blind fashion, both the patient and therapist are led to believe that the patient is receiving medication. It is this belief, not present in the typical practice of psychotherapy, that makes the experimental condition a poor reflection of real life.

Hollon and DeRubeis (1981) present a series of mathematical proofs to demonstrate the nonequivalence of psychotherapy plus placebo and psychotherapy alone. They conclude that

> the only instance in which the tacit assumption of equivalence of placebo plus psychotherapy to psychotherapy alone could prove to be true would be the case in which a negative interaction [of placebo and psychotherapy] precisely offset the expected positive effects associated with pill taking. (p. 470)

A series of examples is presented demonstrating the noncomparability of psychotherapy plus placebo and psychotherapy alone. For instance, in Klerman *et al.*'s (1974) comparison of psychotherapy and amitriptylene, a negative interaction between psychotherapy and placebo occurred. Psychotherapy alone achieved a relapse rate of 16.7%. The addition of placebo raised the rate to 28%. A review of additional studies that included comparisons of psychotherapy plus placebo versus psychotherapy alone revealed a common tendency for psychotherapy plus placebo to *underestimate* the effects of psychotherapy alone.

On the basis of these analyses, Hollon and DeRubeis (1981) conclude that the typical 2 by 2 factorial design is poorly equipped to test the comparative efficacy of psychotherapy and drugs. They propose that investigators instead utilize a 3 by 3 design, crossing active medication, placebo, and no pill with psychotherapy, nonspecific therapy, and no-psychotherapy conditions. This elegant design allows a full analysis of the relative efficacy of the treatments and the mechanisms by which each produces behavior change. However, if pragmatic considerations prohibit the use of the full design and mechanisms of change are not of interest, a no-drug condition may be preferable to a placebo.

CONCLUSION

In this chapter, we have reviewed and critiqued several outstanding examples of comparative outcome research. We have detailed the multitude of problems facing potential investigators and attempted to provide a few solutions. Finally, we reviewed the problems and prospects involved in the use of single-subject designs for COR studies. We have not attempted to review the COR literature comprehensively or boldly to assert the superiority of one approach over another. While other reviewers have done so with some success, we simply do not believe that the COR literature is sufficiently advanced for many such statements to be made. However, the state of this literature easily points to the directions the field should take.

As we stated at the outset, global outcome questions serve little useful purpose. The field will advance only to the extent that outcome questions reflect specific goals and experimental procedures reflect operational precision (Kazdin & Wilson, 1978b). The greatest need is for careful design of studies that can rule out the staggering number of competing hypotheses, accurately measure treatment effects, and specify the mechanisms by which therapeutic change is produced.

Each COR study will present its own unique set of problems in design, subject selection, and so on. Thus, it is not possible to provide a detailed handbook of COR decisions; every new investigator must apply his own creative genius to the idiosyncrasies of the specific research problem. As an aid to this effort, we have outlined four major issues each researcher should consider: issues of treatment, therapists, patients, and outcome measurement. Several common problems in each area have been detailed, but no comprehensive list has been provided. It is our hope, however, that attention to each of these areas will help the researcher to confront his own unique research problems. It is only by such effort that operational precision may be achieved and our knowledge increased.

REFERENCES

Agras, W. S., Kazdin, A. E., & Wilson, G. T. *Behavior therapy: Towards an applied clinical science*. San Francisco: Freeman, 1979.

Becker, J., & Schuckit, M. A. The comparative efficacy of cognitive therapy and pharmacotherapy in the treatment of depressions. *Cognitive Therapy and Research*, 1978, *2*, 193–197.

Bergin, A. E., & Lambert, M. J. The evaluation of therapeutic outcomes. In S. L. Garfield & A. E. Bergin (Eds.), *Handbook of psychotherapy and behavior change: An empirical analysis* (2nd ed.). New York: Wiley, 1978.

Bergin, A. E., & Suinn, R. M. Individual psychotherapy and behavior therapy. *Annual Review of Psychology*, 1975, *26*, 509–556.

Blanchard, E. B., Andrasik, F., Ahles, T. A., Teders, S. J., & O'Keefe, D. Migraine and tension headache: A meta-analytic review. *Behavior Therapy*, 1980, *11*, 613–631.

Borkovec, T. D., & Nau, S. D. Credibility of analogue therapy rationales. *Journal of Behavior Therapy and Experimental Psychiatry*, 1972, *3*, 257–260.

Borkovec, T., & Rachman, S. The utility of analogue research. *Behaviour Research and Therapy*, 1979, *17*, 253–262.

Boy, A. V. A critique by Angelo V. Boy. In A. O. DiLoreto, *Comparative psychotherapy: An experimental analysis*. Chicago: Aldine-Atherton, 1971.

Cross, D. G., Sheehan, P. W., & Khan, J. A. Short- and long-term follow-up of clients receiving insight oriented therapy and behavior therapy. *Journal of Consulting and Clinical Psychology*, 1982, *50*, 102–112.

DiLoreto, A. O. *Comparative psychotherapy: An experimental analysis*. Chicago: Aldine-Atherton, 1971.

Edwards, A. L., & Cronbach, L. J. Experimental design for research in psychotherapy. *Journal of Clinical Psychology*, 1952, *8*, 51–59.

Ellis, A. Outcome of employing three techniques of psychotherapy. *Journal of Clinical Psychology*, 1957, *13*, 344–350.

Ellis, A. A critique by Albert Ellis. In A. O. DiLoreto, *Comparative psychotherapy: An experimental analysis*. Chicago: Aldine-Atherton, 1971.

Eysenck, H. J. The effects of psychotherapy. In H. J. Eysenck (Ed.), *Handbook of abnormal psychology*. New York: Basic Books, 1961.

Eysenck, H. J. An exercise in mega-silliness. *American Psychologist*, 1978, *33*, 517.

Fiske, D. W., Hunt, H. F., Luborsky, L., Orne, M. T., Parloff, M. B., Reiser, M. F., & Tuma, A. H. The planning of research on effectiveness of psychotherapy. *Archives of General Psychiatry*, 1970, *22*, 22–32.

Gallo, P. S., Jr. Meta-analysis—A mixed meta-phor? *American Psychologist*, 1978, *33*, 515–516.

Garfield, S. L. *Psychotherapy: An eclectic approach*. New York: Wiley, 1980.

Gelder, M. G., Marks, I. M., Wolff, H., & Clarke, M. Desensitization and psychotherapy in the treatment of phobic states: A controlled inquiry. *British Journal of Psychiatry*, 1967, *113*, 53–73.

Glass, G. V. Primary, secondary, and meta-analysis of research. *Educational Researcher*, 1976, *5*, 3–8.

Glass, G. V. Reply to Mansfield and Busse. *Educational Researcher*, 1978, *7*, 3.

Glass, G. V., & Smith, M. L. Reply to Eysenck. *American Psychologist,* 1978, *33,* 517–519.

Goldstein, A., & Wolpe, J. A critique by Alan Goldstein and Joseph Wolpe. In A. O. DiLoreto, *Comparative psychotherapy: An experimental analysis.* Chicago: Aldine-Atherton, 1971.

Gottman, J. M., & Markman, H. J. Experimental designs in psychotherapy research. In S. L. Garfield & A. E. Bergin (Eds.), *Handbook of psychotherapy and behavior change: An empirical analysis* (2nd ed.). New York: Wiley, 1978.

Heimberg, R. G., Montgomery, D., Madsen, C. H., Jr., & Heimberg, J. S. Assertion training: A review of the literature. *Behavior Therapy,* 1977, *9,* 953–971.

Heimberg, R. G., Madsen, C. H., Jr., Montgomery, D., & McNabb, C. E. Behavioral treatments for heterosocial problems: Effects on daily self-monitored and roleplayed interactions. *Behavior Modification,* 1980, *4,* 147–172.

Heimberg, R. G., Cunningham, J., Stanley, J., & Blankenberg, R. Preparing unemployed youth for the job interview: A controlled evaluation of social skills training. *Behavior Modification,* 1982, *6,* 299–322.

Hollon, S. D., & Beck, A. T. Psychotherapy and drug therapy: Comparisons and combinations. In S. L. Garfield & A. E. Bergin (Eds.), *Handbook of psychotherapy and behavior change: An empirical analysis* (2nd ed.). New York: Wiley, 1978.

Hollon, S. D., & DeRubeis, R. J. Placebo-psychotherapy combinations: Inappropriate representations of psychotherapy in drug-psychotherapy comparative trials. *Psychological Bulletin,* 1981, *90,* 467–477.

Kazdin, A. E. Acceptability of alternative treatments for deviant child behavior. *Journal of Applied Behavior Analysis,* 1980, *13,* 259–273. (a)

Kazdin, A. E. Acceptability of time out from reinforcement procedures for disruptive child behavior. *Behavior Therapy,* 1980, *11,* 329–344. (b)

Kazdin, A. E. Acceptability of child treatment techniques: The influence of treatment efficacy and adverse side effects. *Behavior Therapy,* 1981, *12,* 493–506.

Kazdin, A. E., & Wilcoxon, L. A. Systematic desensitization and nonspecific treatment effects: A methodological evaluation. *Psychological Bulletin,* 1976, *83,* 729–758.

Kazdin, A. E., & Wilson, G. T. Criteria for evaluating psychotherapy. *Archives of General Psychiatry,* 1978, *35,* 407–416. (a)

Kazdin, A. E., & Wilson, G. T. *Evaluation of behavior therapy: Issues, evidence, and research strategies.* Cambridge, Mass.: Ballinger, 1978. (b)

Kiesler, D. Some myths of psychotherapy research and the search for a paradigm. *Psychological Bulletin,* 1966, *65,* 110–136.

Klerman, G. L., DiMascio, A., Weissman, M., Prusoff, B., & Paykel, E. S. Treatment of depression by drugs and psychotherapy. *American Journal of Psychiatry,* 1974, *131,* 186–191.

Kovacs, M., Rush, A. J., Beck, A. T., & Hollon, S. D. Depressed outpatients treated with cognitive therapy or pharmacotherapy: A one-year follow-up. *Archives of General Psychiatry,* 1981, *38,* 33–39.

Lasky, J. J. The problem of sample attrition in controlled treatment trials. *Journal of Nervous and Mental Disease,* 1962, *135,* 332–337.

Lazarus, A. A. Behaviour rehearsal vs. non-directive therapy vs. advice in effecting behaviour change. *Behaviour Research and Therapy,* 1966, *4,* 209–212.

Ledwidge, B. Cognitive behavior modification: A step in the wrong direction? *Psychological Bulletin,* 1978, *85,* 353–375.

Ledwidge, B. Cognitive behavior modification: A rejoinder to Locke and Meichenbaum. *Cognitive Therapy and Research,* 1979, *3,* 133–139. (a)

Ledwidge, B. Cognitive behavior modification or new ways to change minds: Reply to Mahoney and Kazdin. *Psychological Bulletin,* 1979, *86,* 1050–1053. (b)

Locke, E. A. Behavior modification is not cognitive—and other myths: A reply to Ledwidge. *Cognitive Therapy and Research*, 1979, *3*, 119–125.

Luborsky, L., Singer, B., & Luborsky, L. Comparative studies of psychotherapies: Is it true that "everyone has won and all must have prizes"? *Archives of General Psychiatry*, 1975, *32*, 995–1008.

Mahoney, M., & Kazdin, A. Cognitive behavior modification: Misconceptions and premature evacuation. *Psychological Bulletin*, 1979, *86*, 1044–1049.

Mansfield, R. S., & Busse, T. V. Meta-analysis of research: A rejoinder to Glass. *Educational Researcher*, 1977, *6*(9), 3.

Meichenbaum, D. Cognitive behavior modification: The need for a fairer assessment. *Cognitive Therapy and Research*, 1979, *3*, 127–132.

Meltzoff, J., & Kornreich, M. *Research in psychotherapy*. New York: Atherton, 1970.

Nau, S. D., Caputo, L. A., & Borkovec, T. D. The relationship between credibility of therapy and simulated therapeutic effects. *Journal of Behavior Therapy and Experimental Psychiatry*, 1974, *5*, 129–134.

Overall, J. E., Hollister, L. E., Pokorny, A. D., Casey, J. F., & Katz, G. Drug therapy in depressions. *Clinical Pharmacology and Therapeutics*, 1962, *3*, 16–22.

Paul, G. L. *Insight vs. desensitization in psychotherapy: An experiment in anxiety reduction*. Stanford, Calif.: Stanford University Press, 1966.

Paul, G. L. Strategy of outcome research in psychotherapy. *Journal of Consulting Psychology*, 1967, *31*, 109–118.

Paul, G. L., & Shannon, D. T. Treatment of anxiety through systematic desensitization in therapy groups. *Journal of Abnormal Psychology*, 1966, *71*, 124–135.

Pearl, D. Psychotherapy and ethnocentrism. *Journal of Abnormal and Social Psychology*, 1955, *50*, 227–229.

Pokorny, A. D., & Klett, C. J. Comparison of psychiatric treatments: Problems and pitfalls. *Diseases of the Nervous System*, 1966, *27*, 648–652.

Presby, S. Overly broad categories obscure important differences between therapies. *American Psychologist*, 1978, *33*, 514–515.

Rachman, S. J., & Wilson, G. T. *The effects of psychological therapy* (2nd ed.). Oxford, England: Pergamon, 1980.

Roback, H. B. The comparative influence of insight and non-insight psychotherapies on therapeutic outcome: A review of the experimental literature. *Psychotherapy: Theory, Research, and Practice*, 1971, *8*, 23–25.

Rush, A. J., Beck, A. T., Kovacs, M., & Hollon, S. Comparative efficacy of cognitive therapy and pharmacotherapy in the treatment of depressed outpatients. *Cognitive Therapy and Research*, 1977, *1*, 17–37.

Rush, A. J., Hollon, S. D., Beck, A. T., & Kovacs, M. Depression: Must pharmacotherapy fail for cognitive therapy to succeed? *Cognitive Therapy and Research*, 1978, *2*, 199–206.

Shlien, J. M. Comparison of results with different forms of psychotherapy. *American Journal of Psychotherapy*, 1964, *18*, (Monograph Suppl. 1), 15–22.

Shlien, J. M. Cross-theoretical criteria for the evaluation of psychotherapy. *American Journal of Psychotherapy*, 1966, *20*, 125–134.

Shlien, J. M., Mosak, H. H., & Dreikurs, R. Effects of time limits: A comparison of client centered and Adlerian psychotherapy. *Journal of Counseling Psychology*, 1962, *9*, 31–34.

Sloane, R. B., Staples, F. R., Cristol, A. H., Yorkston, N. J., & Whipple, K. *Psychotherapy versus behavior therapy*. Cambridge, Mass.: Harvard University Press, 1975.

Smith, M. L., & Glass, G. V. Meta-analysis of psychotherapy outcome studies. *American Psychologist*, 1977, *32*, 752–760.

Tucker, J. E. Group psychotherapy with chronic psychotic soiling patients. *Journal of Consulting Psychology,* 1956, *20,* 430.

Weissman, M. M. The psychological treatment of depression: Evidence for the efficacy of psychotherapy alone, in comparison with, and in combination with pharmacotherapy. *Archives of General Psychiatry,* 1979, *36,* 1261–1269.

Weissman, M., Klerman, G. L., Paykel, E. S., Prusoff, B., & Hanson, B. Treatment effects on the social adjustment of depressed patients. *Archives of General Psychiatry,* 1974, *30,* 771–778.

Zitrin, C. M., Klein, D. F., & Woerner, M. G. *Behavior therapy, supportive psychotherapy, imipramine and phobias.* Unpublished paper, Long Island Jewish-Hillside Medical Center, 1976.

Statistical Considerations

ANTHONY J. CONGER

INTRODUCTION

Trends in Statistical Analysis

Psychotherapy outcome studies vary widely in virtually all aspects of design methodology and statistical analysis. Some studies employ simple manipulations on an individual and attempt to show effectiveness via functional analysis in A-B-A type designs, while others may involve complex sampling from diverse populations, multiple treatment and control groups, and sophisticated multivariate analysis on multiple dependent measures and control variables. Behavior therapy studies, historically noted for focusing on the analysis of single individuals and using clinical significance as a criterion of success, increasingly rely on between-group factorial designs and even multivariate analysis using statistical significance as a criterion for success. The sophisticated analyst employs procedures such as generalizability theory to evaluate the quality of measurement, time-series analysis on repeated measures, and multivariate analysis of variance to handle multiple dependent measures. Infrequently, procedures such as factor analysis, discriminant function analysis, canonical correlations, and cluster analysis can be found in treatment outcome studies. Indeed, even second order analysis of data is encouraged as the ultimate evaluation of outcome as in meta-analysis.

The trend away from n of 1 studies has been encouraged by the seminal "how to" and "what to" guidelines for psychotherapy outcome studies (Gottman & Markman, 1978; Paul, 1967). On the one hand, the use of more sophis-

---

ANTHONY J. CONGER • Department of Psychology, Purdue University, West Lafayette, Indiana 47907.

ticated designs and analysis should be applauded for their ability to increase the value of therapy research. On the other hand, recurrent problems in the application and inferences of the more sophisticated procedures need to be remediated before full benefits of these methods can be realized.

General Issues

This chapter has as its goal the evaluation of diverse statistical procedures in terms of their potential contributions to psychotherapy outcome research. In this regard, topics of the choice and evaluation of measures, multivariate and covariance analysis, and clinical versus statistical significance will be examined; the major focus, however, will be on three general issues: appropriateness, comprehensiveness, and power.

Choosing Correct Statistics

An obvious starting point for examining statistics in therapy outcome studies is the correctness of the statistics chosen to evaluate outcome. For this purpose, the concept of appropriateness is introduced. The word *appropriateness* refers to the match between a statistical procedure and the question asked. One must consider the set of operations (sampling, design, procedures, and measures) used to collect information and the statistical theory upon which inferences are reached in an attempt to answer the question. For example, one issue directly involved with appropriateness is what statistics are to be used to measure clinical significance without sacrificing statistical significance.

Using Statistics to Answer Statistical Questions

Comprehensiveness considers the selection of procedures that provide statistical (quantitative, stochastic) answers to statistical questions and therefore vitiate the need to rely on extrastatistical methods (e.g., verbal discussion and integration of partial results) to provide answers. An example of a novel comprehensive approach is to use meta-analysis to integrate results from diverse outcome studies rather than using simple but problematic subjective interpretations of tabulations of diverse studies (cf. Meehl, 1978).

Making the Most of Statistics

The need to rely on statistical inference introduces the possibility of erroneous conclusions. We may inadvertently state that one treatment is better or worse than another when it really is not (a Type 1 error) or that two procedures do not differ when they really do (a Type 2 error). The general concept governing inferential errors is power. Power needs little clarification. It is defined as the probability of correctly rejecting false null hypotheses. However, the role of power, or its absence, is ubiquitous and must be considered as a major statistical issue in outcome research.

Goals

This chapter has several goals. The most manifest level of discourse is directed at clarifying and rectifying several issues or problems currently evident in outcome research. At a less obvious level, the hope is that researchers will become more concerned about their selection of statistical procedures along the lines of choosing statistics appropriate for a particular problem (appropriateness), using statistical procedures that avoid the need for verbal integration of piecemeal, tangential analysis (comprehensiveness), and using designs, operations, measures, and significance tests that maximize power.

In order to accomplish the several goals described above, a more thorough presentation of the three deeper issues is provided. This is followed by an examination of the importance of rigorous measurement and variable specification for outcome studies, including a discussion of problems evident in reliability analyses and by a detailed examination of outcome statistics. The last includes clinical versus statistical significance, the use of multiple dependent measures, and statistical versus methodological controls. The issues of power, appropriateness, and comprehensiveness are illustrated throughout these latter discussions.

THEORETICAL CONSIDERATIONS

The material presented below defines and illustrates for general statistical problems the issues of appropriateness, comprehensiveness, and power. The reader should be informed, however, that while power is a formal statistical concept to be found in all statistical texts, *appropriateness* and *comprehensiveness* are not formal statistical terms and are introduced as conveniences to subsume more formal and fundamental statistical concepts such as sufficiency and efficiency.

Appropriateness

A primary decision underlying the choice of a statistical procedure involves the degree to which the procedure addresses the questions asked. The concept of appropriateness is meant to subsume all aspects of the match between statistical procedures and the question, including population definitions (e.g., finite vs. infinite), sampling from populations of independent variables, dependent variables, and experimental units (e.g., stratified random sampling of clients), specifications of measurement procedures (e.g., categorical vs. interval scale), and so forth. Virtually every consideration made in the conceptualization and operationalization of research has its counterpart in the consideration of an appropriate choice of statistical procedures. All too often, however, research paradigms are used freely, with little concern for the appropriateness of the paradigm for a specific problem. Further, some of the paradigms and associated statistical techniques may be justifiable only on the grounds that they have been used before, despite the fact that they may be inherently incorrect. Unfortunately, there are a variety of questions worth asking that cannot be readily addressed by unambiguously appropriate procedures. Thus, there are often no "correct" answers, and appropriateness can often only be considered in relativistic terms. One example, in which the interplay between questions and procedures is especially evident, is the choice of analysis-of-variance models appropriate to different kinds of inferences and generalizations. For example, if a researcher wishes to show that the number of years of experience of therapists is inversely related to behavioral change in the treatment of depression, it would not be sufficient to include two researcher-selected levels of experience (e.g., up to 5 years vs. 10 to 15 years) as a fixed factor in an analysis of variance. Since the intended generalization is to the entire population of years of experience, levels of this factor should be randomly selected and analyzed as a random factor. The selection of the levels by the experimenter is consistent with a fixed factor design, in which only the specific comparison of the two levels is warranted. The decision to choose one approach versus another is critical, not only from the standpoint of matching a procedure to a question but also from that of whether a statistical test exists. For example, in a two-factor mixed design (one random and one fixed factor), there is no unbiased test of significance for the fixed factor (cf. Hays, 1973).

Other examples of decisions involving the appropriateness of statistics are given in subsequent sections of this chapter.

Comprehensiveness

Comprehensiveness of statistical procedures is viewed as a desirable characteristic to the extent that it provides statistical and quantitative answers to questions requiring such answers. The basic notion is that the piecemeal presentation of results and reliance on verbal and, perhaps, subjective integration and interpretation should be avoided. An example of a statistical fallacy occurring with a noncomprehensive approach is as follows. A treatment and control group are measured pre- and posttreatment and changes from pre to post are investigated via t tests within each group. Perhaps the results show that there is a significant change for the treatment group $(p < .05)$ but not for the control group $(p > .05)$. Based on these results, it might be concluded that treatment works better than no treatment; however, this has not been demonstrated! What was demonstrated was that the treatment group showed a significant gain while the control group did not. If the actual t values for 20 degrees of freedom were $t = 2.10$ for the treatment group and $t = 2.08$ for the control group, there would be no significant difference in the amount of change. For the question of whether the treatment group changes more than the control group, the appropriate test is an analysis of variance in which a significant group-by-occasion interaction is required in order to argue for a significant change. While to some researchers this is an obvious and perhaps trivial example, to others it is not. In a fairly recent article on social skill training in children, the authors argued that because the two treated children showed significant improvement while the untreated children did not, the treatment appeared effective. A closer look at the results showed that the change for treated children just reached significance but change for the untreated just missed. Conceivably, the separate analyses described would be justifiable, but only for a different question. For example, if the researcher wished to show that the treatment is effective for children with moderate interaction deficiencies but not for those with severe deficiencies, separate analyses would be needed. It is only for the relative comparison of effectiveness that the need to demonstrate an interaction exists.

The problem of verbal integration of piecemeal analysis is manifest with a variety of procedures. For example, arguments are often made about differences in the reliabilities of variables or differences in correlations without subjecting the observed differences to statistical tests. That is, what is often tested is whether the correlations differ from zero but not whether they differ from another. More often than not, investigations of factor structures, multiple correlation or regression equations, and canonical correlation equations (which are compared across different groups) are done on the basis of superficial results (i.e., they look different, or they have different numbers of factors). Such comparisons are invidious. The use of comprehensive procedures that simultaneously investigate between-group differences and relationships among variables may well indicate that specific group membership may not be relevant. Alternatively, the prior investigation of differences between groups might indicate an absence of differences, suggesting that one analysis on pooled data would suffice or that differences do exist and separate analyses are required. For example, in an investigation of the profile reliability of the WISC-R (Conger, Conger, Farrell, & Ward, 1979), it was shown that raw correlation matrices did not differ across age groups, thus arguing that their principal component structures did not differ. But the matrices adjusted for unreliability did differ, and thus the reliable profiles would differ. An additional alternative may also exist: results derived on one group can be cross-applied to another to determine group differences. For example, if the relationship between independent variables and a dependent variable is derived by multiple regression procedures separately on a sample of males and a sample of females, they might well look different. But the equation derived on males can be used for females, and the resulting R square can be compared to that derived for females (and of course vice versa). The absence or presence of significant differences then directly tests whether different functions are required, subject to the usual considerations of power and the logic of interpretation.

Power

One recurrent problem in therapy outcome studies is the *a priori* low probability of successfully answering the research questions. This is not to say that the questions *per se* are at fault; rather, the methods used to address them are inadequate. Generally, outcome research questions are formulated along the lines that clients will show more change with one treatment procedure than another. With high probability, the researcher is stating a truism, in that it is inconceivable that intervention will not somehow affect a client (for better or worse) or that different intervention strategies will not show differential effects (cf. Bakan, 1967). This should cause concern to researchers who have not been successful at demonstrating treatment effects, but they might derive solace from the possibility that the results of a more powerful test of their hypothesis would argue that no treatment was better than the hypothesized treatment.

The basic concept underlying successful experimentation (assuming proper methodology) is sufficient power to reject null hypotheses (e.g., no difference between treatment groups) in favor of alternative hypotheses (e.g., the groups are not the same). Power, as previously stated, is the probability of rejecting the null hypothesis when it is false. Since most null hypotheses as typically stated are in fact false, it is the differential power inherent in the research design and choice of statistical procedures that leads to rejecting the null hypotheses.

Power is a relatively complex function of several factors, including sample size, the Type I error rate (alpha), variance of the dependent variable, the statistical test, and the nature of the null hypothesis. All of these factors are to a certain extent under the control of the researcher. In addition, power is also a function of between-group differences associated with the experimental manipulations. Depending on the freedom an experimenter has in choosing the experimental manipulations, this factor may also be controllable.

Power is a function of three general factors: effect size, residual variance, and the statistical decision rule. Each of these factors, in turn, may be broken down into finer components. In general, effect size refers to the difference between groups that is associated with the experimental manipulations. To the extent that the researcher may choose the experimental manipulations, this factor can be controlled (e.g., if an experiment is a replication or an application of prespecified procedures where no freedom exists in choosing the manipulations). Similarly, if a factor is random (as discussed above), there is no choice; however, when fixed factors are used, the comparison of extreme divergent levels will always pay off more than the comparison of virtually identical levels.

The second factor influencing power is residual or within-group variance. The usual test of significance in analysis of variance compares between-group variability (i.e., effect sizes) to within-group variability. Hence, as within-group variance is decreased, power is increased. The usual advice given for increasing power is to use large sample sizes that decrease the residual term. This advice should generally be followed; tables showing the power for different sample sizes for various designs have been carefully prepared by Cohen (1969). Clinical researchers have many problems, however, in realistically increasing sample sizes to the desirable magnitude. Often cited problems include the limited availability of clients of a particular type, of adequately trained therapists to conduct the treatment, and the expense of running large-scale studies. It is easy enough to say that the cost of low power may be at least as high as the cost of the more powerful studies (in the sense of accumulated time and expense across the studies that fail to show anything), not to mention the cost to clients and society due to the absence of well-documented, effective treatment procedures. However, the problems still place practical constraints on frustrated individual researchers. Fortunately, there are, in many cases, feasible alternative ways of reducing within-cell variation. In a typical outcome study there are three sources influencing within-group variance: variability of treatment delivery, nonhomogeneity of experimental units (clients), and error of measurement of the dependent measure. Thus, the use of greater rigor in defining and delivering treatments, more careful selection or classification of clients (including the use of covariates), and reliance on reliable measurement procedures can be used to offset the need for larger sample sizes. This topic is discussed more extensively below.

Finally, the third general factor influencing power is the statistical decision rule. This includes the choice of alpha (Type I error rate), the specification of hypotheses (directional vs. nondirectional, and planned comparisons vs. any and all comparisons), and the selection of statistical tests (e.g., parametric vs. nonparametric). Alpha levels are to a great extent dictated by the general scientific community; little freedom of choice exists. In extreme cases, experimenters might be able to argue that the use of a more "liberal" error rate (e.g., alpha = .10) is justified by the importance of the problem, but generally the Type I error rate cannot be manipulated simply to enhance power. The use of directional hypotheses and/or planned comparisons is, however, a reasonable means of increasing power. Although there is some concern (cf. Bakan, 1967) about the use of directional hypotheses, they are superior to nondirectional hypotheses for a large variety of clinical outcome studies. In particular, for the very basic investigation of whether a proposed new treatment is superior to an existing procedure, the research question is itself directional. There is really no interest in finding that the new treatment is significantly worse. The use of directional hypotheses is infrequent, though, in studies using F tests (e.g., analyses of variance). This author, for example, has had to argue with reviewers that although the F-test is one-tailed, it can be used to test directional hypotheses as follows: conduct the F-test at $p = .10$ and look at the means of the groups. If the means favor the directional hypothesis, then the null hypothesis can be rejected at the $p = .05$ level. Fortunately, this argument works. As researchers become more familiar with the general linear model (cf. Cohen, 1969), they will also be more inclined to use planned comparisons in investigating differences among several groups.

Planned comparisons allow for the increase of power by restricting what differences are investigated in the same way that directional hypotheses increase power for detecting one type of difference. For example, an investigation into the benefits of relaxation versus guided imagery versus a placebo versus no treatment may be subjected to a single three-degree-of-freedom test comparing all groups to one another (requiring *post hoc* analyses to determine what differs from what) or to three one-degree of freedom tests. For example, one might compare relaxation to guided imagery, placebo to control, and the average of relaxation and guided imagery to the average of placebo and control. The last procedure would provide more powerful tests for each of the specified comparisons than would the former. However, it would have zero power for testing, say, whether the placebo differed from the average effect of the remaining conditions. Researchers should definitely consider the advantages of planned comparisons over any and all comparisons, not only from the standpoint of enhanced power for the selected tests but also to avoid the need to do *post hoc* analyses, which typically have low power. In addition, thought should be given to the fact that, in a typical multifactorial design, the tests of main effects and interactions are actually a statistical contrivance for conducting one set of *a priori* tests. Indeed, an overall comparison among all groups could be made just as simply, as could tests that most closely match the questions of interest to the experimenter.

EVALUATION AND DESCRIPTION OF VARIABLES

The Importance of Reliability

Among the major theoretical and practical problems facing researchers are the specification, description, and measurement of independent variables (manipulation or classification), control variables, and dependent variables. Although many of the issues involved in variable specification are theoretical or methodological (i.e., outside the scope of this chapter), there are several aspects that sufficiently interact with or result from statistical problems to warrant their coverage. In general, the oft cited need for therapy outcome studies to determine what works best for whom requires a reasonably precise specification (and delivery) of the "what," reliable measurement and classification to determine the "whom," and specification and reliable measurement of the variable(s) involved in the "works best." All of the above can generally be subsumed under the rubric of reliability and validity or even more generally under the concept of liberalized reliability and validity as discussed in generalizability theory (Cronbach, Gleser, Nanda, & Rajaratnam, 1972).

From a statistical standpoint, reliability is important. Consider a simple two-way ANOVA model:

$$Y_{ijk} = M + M_i + M_j + M_{ij} + E_{ijk}$$

where M_i denotes the effect of Treatment i, M_j denotes an effect for Client type j, M_{ij} represents the interaction of Treatment i and Client type j, and k denotes Subject k. This model assumes consistency of the delivered treatment across all subjects and requires that all subjects of a client type be equivalent. The only variable that is allowed to vary as a function of individual subjects is the residual E_{ijk}, which presumably is random noise (like error of measurement on the dependent variable). To the extent that M_i, M_j, and M_{ij} are also allowed to vary, the research is simultaneously deviating from that which was supposedly being investigated (an issue of validity). Consequently, the residual term is increased, thereby reducing power. The reduction in power can conceivably be made up by increasing sample size, but the threat to construct validity cannot. In addition, weakened effects seriously jeopardize the detection of clinically significant effects that may be independent of sample size.

Research that includes individual difference variables is also threatened by increased imprecision of measurement. Outcome-study analyses that include repeated measures, (e.g., pre, post, and follow-up assessments), covariates, (e.g., pretreatment history), or potential moderator variables (e.g., presence of additional problems) are all limited by the reliability of the individual difference variables. Classical test theory provides a well-known caveat that validity (i.e., the manifest relationship among variables) cannot exceed the product of the square roots of the reliabilities of the variables involved. Thus, attempts to measure change, control for background variables, or determine moderators of treatment may fail for reasons of poor reliability, but the failure may be erroneously attributed to theoretical or substantive factors.

Reliability is also a critical factor in multivariate statistical tests or procedures. The large variety of multivariate procedures, including multiple regression, canonical correlation, principal components analysis, and multivariate analysis of variance involve intercorrelations among the variables. This matrix of intervariable relationships will be differentially attenuated for variables that have low reliability compared with those that have high reliability. The result of this is that the "good" variables are *a priori* more likely to appear as significantly important, while the "poor" variables will be cast aside. For strictly empirical purposes, this does not pose a problem; however, substantive interpretations of results would incorrectly favor variables with the higher reliabilities. In brief, reliable specification, operationalization, and measurement enhance all aspects of statistical tests and inferences. With an understanding that reliable independent variables are as important as reliable dependent variables, the remainder of this section will be devoted to a consideration of reliability issues focusing primarily on dependent variables.

What Should Be Measured?

The selection of dependent measures in therapy outcome studies involves three subissues: the selection of substantive constructs relating to the clients' problems and their remediation, the operationalization and measurement of those constructs, and the analysis of the measures in the context of the overall design.

In general, research designed to demonstrate therapy effectiveness for "real" clients with "real" problems should probably include multiple dependent outcome measures, including measures of client (or significant other) reports of subjective feelings and beliefs, measures of clinicians' evaluations, and objective behavioral criteria (Paul, 1967). Within each of these domains there are multiple alternatives from which to select, depending on the specific problem and the researchers' theoretical and personal preferences. Each domain, typically, has its own special set of problems, and the combination of domains raises still more.

Client report, for example, possesses a degree of face validity to the extent that it relates directly to the client's experiences; clinicians' evaluations of improvement provide information from the viewpoint of the service provider and/or gatekeeper; behavioral and physiological measures index more objective improvement. Problems with client and clinician reports are their subjectivity and possible confounding by demand characteristics as well as their relatively low reliability. The more objective measures can typically be collected so as to have high reliability (of a sort), but they come under attack because they may be less generalizable to the real world and may not be highly correlated with other criteria for improvement.

Appropriate Indices of Reliability

At what level should reliability be assessed? Consider a therapy outcome study focusing on decreasing the frequency of out-of-seat behavior of individual children in a classroom. The researcher is probably interested only in a summary measure of frequency for each session, although each instance of out-of-seat behavior will be recorded. In this situation there is a choice of levels of analysis: reliability could be measured only for the summary scores (the score to be used in the outcome analyses) or at the finer level of each occurrence/nonoccurrence of the behavior (the level at which data are collected).

For subsequent data-analytic purposes, demonstration of interobserver agreement on total frequencies is adequate. However, the analysis of reliability at the finer level provides something extra: confidence that the sum score is a valid measure of the behavior. That is, agreement for sum scores could be obtained for invalid or artifactual reasons. For example, if observers were not blind to subject classifications (e.g., treatment group vs. control group), they could misperceive frequencies in accord with their expectations. In fact, in a multigroup design, observers could not only disagree at the finer event level but also for all children within a group. But as long as they agreed for the group means and the group means differed across groups, high reliability could be shown. For example, if children were observed in different classrooms at different times for inadvertently different lengths of time, the calculation of reliability on a pooled distribution of these scores would likely yield "good" but invalid results. On the other hand, sum score reliability as typically measured could be virtually zero despite perfect interobserver agreement at the finer level. Indeed, the best possible outcome would be where all subjects of a designated type have identical scores but subjects of different types have different scores (i.e., minimum within-group variance and maximum between-group variance). By analyzing the reliability of the data at the finer level (i.e., the level at which it is collected), the researcher could immediately ascertain whether the observers were seeing the events at the same time. If so, then unreliability of total scores merely signifies intersubject homogeneity (possibly a good thing) if the observers do not agree. Thus, high interobserver reliability across subject total scores should definitely be suspect.

The choice of level of analysis is clearest when there are opportunities for simultaneous and independent measurements (indeed, this is the essence of reliability). But many measures do not lend themselves to either simultaneous or independent measurements. For example, a client's self-reported improvement can be neither simultaneously nor independently assessed. For such measures, the term *reliability* is still commonly used, but its meaning is clearly different. The separation of different kinds of reliability is critical, however, and warrants further discussion (see below).

A second area of concern in describing the quality of data is the choice of a statistic. Traditionally, reliability analyses focused on evaluation of interval level data and were equated with Pearson product moment correlation coefficients (cf. Lord & Novick, 1968). There are, however, a variety of problems in which the data are categorical (nominal) or in which something other than mere agreement among arbitrarily standardized scores is required.

Cohen (1960) introduced the kappa coefficient as a measure of reliability for categorical data. A variety of extended uses or variants on its use have been proposed as well (Conger, 1980). But despite the excellent acceptance of kappa in many quarters, behavioral psychologists, among others, often resort to simple percent agreement among observers or coders. Percent agreement is easily calculated but inappropriate as a measure of data quality. The basic problem with percent agreement is that it does not take into account the proportion of times agreement would occur by chance alone. The problem is especially evident when one event class (e.g., presence of behavior X) has a high base rate. For example, if in-seat behavior occurs 95% of the time according to each of two independent observers, the minimal agreement possible between the two observers is 90%. All too often, such a high level of agreement would be taken as evidence of excellent reliability when, in fact, it is evidence of extremely poor agreement. If for example, in-seat behavior was being recorded by depressing a button each time it occurred and releasing the button each time it terminated, even better agreement would be obtained between the two observers if one of them fell asleep on his or her button, holding it down for the entire session. In this case, the percent agreement would be 95%! To avoid this obvious problem, various machinations have been used (e.g., calculating agreement only for "presence" or "absence"). But this results in noncomprehensive statistics in which the reliability for event presence is unrelated to the reliability for event absence (although when there are only two possible outcomes, the outcomes are necessarily correlated -1). The use of a statistic that avoids chance inflation or fractionates a single reliability analysis into multiple analyses would appear desirable. Cohen's kappa possesses all the necessary characteristics of an appropriate and comprehensive measure of reliability. It has also been shown that when chance agreement is taken into account, the various alternatives to kappa either become equivalent to kappa (Fleiss, 1975) or are equivalent except for scaling (Conger & Ward, in press). A different problem is the selection of an appropriate index of agreement when interval-level data (e.g., ratings of improvement) are available. In the above discussion on categorical data, agreement occurs when observers provide identical values. However, it has been traditional in test theory to measure agreement for interval-level data by the relative equivalence of standardized scores. Conceivably, perfect reliability can be claimed even though two raters never give identical scores to the same subject. A cogent discussion of the appropriate statistic to use to measure reliability has been provided by Mitchell (1979) and is generally available in texts covering the intraclass correlations. In short, the intraclass approach is to be preferred to the more traditional Pearson interclass procedure because it allows for selection of the most appropriate index of reliability. For example, if session summary scores on out-of-seat behavior are available from each of two independent observers, three different indices could be used to measure reliability. Selection of the most appropriate measure depends on the intended use of scores obtained from the observers. Assuming that two observers have independently provided data on a sample of subjects, the linear model for this design would be:

$$Y_{ij} = M_i + M_j + M_{ij} + E_{ij}$$

where Y_{ij} is the obtained score, M_i is the "true" score for Subject i (i.e., an average score for a population of observers), M_j is the population mean for Observer j, M_{ij} is the interaction effect, and E_{ij} corresponds to a residual that is confounded with the interaction. A random-effects analysis of variance of these data will yield sums of squares and mean squares for persons (based on the variance of the M_i's), for observers (based on the variance of the M_j's), and for the interaction of persons and observers confounded with the residuals. From the mean square values, variance estimates for persons (V_p), observers (V_o), and the interaction $(V_{po,e})$ may be derived.

The most rigorous measure of reliability compares the estimated person variance to the sum of person, observer, and interaction variances:

$$R_2(M_i, Y_{ij}) = V_p/(V_p + V_o + V_{po,e})$$

This coefficient is necessary for situations in which scores from two (or more) observers are to be pooled as if a single data collection instrument were used for all subjects. For example, if each of two observers will observe half of the sessions, this statistic, which takes into account between-observer mean differences (V_o), is the required measure of reliability. If, however, all data will come from a single observer (except for the reliability substudy), then the appropriate measure need not include differences in observer means. The reliability coefficient could thus be:

$$R_2(M_i, Y_i) = V_p/(V_p + V_{po,e})$$

This coefficient describes reliability for scores obtained from a randomly selected observer whose mean is irrelevant but whose standard deviation of scores is not. Hence, the residual term penalizes for differences in observer standard deviations by inclusion of the interaction V_{po}. Otherwise, this intraclass coefficient could be obtained by converting the Y_{ij}'s to deviation scores $(Y_{ij} - MY_j)$ and using the first reliability equation.

The third reliability measure is useful only for those situations in which neither the mean nor the standard deviation of individual observers is relevant. For example, in correlational studies, all variables are arbitrarily standardized and relationships are investigated via Pearson interclass coefficients. The reliability coefficient in this case can be obtained from the above data set by converting all scores to standard Z scores: $Z_{ij} = (Y_{ij} - MY_j)/SDY_j$. Use of Z scores in the first reliability equation provides a value equivalent to the Pearson product moment correlation for interobserver reliability. Except in unusual cases, the largest value will be for the Pearson and the smallest will be for the coefficient that penalizes for interobserver mean differences. But larger magnitudes should not determine the choice of a statistic; rather, the intended use of scores should. In the majority of situations, the probable choice is the second statistic corresponding to an intraclass reliability coefficient, which includes a penalty for differences in standard deviations of scores obtained from different observers.

Generalizability Theory and Comprehensiveness

Unfortunately, reliability is not measured as simply as described above, and concepts like the stability of scores over time, or in different settings, or on different forms become confused with the replicability of scores collected simultaneously and independently. Classical test theory (Lord & Novick, 1968) perpetuated the myth that different measures of reliability could be used interchangeably by making seemingly innocuous assumptions about the constancy of a mythical "true" score. The best alternative is provided by generalizability theory (Cronbach *et al.*, 1972), which allows for the assessment of different sources of error in determining reliability. Thus, rather than assessing reliability for different observers and then for different situations and occasions in the hope that all three indices are roughly equivalent, a single comprehensive statistical procedure could be used to assess all three sources of variance simultaneously. That is, generalizability theory or, more generally, multifacet intra-class reliability statistics, can be used to estimate components of variance for each identified potential source of variation in the context of a single design. This procedure not only controls for the sources in terms of main effects but also provides information about joint effects in the form of interaction components. From the estimated variance components (in this case 15 components can be identified, corresponding to person, observer, situation, and setting and the various two-, three-, and four-way interactions), reliability coefficients can be found for a variety of observed score uses depending on whether scores are pooled across different observers, settings, situations, or combinations of them or whether certain sources are held constant. Details of the various possibilities are provided by Cronbach *et al.* (1972).

In summary, reliable measurement is an extremely critical aspect of clinical outcome studies that hope to do more than merely make a claim that some treatment is significantly better than none. To the extent that we wish to show what treatment works best for which problems on certain individuals in given settings, reliable measurement is absolutely necessary. Without it, we are bound to be plagued by inconsistent findings and erroneous inferences.

INFERENTIAL STATISTICS

Since psychotherapy outcome studies use a broad range of designs and statistics, a thorough evaluation of all procedures will not be attempted. Instead, detailed coverage of a limited set of problems is provided in the hope that others will address more fully those issues not covered by this author. The issues chosen include clinical versus statistical significance, the use of multiple dependent variables, and statistical versus methodological control. These topics do not exhaust statistical issues in outcome studies, but they do represent a reasonably coherent set of problems in which issues of appropriateness, comprehensiveness, and power are evident. Other procedures like time series (Glass, Willson, & Gottman, 1975), meta-analysis (Smith & Glass, 1977), and the more traditional "alphabet soup" designs (i.e., A-B-A–type designs) involve similar issues, but their evaluation will be left to others.

As previously discussed, clinical outcome studies require decisions about the number and nature of dependent and control measures. Generally, if the advice of experts on psychotherapy research is followed, outcome studies will include repeated observations of multiple dependent measures selected from different sources (therapists and clients) and different response domains (self-report, physiological measures, and behavioral measures). In addition, several control or moderator variables may be included in this set of variables, all of which are then to be analyzed for between-group and across-occasion differences in the context of a factorial analysis of variance (ANOVA) design. But this often leaves the researcher with a bewildering data analysis problem that includes questions of how the variables or results of statistical tests relate to the improvement of individual clients, which variables actually show differences and which do not, and whether certain individuals benefit more from a particular type of treatment than from another. All too often, answers to these questions are expected from studies that have a limited client population on which to draw or that attempt to make up for inadequate sample sizes by drawing upon diverse populations. It is from this matrix of problems that the issues discussed below were derived.

Clinical versus Statistical Significance

One of the advantages of limiting outcome studies to functional analyses of single individuals is that there is usually less ambiguity about the clinically relevant improvement of the client. However, such studies are greatly limited in the generalization to other clients and do not lend themselves to detailed analyses involving multiple treatment comparisons on multiple outcome measures. Scientific evaluations of treatments require the possibility of replication by others on roughly equivalent experimental units (i.e., clients). As a consequence, for better or worse, therapy outcome studies have moved toward the use of standardized diagnostic and outcome measures. A major difficulty encountered in this tendency of using standardized measures of treatment outcome in experimental designs on groups is that the benefits of treatment for individual clients is lost. That is, the goal of therapy—to produce meaningful changes in the behavior of an individual—may not be attainable by a treatment program that significantly changes the average performance of one group relative to the average performance of a control group. In fact, if a significant group-by-occasion interaction is found in a standard treatment versus control group by pre- versus postevaluation, there is no guarantee that anyone improved in terms of absolute performance (i.e., perhaps the control subjects deteriorated). Even if pre- to postscores show a significant improvement for the treatment group, questions of the magnitude of change and universality of change still exist. Thus, a significant change in average performance may indicate a relatively constant but small relative change for all group members or some substantial change for some individuals with possible losses for others. In neither of the cases would an inference of effective therapy be warranted.

To complicate matters further, the advice of statisticians calling for larger sample sizes and more precise measures in well-controlled studies so as to enhance power simply increases the likelihood of detecting clinically irrelevant differences. The issue, however, is one of specification of appropriate measures of clinically relevant improvement. From a statistical standpoint, significance is the *sine qua non* of confident inference; no matter how large a difference is shown to be associated with a therapeutic procedure, confidence in the repeatability of such a difference is what is being measured by the statistical tests. The issue, therefore, should not be considered as one of choosing between statistical versus clinical significance; rather, the issue is one of choosing criteria for clinical significance that can be assessed for statistical significance.

The criteria for clinically relevant improvement can be established in several different ways, but two methods seem most appropriate. First, the null hypothesis can be set up so that differences of a prior magnitude must be achieved in order to reject it. For example, in a design where subjects are randomly assigned to either a treatment or control group and only posttreatment measures are collected, the usual null and alternative hypotheses are:

$$Ho: \mu(T) = \mu(C)$$
$$Ha: \mu(T) < \mu(C)$$

In this context, any difference between the groups is deemed acceptable; however, if on the dependent variable a clinically relevant difference is at least 3 score units, then the hypotheses could be given as:

$$Ho: \mu(T) - \mu(C) = 3 \text{ score units}$$
$$Ha: \mu(T) - \mu(C) < 3 \text{ score units}$$

Rejecting the null hypothesis in favor of the alternative in the latter case is equivalent to asserting that on the average, the treatment group improved by a clinically relevant amount compared to the controls. The latter test is obviously more meaningful, but since the difference required for statistical significance is larger than for the standard test, greater power is needed to detect a difference.

As an alternative to the *a priori* setting of a criterion difference in significance tests, *post hoc* evaluations of the magnitude of change could be evaluated via confidence intervals. Thus, if the confidence interval for between-group differences could be shown to include mean differences between 2.5 and 4.5 score units, a claim of clinical significance could be offered; if the differences range from 1.1 to 2.3 score units, the researcher could be confident that the treatment is better than no treatment but that it also does not result in a clinically meaningful change.

In these procedures, there is still the problem of the degree to which improvement is a function of large changes for some individuals versus relatively uniform changes for all individuals. The distribution of scores or changes could be investigated to assess the degree to which all versus just some clients benefited from treatment. In the presence of differential change, a claim of clinical benefit would necessarily be tenuous, but follow-up investigations of "who" benefits from this treatment could be done.

But a better alternative exists. Clinically relevant performance could be set *a priori* for each outcome variable or combination of outcome variables, and each individual in the treatment and control groups could be assigned a score of 1 if they reached criteria or 0 if they did not. The statistical question would then be: Did a greater proportion of subjects in the treatment group reach criteria? Such procedures have been employed in treatment outcome studies (cf. Eyberg & Johnson, 1974; Paul, 1966), but too infrequently. However, there are procedures for assessing benefits to single individuals, and only the failure of researchers to specify clinical improvement keeps the issue of clinical versus statistical improvement alive.

Multiple Dependent Measures

When several dependent measures are collected in the context of a factorial design, the recommended statistical procedure is a multivariate analysis of variance (MANOVA). This procedure is preferred to multiple univariate analyses of variance because it is more powerful and comprehensive. However, its increased power and comprehensiveness are relative to the questions asked. MANOVA is singularly most appropriate for addressing the issue of whether there are differences, in the form of weighted linear composites, among groups. It does not specifically test whether designated groups differ or whether groups differ on the simple univariate variables. It does, however, allow for experimentwise control of Type I errors and thus minimizes capitalization on chance differences. MANOVA also takes into account the relationships among the dependent measures and to this extent avoids the need for later verbal integration of univariate findings. Unfortunately, this procedure does not always provide researchers with answers to the questions asked (i.e., on which variables do the groups differ?). The results of a MANOVA are made even less relevant to clinicians when the variables are not put in the form of clinically relevant improvement, as discussed above.

Multivariate analysis of variance can be used for virtually all the same designs appropriate for univariate analyses. Proper application requires that all univariate assumptions about distributions and independence be met and that certain multivariate assumptions be met as well. In particular, just as univariate ANOVA assumes normal distributions within cells, MANOVA requires multivariate normal distributions. And as ANOVA requires homogeneity of variances among cells, MANOVA requires homogeneity of variance–covariance matrices among dependent measures among cells. That is, the pattern of relationships among variables should not differ across different groups or conditions of the design. To a limited degree, deviations from assumptions can be tolerated in MANOVA as for most statistical tests. However, the degree of robustness tends to decrease as complexity increases, and multivariate, multifactorial designs are decidedly less robust than simple t tests. This issue is raised because assumptions underlying the MANOVA are frequently ignored and hardly ever tested. Testing the assumptions is furthermore made difficult when small sample sizes are used (i.e., there is too little power to determine deviations from the null hypothesis that the assumptions are met). The possibility that MANOVA assumptions are not met is one reason for considering less complex alternatives, despite the comprehensiveness of the MANOVA tests.

MANOVA is designed to weight the dependent variables so as to maximize (on the weighted composite) the F ratio comparing effects to error. For example, a simple design comparing a treatment group to a placebo and a control group for which there are pre, post, and follow-up measures would provide independently maximized composites for the treatment effect, for the occasion effect, and for the interaction of treatment by occasion. The weighting simultaneously takes into account the effects for each dependent measure and the dependencies among the measures, thus avoiding the need to do this later. The test of significance of the multivariate composite considers the number of dependent measures used and thereby avoids capitalizing on chance.

Depending on the number of variables and degrees of freedom for effects being investigated, several uncorrelated composites may be found to differentiate among groups in different ways. The number of potentially significant composites is the lesser of the number of dependent measures or degrees of freedom. In the groups-by-occasions design mentioned above, two composites might differentiate among the treatment, placebo, and control group or among the pre, post, and follow-up occasions. However, the interaction of groups by occasions could yield four different composites if at least four dependent measures were used. Only a thorough (and perhaps highly creative) investigation of univariate results would yield comparable findings.

Perhaps the major drawback to doing a MANOVA is that the weighted composite that maximizes the effect under investigation is not readily interpretable. The composite resembles a multiple regression equation by assigning a weight to each variable. The weights specify how the composite is to be formed, but they do not specify the meaning that can be assigned to the composite (i.e., how it is to be interpreted). Obtaining a highly significant difference between groups on a strange-looking composite would provide little solace to an experimenter who wanted to know on which variables the groups differed. Even if the researcher were content to put aside the question of which variables the groups differ on, interest as to what the composite meant would probably remain.

Some researchers have attempted to avoid the interpretational problem while maintaining control of Type I errors by a well-intentioned but incorrect use of MANOVA. The MANOVA is done to determine if overall (experiment-wise) significance exists. In the presence of a nonchance multivariate relationship, significance is evaluated for each dependent measure by standard univariate tests. This procedure is tempting in that it avoids the need to deal with the multivariate composites and also promises to answer on which variables differences do exist. However, the investigation of univariate relationships in this fashion is not a statistically logical follow-up. Often, no serious consequences result, but in some cases the multivariate test may be significant because of variables that are not themselves significant by univariate tests. The results of a MANOVA using six dependent variables to evaluate differences in interaction behaviors of three dyad types (two popular children, two unpopular children, or one popular and one unpopular child) crossed with sex and repeated in each of three segments of an interaction (waiting period, puzzle task, and conversation period) demonstrated highly significant differences for a variety of main effects and interactions, but two types of findings are relevant for current purposes. With only one exception, the variables that contributed to significance of the multivariate composite were themselves significant by univariate tests. The exception occurred for the main effect of dyad type, for which the only significant univariate effect occurred for "Laughs" (more frequent for popular child dyads and less so for the unpopular dyads). The significant multivariate composite weighted the variables as follows:

$$Y = .94*\text{Talk} - .93*\text{Orientation}$$

with all other weights being quite small (less than .35 in absolute value). Interestingly, Laughs did not contribute to the multivariate composite (the weight was .33) and neither Talk nor Orientation were significant by themselves (both p's > .15). The means indicated (nonsignificantly) less talking among popular than unpopular children and (nonsignificantly) more orientation toward the partner. The basis of the significant contribution of the difference was not clear, however, until the relationship between Talk and Orientation was investigated within groups. For any given group, these two variables were positively correlated (e.g., popular dyads which talked more also showed more inter-partner orientation). Thus, the multivariate composite was significant because of a difference in the pattern of relationship between these two variables within groups compared to across groups. In this situation it is quite clear that significance of the multivariate test has no bearing on interpreting univariate tests and vice versa. But this is also the general issue, namely, that a MANOVA test is not specifically related to multiple univariate tests nor does it control for them.

If a researcher is not interested in multivariate composites, then MANOVA should not be used; instead, the univariate ANOVAs should be investigated directly. Of course, there is still a need to avoid capitalizing on chance relationships, and the Bonferroni method of experimentwise Type I error control should be employed (i.e., divide alpha, the Type I error rate, by the number of dependent measures). This may be costly if the number of variables is large, and this cost carries over into a lowered power as well. It should also be noted that this procedure will merely indicate which variables are significant and which are not. It does not test whether one variable is more significantly related to the independent variable than another.

Often the researcher has *a priori* hypotheses about which variables are related and which are not. This provides still another alternative to the standard use of MANOVA: namely, the construction of composites based on prior hypotheses that are then tested for significance. The tests take the form of those recommended by Cohen and Cohen (1975) for multiple regression analyses. That is, the experimenter-constructed composite is viewed as a model, and the multivariate composites are investigated to see whether the optimal weighting is significantly better than the model. Unless the multivariate composite is significantly better than the (univariate) composite, the experimenter can conclude that the *a priori* weighting contains the basis of significant differences for the effects under investigation. This type of statistical procedure can be used to investigate individual variables as well. If the elimination of a variable from a significant composite does not result in a significantly weaker relationship, it is not needed. Of course, as is the case in multiple regression, different combinations may result in equally good or poor relationships, but this procedure does allow the investigation of which variables differentiate and which do not without sacrificing power, comprehension, or appropriateness.

Finally, if the MANOVA composite is the best or if there are no prior models to investigate, procedures do exist to facilitate the interpretation of the composites. First of all, the interpretation of any composite is dependent on the definition of ascribed meaning. With complexly determined variables like those under consideration, two factors need to be taken into account: the weighting used to form the composite and the relationship that the composite has to the variables used to form it. The former is manifest and is typically provided as output from computer programs. However, the latter is viewed as more important but is available only in some computer programs. The example used above to demonstrate lack of correspondence between multivariate and univariate tests was done via the MANOVA program available in SPSS. This program gives both multivariate and univariate tests, weights for significant multivariate composites, and correlations of composites with the dependent measures. Since this program also allows for user-defined planned comparisons, it can do virtually all of the necessary statistical analyses required for an appropriate, comprehensive analysis that also maximizes power.

Statistical versus Methodological Control

Based on formal statistical assumptions for the analysis of covariance (ANACOVA), this procedure may be the most widely abused. Covariance controls are used directly in analysis of variance situations and explicitly or implicitly in multiple regression designs. The intent of covariance analysis is usually to hold constant, by statistical means, individual difference variables that mask differences due to independent variables. By such adjustments, greater power can be obtained because the "error" variance is reduced. The formal assumptions governing the use of ANACOVA include normal distributions, homogeneity of variance and covariance across groups, relatively error-free measurement, and, most importantly, randomly constituted groups. Under these conditions, adjustments made by holding the covariates constant can be interpreted along the same lines as if methodological control were used. Thus, if between-group differences are enhanced by the ANACOVA, so much the better. As with the analysis of variance, deviations from distributional assumptions will jeopardize the meaning of statistical tests. However, deviations from precise measurement of the covariate or from the requirement of random assignment will jeopardize the interpretation of results, with potentially severe consequences. Lord (1960), for example, showed that covariance adjustments with fallible measures can result in relationships that are actually opposite in direction to the true underlying relationship—an obviously undesirable result.

Deviation from random assignment is the more likely problem in psychotherapy outcome studies, however, and usually arises in the context of "selection" factor confounds (Campbell & Stanley, 1966). All too often, investigations of treatment effectiveness include the comparison of groups that differ *a priori* on some factor in addition to the independent variables under investigation. For example, to investigate the effectiveness of marital therapy for spouse-abuse cases, three groups might be required: a treatment group for couples seeking therapy, a waiting-list control group of couples seeking therapy, and a group of couples to be treated by individual therapy. If the entire sample of couples is randomly assigned to the different groups, then covariance controls can be instituted for variables such as severity of problem, number of years married, SES, and so on. Unfortunately, we might find that the researcher has limited access to couples who can be treated by marital therapy but can find other spouse-abuse couples at two other mental health centers, one of which is quite willing to withhold treatment because of limited resources and the other of which normally treats such cases by individual therapy. These three groups may differ on the basis of pretest data on variables such as severity of problem, number of years married, SES, and so on. Obviously, there is a need to control for these extraneous differences, and ANACOVA would appear appropriate. Indeed, an ANACOVA could be done to control for these other variables, but according to the logic of the control, the inferences based on adjusted data are limited to the following: if between-group differences disappear, one may conclude that they were due to the extraneous variables; however, if they do not disappear or are enhanced, then one can only conclude that the set of control variables chosen did not account for between-group differences. It is not legitimate to conclude that treatment effects have been demonstrated under any circumstances (for an opposing view, see Overall & Woodward, 1977). The basis of such a conservative approach is straightforward. Since the groups differ *a priori* by a selection factor, they may differ on the dependent variable only because of factors correlated with the selection. Failure to find such correlates may be due to imprecise measurement, low power for the overall design, or a lack of knowledge about the factors on which they do differ. On the other hand, if a covariate is found that removes between-group differences, definite information has been provided. Typically, however, this is not the sort of information that is being sought. In general, reliance on statistical controls that are methodologically confounded with between-group differences cannot be justified as a legitimate procedure. Before resorting to such "queasy" experimental designs (Campbell, cited in Margolies, 1982), researchers should go back to the drawing board to redesign their studies.

CONCLUSION

The introduction to this chapter delineated several problems or issues that permeate the arena of psychotherapy outcome studies. These problems are by no means the sum total of statistical crises evident in this field or in evaluation studies in general. Specific focus was placed on the traditional experimental designs; quasiexperimental procedures were almost totally ignored. The rationale for this decision is based on the relatively weaker inferences allowable from the latter and the relatively greater complexity of the statistical procedures needed.

The issues selected to be addressed were the appropriate choice of statistics, reliance on comprehensive statistical procedures, and the need for greater power in research designs. The format chosen as a vehicle for discussing these issues included a moderate degree of general discussion, accompanied by more detailed examination of the importance of reliable measurement, clinical versus statistical signficiance, multiple dependent measures, and statistical versus metholodogical control. It is hoped that the examples were plentiful and clear enough to demonstrate the issues involved.

Attention was drawn to the issues of appropriateness, comprehensiveness, and power, because the exercise of greater care in selecting statistics according to these principles should result in better (replicable and valid) psychotherapy outcome studies. Research on "real" clients is often limited by practical constraints; it is with this understanding that much of the advice was offered. That is, practical problems impose limitations on sample sizes and manipulations, with a general consequence of a reduction of power and the likelihood of demonstrating between-treatment group differences. But the reduction in power can, to a degree, be compensated for by the use of more rigorous client classification, homogeneous delivery of treatments, and reliable outcome measures. Additional gains in power are available by appropriately using planned comparisons, including directional hypotheses.

The best advice on how to conduct meaningful outcome studies was, perhaps, provided by Paul (1967), and he has also provided the field with an excellent model of rigorous research (Paul, 1966). Despite the availability of these resources, therapy outcome studies during the 15 years since Paul's studies have not borne witness to an upsurge of well-conducted research (see, for example, Conger & Keane, 1981, for a review of problems in social skills interventions with children). While practical problems in the design and execution of research do exist, there are a number of other factors that may be embarrasingly more important. The oft cited "publish or perish" rule governing the admittance of novices to the ranks of the truly worthy (i.e., tenured faculty) surely plays a role in the selection of research topics and methods. In many settings researchers are expected to be not only prolific but to demonstrate the ability to publish as sole authors. The need to publish many studies cannot help but to discourage larger multigroup, longitudinal designs and to encourage the taking of "shortcuts" in the conduct of the more modest designs. Actual or implied prohibitions against collaborative research impose further limits by eliminating incentives (e.g., coauthorship), thereby requiring each individual investigator to be a substantive, methodological, and statistical expert or to fall into traps because of some deficiency in one area. The perpetuation of problems in methodology and statistics may continue for another 15 years (or more) unless there are changes in the sociopolitical environment of research.

REFERENCES

Bakan, D. The test of significance in psychological research. In D. Bakan (Ed.), *On method: Toward a reconstruction of psychological investigation.* San Francisco: Jossey-Bass, 1967.

Campbell, D. T., & Stanley, J. C. *Experimental and quasi-experimental designs for research.* Chicago: Rand-McNally, 1966.

Cohen, J. A coefficient of agreement for nominal scales. *Educational and Psychological Measurement,* 1960, *20,* 37–46.

Cohen, J. *Statistical power analysis for the behavioral sciences.* New York: Academic Press, 1969.

Cohen, J., & Cohen, P. *Applied multiple regression/correlation analysis for the behavioral sciences.* Hillsdale, N.J.: Lawrence Erlbaum Associates, 1975.

Conger, A. J. Integration and generalization of kappas for multiple raters. *Psychological Bulletin,* 1980, *88,* 322–328.

Conger, A. J., & Ward, D. Agreement among 2 × 2 agreement indices. *Educational and Psychological Measurement,* in press.

Conger, A. J., Conger, J. C., Farrell, A. D., & Ward, D. What can the WISC-R measure? *Applied Psychological Measurement,* 1979, *3,* 421–436.

Conger, J. C., & Keane, S. P. Social skills intervention in the treatment of isolated or withdrawn children. *Psychological Bulletin,* 1981, *90,* 478–495.

Cronbach, L. J., Gleser, G. C., Nanda, N., & Rajaratnam, N. *The dependability of behavioral measurements.* New York: Wiley, 1972.

Eyberg, S. M., & Johnson, S. M. Multiple assessment of behavior modification with families: Effects of contingent contracting and order of treatment problems. *Journal of Consulting and Clinical Psychology,* 1974, *42,* 594–606.

Fleiss, J. L. Measuring agreement between two judges on the presence or absence of a trait. *Biometrics,* 1975, *31,* 651–659.

Glass, G. V., Willson, V. L., & Gottman, J. M. *Design and analysis of time series experiments.* Boulder: Colorado Associated University Press, 1975.

Gottman, J. M., & Markman, H. J. Experimental designs in psychotherapy research. In S. L. Garfield & A. E. Bergin (Eds.), *Handbook of psychotherapy and behavior change* (2nd ed.). New York: Wiley, 1978.

Hays, W. L. *Statistics for the social sciences* (2nd ed.). New York: Holt, 1973.

Lord, F. M. Large-sample covariance analysis when the control variable is fallible. *Journal of the American Statistical Association,* 1960, *55,* 307–321.

Lord, F. M., & Novick, M. R. *Statistical theories of mental test scores.* Reading, Mass.: Addison Wesley, 1968.

Meehl, P. E. Theoretical risks and tabular asterisks: Sir Karl, Sir Ronald, and the slow progress of soft psychology. *Journal of Consulting and Clinical Psychology,* 1978, *46,* 806–834.

Mitchell, S. K. Interobserver agreement, reliability, and generalizability of data collected in observational studies. *Psychological Bulletin,* 1979, *86,* 376–390.

Overall, J. E., & Woodward, J. A. Nonrandom assignment and the analysis of covariance. *Psychological Bulletin,* 1977, *84,* 588–594.

Paul, G. *Insight vs. desensitization in psychotherapy.* Stanford, Calif.: Stanford University Press, 1966.

Paul, G. Strategy of outcome research in psychotherapy. *Journal of Consulting Psychology,* 1967, *31,* 109–118.

Smith, M. L., & Glass, G. V. Meta-analysis of psychotherapy outcome studies. *American Psychologist,* 1977, *32,* 752–760.

IV

General Issues

10

Patient Characteristics and Their Relationship to Psychotherapy Outcome

MICHAEL J. LAMBERT and TED P. ASAY

INTRODUCTION

An important variable relating to both the course and outcome of psychotherapy involves the patient and the characteristics he or she brings into the therapeutic situation. Past research in this area has attempted to relate patient variables such as motivation, expectancies, demographic characteristics, diagnosis, severity of maladjustment, and personality traits to continuation in therapy, therapy outcome, and in-therapy patient behavior.

Results of research dating back to the 1940s have shown the importance of patient characteristics in the course and outcome of psychotherapy (Garfield, 1978; Lambert, 1979; Luborsky, Chandler, Auerbach, Cohen, & Bachrach, 1971; Meltzoff & Kornreich, 1970). More recent research has shown not only that the patient's characteristics in psychotherapy are important but also that what the patient brings into the therapy situation is the single most important and influential factor relating to outcome (Lambert, 1979; Luborsky, Mintz, Auerbach, Christoph, Bachrach, Todd, Johnson, Cohen, & O'Brien, 1980; Strupp, 1980a). In this chapter, past research on patient characteristics in psychotherapy will be reviewed and conclusions drawn that seem to be supported by the accumulated research reports. There will also be a discussion centering around the patient's ability to form a relationship with the therapist.

MICHAEL J. LAMBERT and TED P. ASAY ● Department of Psychology, Brigham Young University, Provo, Utah 84602.

This patient trait seems to be a precondition for much of the improvement in psychotherapy. Throughout the chapter, current issues related to patient variables in both research and practice will be discussed.

MOTIVATION

Since the nature of psychotherapy necessarily requires the active involvement of the patient, it is not surprising that there is considerable agreement among clinicians that the well-motivated patient has a much better chance of improving as a result of the therapeutic process (Lambert & Utic, 1979; Luborsky *et al.* 1971; Meltzoff & Kornreich, 1970; Wolberg, 1977). A lack of motivation at the beginning of therapy or the patient's inability to develop it over the course of treatment has been used both as an after-the-fact explanation for failure in psychotherapy as well as a means of predicting response to treatment before it has been offered. Motivation implies not only a desire to change but also a desire to change in ways that are congruent with the goals and values of the therapist. For example, Wolberg (1977) pointed out numerous types of motives that a patient may have. A patient may want to use the therapeutic relationship

> as a means to power, success or perfectionism . . . as a social experience because he is lonesome or frustrated in his personal life . . . he may desire to convert the therapist into a parental figure to satisfy a dependency need. Or he may search for an idealized image of himself in the therapist with which he can identify. (p. 424)

All of these desires on the part of the patient would probably not be in line with the goals of the therapist and if not given up would prove to be a hindrance in therapy. Motivation, therefore, is a complex variable that is very similar to and may at times overlap with the expectations the patient has for therapy.

Motivation has proven to be difficult to study for two reasons. First, motivation is not fixed. Some patients who are well motivated at the outset quickly lose their motivation for therapy, while others who are perceived as initially unmotivated become much more motivated after a few therapy sessions. Second, the term *motivation* is vague. It has been defined as the voluntary and willing effort of the patient to engage in therapy without being compelled to do so. Motivation has also been referred to as "need to change" in connection with the felt disturbance or felt anxiety of the patient. Thus, research studies on motivation are often based on differing definitions of this variable, making the interpretation of their results somewhat difficult.

Perhaps as a result, past research has yielded equivocal results as to the contribution of patient motivation in obtaining positive therapeutic outcomes. While some studies report a high relationship between motivation and improvement, other studies show little or no relationship between the two. In a study by Cartwright and Lerner (1963), for example, it was reported that subjects who were high on the "need to change" variable improved significantly more in therapy than those who were low on this variable. Strupp, Wallach, Wogan, and Jenkins (1963) found that patient motivation for therapy (before therapy) correlated significantly with overall success in therapy as rated by the treating psychotherapist. In their extensive review of factors influencing psychotherapy, Luborsky *et al.* (1971) reviewed five studies dealing with patient motivation and concluded that good motivation for therapy tends to be positively related to outcome. They also concluded that while amount of motivation tends to be positively related to outcome, type of motivation is not. Thus, patients with congruent motives for treatment did not fare better than those with noncongruent motives (e.g., treatment should change their life situation rather than themselves). Schroeder (1960) conducted a study wherein she hypothesized that clients with both very high and very low acceptance of responsibility, a variable closely related to motivation, would present more difficulty for therapy than clients who accepted a moderate amount of responsibility. Although the results of the study did not support the hypothesized curvilinear relationship, the results did show that high responsibility was associated with difficult therapy and high movement in therapy. Low responsibility was associated with low movement and nondifficult therapy. Although these results may have been confounded by the tendency of therapists to rate those staying longer in therapy as more improved, it may be inferred that there was a positive relationship between responsibility and motivation and that a positive relationship also existed between these two variables and patient improvement.

In a more recent study by Gomes-Schwartz (1978), outcome was predicted from psychotherapy process variables. It was reported that the dimension "patient involvement" was consistently the best predictor of outcome and appeared to be independent of therapist technique and the level of relationship variables offered by the therapist. It seems that patients enter therapy with this capacity, which is at least in part a motivational trend, and that it is central to improvement in therapy. Heilbrun (1978) studied the usefulness of the Counseling Readiness Scale for predicting length of stay in brief psychotherapy. This scale, which is a measure of motivation, correlated with number of sessions in treatment which, in turn, correlated with treatment outcome. Other studies (Lothstein, 1978; Perry, Gelfand, & Marcovitch, 1979; Pomerleau, Adkins, & Pertschuk, 1978), which will not be discussed here, have also shown a positive relationship between patient motivation and change in therapy.

In contrast to the above mentioned studies, there are many investigators who report little or no relationship between motivation and outcome. Siegel and Fink (1962) reported a study where patients were placed in high- or low-motivation groups based on ratings by a psychiatric team using data obtained from case records. Results of this study indicated that patient motivation was unrelated to outcome. In a study with college homosexuals, Ross and Mendelsohn (1958) found that initial strong motivation or desire to change a homosexual orientation to a heterosexual one was not always necessary for a change in sexual orientation. They concluded also that individuals benefiting from therapy did not ordinarily possess a high degree of motivation. Rosenthal and Frank (1958) report an investigation with Phipps Clinic outpatients in which improvement was found in 54% of those with moderate motivation and 43% of those with high motivation. These results also indicate that high motivation was not a necessary condition for improvement in therapy. In a study involving school counseling, Volsky, Magoon, Norman, and Hoyt (1965) were unable to demonstrate that motivation was related to outcome.

A possible explanation of the importance of motivation as a factor in patient improvement in some studies and not in others may have to do with the fluctuation and change in patient motivational states. As mentioned earlier, motivation is not fixed but may change over the course of therapy. Some patients who are highly motivated at the beginning of therapy may lose their motivation over time, whereas patients who are initially unmotivated may become much more motivated during treatment. In this line of reasoning, which is congruent with the observations of other writers (Ends & Page, 1957; Lambert & Utic, 1979; Meltzoff & Kornreich, 1970), it is assumed that, if not initially, then at some time during the therapy process, the unmotivated patient must become motivated if any improvements are to be made. What specific factors are involved in the patient's movement from one motivational state to another is a question worthy of future research. Some investigators (Gendlin, 1961; Vriend & Dyer, 1973) have argued that patients who are in therapy involuntarily or who are skeptical, pessimistic, or otherwise unmotivated can all benefit from psychotherapy if the therapist is skillful enough to get them involved in the treatment process.

It seems clear that patient motivation is a complex therapeutic variable that needs to be studied in much more detail. As mentioned earlier, part of the difficulty in studying motivation has to do with the vague, nonspecific ways this variable has been defined. Once an agreed-upon standard definition of motivation is obtained, research involving this variable will become more specific and useful. It may even be found that motivation is essential in the treatment and prognosis of some and not essential in others. The changing nature of patient motivational states also makes research difficult and may lead to incorrect conclusions about the presence and value of motivation in the treatment of various types of patients.

As a prognostic variable, motivation generally appears to have modest predictive value, although, as of yet, there has been no delineation of specific patient motivational attributes that would provide information to clinicians as to whether or not a particular patient could benefit from the psychotherapeutic process. The delineation and elaboration of these specific patient attributes is a topic that could profitably be explored by future researchers. At the present time, clinicians must rely mostly on subjective, intuitive judgments in determining patient motivational states. In addition, if therapy is to be effective with patients judged to be low in initial motivation, the clinician must try to alter the existing motivational state during the course of therapy.

EXPECTATIONS

An area that has received a considerable amount of attention from clinician and researcher alike involves the patient's expectations regarding psychotherapy. Patient expectations can be conceptualized as having two distinct aspects. The first is the expectation for change, the prognostic expectancy, or the optimism the patient has for benefiting from treatment. The second involves the expected role performance of the therapist and patient as well as the procedures to be followed in therapy.

Both the therapist and patient have expectations about psychotherapy that may complement or contradict each other. It is also clear that patient expectancies are not fixed but are subject to change over time and may change significantly over the course of therapy. The problems associated with changing patient expectations are similar to those discussed in connection with patient motivation and will not be reviewed here. Most investigations of patient expectations have been concerned with how expectancies influence either the outcome of or continuation in therapy.

Research on patient expectations has suggested a positive relationship between expectation for improvement and later improvement, a curvilinear relationship between the two, and no relationship between expectancies and outcome. One of the pioneers of research on client expectancies and their relationship to outcome was Jerome Frank. In his classic work *Persuasion and Healing,* Frank (1961) argued that the therapeutic enterprise carries with it the strong expectation that the patient will, in fact, be helped. Frank (1961) suggests that it is the offering of hope that something can be done that represents the underlying factor in different schools of psychotherapy and in other forms of healing such as the placebo effect in medicine and various types of religious cures. In earlier studies, Frank and his colleagues (Frank, 1959; Frank, Gliedman, Imber, Stone, & Nash, 1959) produced evidence indicating that the expectations the patient brings into therapy have an important influence on the outcome of therapy, and the greater the felt distress, the greater the likelihood of improvement. These findings were supported by other investigations (Goldstein, 1960a; Friedman, 1963; Lipkin, 1954), which also demonstrated a positive relationship between expectations and change in therapy. At times, the relationship between expectancies and outcome has appeared more complex. In a study by Goldstein and Shipman (1961) for example, it was reported that a curvilinear relationship existed between patient expectancies and improvement, with those having moderately positive expectancies improving the most. In contrast to these findings, other investigators (e.g., Brady, Reznikoff, & Zeller, 1960; Goldstein, 1960b;) have reported no relationship between expectancies and symptom improvement.

Results from more recent studies in this area have also yielded conflicting results. Martin, Sterne, and Hunter (1976), for example, studied the expectations of hospitalized schizophrenics and their therapists in relation to short- and long-term outcomes. Both these outcomes, as measured by improvement on the MMPI and the Brief Psychiatric Rating Scale, were positively correlated with prognostic expectancies. Tollinton (1973) also found that initial expectations about psychotherapy were related to outcome in the initial stages of treatment with neurotics, but as therapy progressed this relationship diminished. A positive relationship between patient expectancies and outcome was also found in studies by Halperin and Snyder (1979), Borkovec, Grayson, and Cooper (1978), Uhlenhuth and Duncan (1968), and Garrison (1978).

In contrast to the studies mentioned above, Greer (1980) reported that lower expectations of improvement were associated with more favorable outcomes in therapy. Imber, Pande, Frank, Hoehn-Saric, Stone, and Wargo (1970) also found that negative results were obtained when positive expectancies for improvement were induced in subjects during a 4-week treatment period. Similar results were reported by Piper and Wogan (1970).

In addition to studies on therapy outcome, investigators have addressed the issues of the relationship between client expectancies and continuation in therapy versus premature termination. Most studies in this area have involved patient perceptions of expected patient–therapist role performance in therapy. Heine and Trosman (1960), for example, conducted an investigation in which premature terminators were shown to be more likely to expect to give passive cooperation as a means of reaching their goals. Terminators were also more likely to seek medication or diagnostic information and to assume that this constituted proper treatment. Those who stayed in therapy showed expectations that were congruent with the therapist's expectation that they engage in active problem solving and active participation in the change process. In another study, Heine (1962) reported that patients who came in for 12 or fewer sessions had significantly different expectations about the therapist than patients who remained for 13 or more therapeutic sessions. Investigations by Overall and Aronson (1962), and Clemes and D'Andrea (1965) tend to support these findings.

A general conclusion that can be drawn about patient expectancies and continuation in therapy is that patients are less likely to drop out of therapy when their expectancies are compatible with the therapist's expectancies (Garfield, 1978; Lambert & Utic, 1979). It is also interesting to note that no conclusive evidence exists to indicate whether failure to meet the client's expectations of therapist behavior has a negative effect on therapy outcome for those patients who remain in treatment (Duckro, Beal, & George, 1979).

Although initial research on patient expectations and improvement in therapy did indicate a positive relationship between the two, the equivocal results obtained in subsequent studies have led recent reviewers (Goldstein, 1962; Lick & Bootzin, 1975; Morgan, 1973; Perotti & Hopewall, 1976; Wilkins, 1973) to take a more cautious stand. Criticisms of studies showing a positive relationship between expectations and outcome have been offered by Perotti and Hopewall (1976) and Wilkins (1973). Among the most important drawbacks cited are that expectancies have been inferred instead of actually being measured, and that expectancies and outcome ratings are both based on patient self-report data.

Wilkins (1973) also discussed several conceptual and methodological considerations that "call into serious question the legitimacy and utility of employing expectancy as an explanatory construct to account for gain in psychotherapy" (p. 70). These considerations include the conditions under which expectancy effects have been demonstrated, the circular definition of experimentally induced expectancy, the attribution of causality to expectancy, the issue of expectancy versus prediction, and the methodological confounding of expectancy and feedback effects. For further elaboration of these and other issues such as expectancy traits and states and the meaning of the term "expectancy," the interested reader is referred to the reviews by Wilkins (1973), Lick and Bootzin (1975), Perotti and Hopewall (1976), Kazdin and Wilcoxon (1976), and Garfield (1978).

At the present time there is significant controversy over the conclusions that have been offered in the area of patient expectations. The discrepancy in conclusions as well as the significant methodological and conceptual drawbacks of current research ultimately point to the conclusion that the place of the variety of concepts and behavior loosely referred to as "expectations" in psychotherapy outcome is yet to be determined.

DEMOGRAPHIC VARIABLES

Social Class

A patient variable that has received a great amount of attention as it relates to various aspects of psychotherapy is the socioeconomic status or social class of the patient. Most studies on social class have concerned themselves with the relationship between social class and selection for as well as continuation in therapy. A relatively small number of studies have considered the effects of social class on therapy outcome itself.

As to the selection of patients for psychotherapy, most research evidence indicates a positive relationship between social class and referral into psychotherapy. Myers and Schaffer (1954), for example, found a significant relationship between social class and referral to long and intensive therapies offered in a community clinic. In this study, insight-oriented psychotherapies were not recommended for 64% of the lowest-class patients. Ryan (1969) also reported that of patients treated in private practice by psychiatrists in Boston, four out of five had gone to college or were in college and were considered to be middle- or upper-class individuals. Other studies that support these reports have been offered by Rosenthal and Frank (1958), Crowley (1950), Gallagher, Levinson, and Erlich (1957), Brill and Storrow (1960), and Cole, Branch, and Allison (1962).

Social class influences not only which individuals are selected for therapy but also seems to be important in the patient's acceptance of psychotherapy. Most evidence in this area indicates that lower-class individuals are less likely to accept traditional psychotherapy once it is made available. In a study of this phenomenon, for example, Imber, Nash, and Stone (1955) found that almost all the patients in their sample who failed to return after initial screening were members of the lower class. Rosenthal and Frank (1958) reported a significant negative relationship between level of income and acceptance of treatment by patients; Yamamoto and Goin (1966) also found that lower-social-class patients were less likely to keep their initial appointment.

For those individuals who do return for treatment, the research tends to show that there is a relationship between social class and type of therapy received as well as the type of therapist assigned (e.g., private practice, level of experience). In an important study by Hollingshead and Redlich (1958), for example, results indicated that different social classes received different kinds of treatment, with long-term psychoanalytic treatment being given most often to upper- and middle-class patients. These patients tended also to be treated most often by private practitioners. Lubin, Hornstra, Lewis, and Bechtel (1973) also found that patients with less than 12 years of education and lower occupational ratings were assigned much more frequently to inpatient treatments that did not include individual therapy and less frequently to individual outpatient psychotherapy. Evidence supporting these findings was also reported by Ryan (1969).

In a study by Schaffer and Myers (1954) it was found that a relationship existed between the rank of the therapist and the social class of assigned patients. Senior staff members treated more of the higher-social-class patients while residents and medical students treated more lower-class patients. Subsequent findings have supported the results of the abovementioned study for psychiatrists (Baker & Wagner, 1966) and psychoanalytic therapists (Kadushin, 1969), among others. However, in some studies (e.g., Cole, Branch, & Allison, 1962), no significant trend was shown for more experienced therapists to treat higher-social-class clients. In summary, the above evidence indicates that the higher the social class of the patient, the more likely it is that he or she will be accepted into treatments that rely heavily on insight, verbal fluency, and middle-class values (e.g., psychoanalysis). Further, higher-social-class patients are usually treated by more expert and experienced therapists.

An aspect of psychotherapy in which the social class of the patient plays an especially important role has to do with how long the patient stays in therapy. Most research in this area supports the view that lower-social-class patients tend to drop out of therapy earlier. In studying this issue, Imber *et al.* (1955) reported that only 57.1% of lower-class patients stayed beyond the fourth interview whereas 88.9% of middle-class patients went beyond the fourth interview. In another study, it was shown that after 30 interviews, 40% of the upper- and 12% of the lower-class patients remained in therapy (Cole *et al.,* 1962). Although most studies in this area support the abovementioned findings (Frank, Gliedman, Imber, Nash, & Stone, 1957; Kamin & Caughlan, 1963), other investigators (Baker & Wagner, 1966; Ross & Lacey, 1961; Strickland & Crowne, 1963) have found no relationship between social class and duration of treatment. However, most of these studies were conducted in psychoanalytically oriented clinics; one used schizophrenics as the patient sample, another used children.

A specific patient variable that is closely related to social class is education. Most studies tend to show a positive relationship between education and continuance in therapy (Bailey, Warshaw, & Eichler, 1959; Katz & Solomon, 1958; Rosenthal & Frank, 1958; Sullivan, Miller, & Smelzer, 1958), although some investigators (Affleck & Garfield, 1961; Garfield & Affleck, 1959) have not found a relationship between the two. It is likely that education becomes a factor in continuation in therapy when it is below a certain level. Also, education probably interacts with other variables, such as the values of the therapist and the goals of the therapy, in a complex fashion.

The relationships between social class and continuation in therapy are probably the result of a variety of "process variables" tapped by these static traits. Fortunately, some of these variables can be modified without a great deal of time and expense. The expectations and attitudes of the patient as well as the personality, attitudes, and values of the therapist seem to be specifically involved. For example, Brill and Storrow (1960) found an interaction between the "psychological mindedness" of lower-class patients and attitudes of middle-class therapists that may have played a role in continuation in therapy. Lower-class patients were found to be less intelligent, showed a tendency to see the problem more as physical than psychological, did not understand the psychotherapeutic process, and showed a lack of desire for psychotherapy. The intake interviewer also had less positive feelings toward lower-class patients and considered them to be poor candidates for psychotherapy. Also, Pettit, Pettit, and Welkowitz (1974) found an interaction between social class and the discrepancy between patient and therapist values. This discrepency was, in turn, related to continuance in therapy.

The problem created by the premature termination of lower-social-class patients has prompted attempts to find ways of helping these types of patients to remain in therapy. One approach, mentioned by Garfield (1978), is to employ more careful screening of patients who are assigned for therapy. This approach assumes that the patient traits (e.g., education) are too immutable to be quickly modified and emphasizes the selection of more educated, intelligent, and psychologically sophisticated patients. It also assumes that the treatments offered cannot easily be altered.

Another approach that has received considerable attention in the research literature involves some type of pretherapy training to prepare the patient and/or the therapist for psychotherapy. For example, Hoehn-Saric, Frank, Imber, Nash, Stone, and Battle (1964) used a role-induction interview to help clients develop more appropriate expectations about the process of psychotherapy. Results of the study indicated that those patients who received the interview exceeded the control patients on 6 of the 16 criteria, including attendance at therapy. Positive results have also been reported using other types of "socialization" interviews (Heitler, 1973), role-induction films (Strupp & Bloxom, 1973), and related procedures (Warren & Rice, 1972). Lambert and Lambert (in press), for example, tested the hypothesis that a standard role-induction interview could be used to decrease the unusually high dropout rate of Polynesian immigrants seeking help at a community mental health center. Thirty patients were randomly assigned to either a standard role induction (presented via audiotape) or to a placebo role-induction control group. Dependent measures included treatment dropout rates, expectancies for therapist role performance, attendance, and satisfaction with treatment. In general, the study showed that this simple procedure had a statistically significant and clinically meaningful impact on expectations, satisfaction, and premature termination from therapy. The authors suggest that the role-induction procedure is an effective method for dealing with patients who are culturally unique, at high risk for having problems, and failing to be served in community agencies.

The methods of dealing with patient attrition discussed above take the viewpoint that the patient is somehow unsuitable and must be changed in order to do well in therapy. Another point of view would suggest that instead of changing the patient, the better approach may be to change the type of therapy utilized with lower-class patients. Attempts in this area have not been abundant, although some have shown promise. In one study, for example, Goldstein

(1973) developed a structural learning therapy for the poor. Also, Szapocznik, Scopetta, and King (1978) reported on the use of a family therapy approach with Cuban immigrants. For further discussion of special therapies for lower-social-class patients, the reader is referred to the review by Heitler (1976).

In general, the research pertaining to social class and continuance in therapy shows that lower-social-class patients tend to drop out of therapy earlier than middle- or upper-social-class patients. The higher attrition rate of lower-social-class patients has been shown to be modifiable when patients and/or therapists receive some type of pretherapy training or when special treatment approaches are used. Although the relationship between social class and continuation in therapy has received considerable empirical support, it should be kept in mind that this relationship probably involves a number of variables that are frequently obscured but often interact with each other. Also, many of the findings in the cited studies have not been cross-validated on new samples, making the results less conclusive. It would be helpful, too, for future studies in this area to define the types of psychotherapies under consideration in operational terms rather than simply as "dynamic" or "traditional."

The final area of importance in a consideration of social class as a patient variable involves the relationship between social class and outcome of psychotherapy. Although there have not been as many studies on this topic as on the areas discussed above, the available research has yielded some meaningful results. For example, Rosenbaum, Friedlander, and Kaplan (1956) reported more upper-class patients than lower-class patients in the "much improved" category. Also, Friedman (1963) found that only 14% of those in his sample receiving psychoanalytic psychotherapy who were rated as improved or recovered were lower-class patients.

In their review of factors influencing outcomes in psychotherapy, Luborsky *et al.* (1971) reviewed five studies dealing with social class. Two investigations reported a positive relationship between social class and outcome (McNair, Lorr, Young, Roth, & Boyd, 1964; Rosenbaum *et al.*, 1956), two found no relationship (Brill & Storrow, 1960; Katz, Lorr, & Rubinstein, 1958), and one (Gottschalck, Mayerson, & Gottlieb, 1967) found that lower socioeconomic status was associated with greater improvement in brief psychotherapy. Studies by Rosenthal and Frank (1958), Cole, Branch, and Allison (1962), Baker and Wagner (1966), and Lorion (1973) also did not report any relationship between social class and outcome.

Studies that have examined educational status in relation to outcome have shown more positive results. In the studies reviewed by Luborsky *et al.* (1971), five reported a positive relationship (Bloom, 1956; Casner, 1950; Hamburg, Bibring, Fisher, Stanton, Wallerstein, Weinstock, & Haggard, 1967; McNair *et al.*, 1964; Sullivan *et al.*, 1958) and two found no significant relationship (Knapp, Levin, McCarter, Wermer, & Zetzel, 1960; Rosenbaum, Friedlander,

& Kaplan, 1956). In another study not listed by Luborsky *et al.* (1971), Frank, Gliedman, Imber, Nash, and Stone (1957) found no relationship between education and outcome. Unfortunately, in many of these studies the results may have been confounded because nonindependent therapist ratings constituted the main criteria of outcome.

In general, patient social class appears to be an important variable in the selection for and continuation in psychotherapy, but it does not relate specifically to therapeutic outcomes among those patients who are adequately prepared for therapy or those who are selected for treatment and eventually complete it. Practically speaking, while lower-social-class patients are more likely to be denied treatment and run a greater risk of dropping out of therapy, lower-class patients who remain in therapy do as well as other patients with respect to outcome. Future research on social class as a patient variable would probably be most beneficial if the issue of how to help lower-social-class patients benefit from psychotherapy was of prime concern. Further testing of pretherapy training procedures and, perhaps most importantly, the testing of modifications of existing therapeutic approaches are areas that deserve careful consideration.

Race

A variable that has generally been correlated with social class, although it has also been studied separately and in connection with patient–therapist interactions, is the race of the patient. As with social class, the underlying factors that seem to affect the influence of race in therapy have to do with the values, expectations, and motivations of the members of various racial groups. Also, prejudice of one racial group toward another has a significant influence on interracial interactions in psychotherapy. Research on race has generally shown that it is an important factor in the acceptance and continuance of the patient in therapy, but this variable does not appear to be reliably related to outcome.

Typical of research in this area was a study of adult outpatients by Krebs (1971). In this investigation, the type of therapy assignment and number of appointments kept were analyzed in terms of race and sex. Findings showed that black females were assigned to crisis-oriented brief therapy more often as compared with whites and with black males. Black females were also reported to have missed a larger number of appointments as compared with the other groups. In a study of 17 community mental health clinics, Sue, McKinney, Allen, and Hall (1974) found that black patients attended significantly fewer sessions than whites and also tended to terminate therapy more often after the first session. Other investigations have also reported a tendency for blacks to drop out of therapy early (Rosenthal & Frank, 1958; Salzman, Shader, Scott, & Binstock, 1970; Yamamoto, James, & Palley, 1968.

Past research on race in psychotherapy has also concerned itself with maximizing positive outcomes through patient–therapist interracial matching. The presence of possible negative effects due to therapist biases, stereotypes, and the lack of shared experiences and values that arise from racial differences has been hypothesized to result in poorer patient outcome. In general, the idea of racial matching, which has usually involved eliminating interracial pairs, has received little support in empirical research. Ewing (1974), for example, assessed the effects of black and white counselors who worked with both black and white students. Each counselor saw 13 black and 13 white students; although both groups of counselors favored black students, the racial similarity of client and counselor was not an important factor in the counseling situation. Similarly, Cimbolic (1972) paired black and white counselors with black and white clients and failed to show any significant racial preference.

In a study involving Mexican-American and Anglo-American therapists and patients, Acosta and Sheehan (1976) reported that both patient groups preferred Anglo-American therapists. Costello, Baillargeon, Biever, and Bennett (1979) reported on a 2-year follow up of Anglo- and Mexican-American patients treated for alcoholism with a multifaceted inpatient/outpatient program offered mainly by Anglo-therapists. The overall recovery rate was estimated to be 39% and, in general, the Mexican-American patients fared as well as or better than the Anglo-American patients. Jones (1978) studied the influence of the race of the therapist and client on process and outcome in short-term psychotherapy. No differences in outcome were reported as a result of racial matching, but differences in process were noted. For example, the importance of race as a psychological factor was emphasized by the black therapists and patients, as were issues of understanding and trust. White therapists and patients, on the other hand, were rated higher on the rapport/liking dimension. Finally, Szapocznik *et al.* (1978) reported success in treating Cuban immigrants with an "ecological structured family therapy." This type of therapy was used because of the emphasis of these immigrants on family orientation, respect for parental authority, external control, and lack of incentive to attain psychological growth.

With regard to the question of interracial pairings in therapy, it seems clear that this has not been shown to be a significant factor in terms of outcome, although it may influence the duration and process of therapy. Research has demonstrated that whites can be helpful to blacks, blacks can be helpful to whites, and Anglo-Americans can be helpful to Mexican-Americans. Relationships between other racial pairs do not seem to be as clear and should be considered in future research. Although a number of studies show that blacks leave therapy with whites at higher rates than would be expected, there is support for the notion that past experience with a unique population by the therapist is helpful in reducing premature termination of therapy. More systematic research on racial pairing is needed before sound conclusions on various types of racial pairings can be obtained.

Intelligence

Another factor that seems to have a bearing on the effectiveness of psychotherapy is the intelligence of the patient. At this point, the relationship between patient IQ and success in psychotherapy has not been clearly agreed on. While most therapists would probably prefer patients with higher intelligence—who also tend to be more verbal, introspective, and better able to learn from therapy—the relationship between intelligence and improvement as well as the level of intelligence required for therapy are important questions that deserve close consideration.

Luborsky *et al.* (1971) reviewed 13 studies dealing with IQ and improvement in therapy. Of these 13 studies, 10 showed a positive relationship between outcome and intelligence. Some of the studies reviewed reported significant differences in intelligence between improved and unimproved groups, while others listed correlations between .24 and .46. Meltzoff and Kornreich (1970), on the other hand, reported 15 studies, 7 of which showed a positive relationship between IQ and outcome and the remaining 8 no relationship. These investigators concluded that high intelligence is not a necessary condition for improvement in psychotherapy. In an interesting study relating to this conclusion, Garfield (1963) found that successful psychotherapy has been carried out with individuals classified as mentally retarded.

Amid the conflicting results of the studies reported above, the general conclusion may be drawn that some minimal amount of intelligence is required for successful psychotherapy, but what that level is remains to be seen. One possible factor is that a minimum level of intellectual ability is required to usefully understand what is happening in therapy and, also, that intelligence seems to be related to the patient's ability to cope with his or her problem and the associated stress and ambiguity that may be involved in overcoming it. Of course, other aspects of the individual are also important and perhaps of greater significance than intelligence. As Garfield (1978) points out, if the correlation between intelligence and outcome is about .30, for example, this would still account for less than 10% of the variance needed to account for outcome ratings. It should further be noted that the type of therapy being used has a direct bearing on the required intelligence of the patient. Patients treated in psychoanalysis are necessarily required to be of above-average intelligence, while in behavior therapy this requirement does not seem to hold.

From a research perspective, it is clear that more studies evaluating the importance of intelligence in psychotherapy are needed. It would be well for future studies in this area to follow the guidelines of Meltzoff and Kornreich (1970), who suggest that studies should evaluate different types of therapy and therapist goals across the range of intellectual abilities of patients. In this format, patient IQ would be the independent variable and outcome the dependent variable. This approach would not only yield results as to the effects of intelligence in psychotherapy but also help clarify the limits of effectiveness of different approaches with patients of various intellectual levels.

Gender

Another demographic variable that should be considered in our discussion is the sex of the patient. In general, research regarding the influence of patient gender has shown this variable to be relatively unimportant in psychotherapy, although it may be a factor in some specific instances (e.g., rape victims). This conclusion is limited by the fact that the findings relating to the sex of the patient in psychotherapy are usually incidental, since this variable has rarely received exclusive focus in research.

Many studies have considered the influence of patient sex on attrition, with most showing no significant differences between males and females in premature termination (Affleck & Garfield, 1961; Craig & Huffine, 1976; Garfield & Affleck, 1959; Grotjahn, 1972; Koran & Costell, 1973). A host of studies have looked at the differential success rate of male and female patients without controlling therapist gender. Cartwright (1955), Rosenbaum, Friedlander, and Kaplan (1956), and Rosenthal and Frank (1958) found no differences in outcome as a function of sex in adult outpatient clinic patients treated with conventional psychotherapy. Also, Lazarus (1963), in treating neurotics with behavior therapy, found no significant results due to patient sex, nor did Katahn, Strenger, and Cherry (1966) while using group therapy and desensitization. Other investigations in this area have reported similar findings (Gaylin, 1966; Hamburg *et al.*, 1967; Jeffrey, Vender, & Wing, 1978; Knapp *et al.*, 1960). Some studies have reported more success with women (Ellis, 1956; Kirshner, 1978; Mintz, Luborsky, & Auerbach, 1971; Seeman, 1954) whereas others report better outcomes for men (Rachman, 1965; Schmidt, Castell, & Brown, 1965).

More recent investigations have addressed the issue of patient–therapist sex pairing in psychotherapy. Presently, there is no evidence to suggest superior effectivenss of identical or complementary sex pairing in therapeutic outcomes. Although there may be subtle types of bias that may occur between opposite-sex and same-sex matches (e.g., males having a more negative view of females), research does not indicate negative effects or superior outcomes resulting from interactions between therapist and patient gender. Some empirical studies have reported greater patient satisfaction with same-sex pairings (e.g., Liljestrand, Gerling, & Saliba, 1978), while others (e.g., Willer & Miller, 1978) show opposite-sex preferences. In a general overview of gender effects in psychotherapy, Kirshner (1978) concluded, among other things, that a "significant minority" of female patients prefer a male therapist over a female therapist. This finding was hypothesized to be a result of a tendency for females to trust men more than women as authorities. In a recent study that examined therapist and patient gender preferences as well as outcome, partial support was found for the preceding conclusions. Jones and Zoppel (1982) reported two related studies investigating the impact of client and therapist gender on the process and outcome of therapy. This retrospective study measured therapist and patient views of outcome following treatment. In the first study it was found that women therapists rated themselves as more successful, particularly with female clients. In the second study it was found that patients regardless of gender agreed that women therapists formed more effective therapeutic alliances than did male therapists. Even so, both male and female clients of male therapists reported significant improvement. Overall, it was apparent that gender effects (based on matching or viewed as a client variable only) did not have a strong influence on outcome. Additionally, patient gender was confounded with diagnostic classification (males received more enduring labels, such as personality disorder).

Based on the above research evidence, it seems reasonable to conclude that even though some patients and therapists may occasionally prefer to work with one sex or the other in therapy, the sex of the patient does not significantly relate to selection for, continuation in, or outcome of psychotherapy. For a further discussion of sex-role issues in therapy, the interested reader is referred to the work of Kaplan (1979).

Age

Studies on patient variables in psychotherapy have also considered the importance of the age of the patient. For the most part, age appears to be most important in the selection of patients. It seems that most psychotherapists hold to the belief espoused by Freud that younger individuals can benefit more from psychotherapy (or psychoanalysis) than older ones. Consequently, most research indicates that, on the average, younger patients are selected more often for treatment than are older patients, particularly when psychoanalytic treatment is involved. Crowley (1950) surveyed the age of patients selected at a low-cost psychoanalytic institute. Accepted patients averaged 28 years of age against 32.6 years for those rejected; it was also found that more applicants over age 40 were rejected. Similar findings were also reported in another study involving a psychoanalytic treatment (Knapp *et al.,* 1960). Also, Bailey *et al.* (1960) surveyed 549 patients at the VA Mental Hygiene Clinic in New York City and reported that the mean age of patients assigned to psychotherapy was significantly lower than that of those assigned to other types of treatments. Other investigations supporting these findings have been reported by Gallagher, Sharaf, and Levinson (1965) and Rosenthal and Frank (1958).

After the patient has been accepted into psychotherapy, most studies indicate that age does not appear to be an important variable in whether or not the patient continues to receive treatment (Affleck & Garfield, 1961; Beck, Kantor, & Gelineau, 1963; Frank *et al.,* 1957; Garfield & Affleck, 1959; Rosenthal & Frank, 1958; Sullivan *et al.,* 1958). In a more recent study, however, Karasu, Stein, and Charles (1979) found that the age of the therapist in relation to the age of the patient was a significant variable in therapy. Specifically, results indicated that therapists more readily developed a treatment relationship with patients who were approximately the same age as the therapist; patients who were younger or older than the therapist were less likely to remain in treatment than were same-age patients. It may be that in some cases therapists prefer to work with patients who are most like themselves in terms of age. This, of course, would have a direct bearing on which patients are selected for treatment and how long they stay in treatment as well as the eventual outcome of therapy.

The relationship of age to the actual outcome of psychotherapy has been the focus of considerable research. While some studies tend to support Freud's notion that younger patients do better in therapy, other studies show no relationship between age and outcome and some have even indicated an inverse relationship. In a study by Casner (1950), for example, it was reported that patients under 30 years showed more improvement than older patients. Hamburg *et al.* (1967) also reported that patients who were 46 and older improved less in therapy. Studies by Feldman and MacCulloch (1965); Stone, Frank, Nash, and Imber (1961); and Warren (1965) also support these mentioned findings in regard to age and outcome.

There have also been several studies that reported no relationship between age and outcome. Of the 11 studies dealing with age reviewed by Luborsky *et al.* (1971), 5 reported that age was not related to outcome (Bloom, 1956; Cartwright, 1955; Gaylin, 1966; Gottschalk *et al.* 1967; Seeman, 1954). Typical of research on this dimension was a recent study by Hall, Bass, and Monroe (1978), who reported that neither age nor socioeconomic status nor chronicity was related to outcome in obese patients.

In summary, age seems to be relatively unimportant in predicting therapeutic outcomes or whether or not the patient will remain in therapy, although in some cases the therapist may prefer to see patients closer to his or her age. In the selection of patients for treatment, however, age seems to play an important role. Most studies correlating age and outcome involve the problem that age is often confounded with other variables in the therapeutic situation. This is usually the result of selection bias on the part of experimenters, who select only older patients who have some special ability that will help them, in therapy, to compensate for their age. On the other hand, younger patients with a poorer prognosis may be selected chiefly because of their age. These are problems to which future researchers should give serious consideration. And, of course, the interaction between specific types of treatment and patient age is in need of future study.

Marital Status

The marital status of the patient has also been considered as a factor in psychotherapy, although studies in this area have been relatively few. In relation to early termination or therapy attrition, marital status does not seem to be an important factor (Frank *et al.,* 1957; Yalom, 1966), although one study (Katz & Solomon, 1958) reported that patients who were married or single remained longer than those who were divorced or separated.

Most studies on outcome have also shown that marital status is a relatively unimportant variable. In studies by Gottschalk *et al.* (1967), Miles, Barrabee, and Finesinger (1951), Rosenbaum *et al.* (1956), and Tolman and Meyer (1957), marital status was reported to be uncorrelated to outcome. On the other hand, two recent investigations have shown marital status and marital adjustment to have a significant influence on outcome. Milton and Hafner (1979) studied 18 couples where the females were being treated for agoraphobia. The females in the study had been agoraphobic for an average of 8.6 years and had been married for 13.7 years on average. Findings showed that patients whose marriages were rated as unsatisfactory before treatment improved less during treatment and were significantly more likely to relapse during the follow-up period than patients with more satisfactory relationships. Also, Luborsky, Mintz, and Christoph (1979) found an interesting correlation between marital status and outcome that withstood replication. In their analysis of data from two large projects, they note that patients treated by therapists with similar marital status (e.g., neither patient nor therapist had ever been married, or patient and therapist were both married) tended to have better outcomes.

Although the bulk of research in this area shows marital status to be unrelated to outcome, the two studies just mentioned suggest that marital status and marital satisfaction both play a role in therapeutic outcomes. The study by Luborsky *et al.* (1979) has implications for future research on patient–therapist interactions as well as patient–therapist matching. Another consideration with regard to marital status is that with some populations, such as schizophrenics and alcoholics, marital status may be a meaningful predictor of outcome because it is indicative of the patient's adaptability and ego strength, both of which are associated with positive therapeutic outcomes. A detailed discussion of the role of ego strength in psychotherapy outcome will be presented later in this chapter.

MANNER OF PATIENT PARTICIPATION IN THE THERAPEUTIC RELATIONSHIP

We now shift our attention from a consideration of specific patient variables to a discussion of what we believe to be of central importance in the process and outcome of psychotherapy—the relationship between the patient and the therapist. Specifically, this relationship will be discussed in terms of the manner of patient participation.

The importance of the patient–therapist relationship in psychotherapy outcome has been discussed and debated by several investigators with varying points of view. For example, Telch (1981) exemplifies the position of those who believe that the major determiners of psychotherapy outcome are technique variables. It is his position that the influence of client–therapist variables is inversely related to the potency of the therapeutic intervention. According to this view, future research would necessarily focus on further development of specific intervention procedures as opposed to the identification of relevant patient–therapist attributes and their interaction.

On the other hand, Frank (1981) has argued that specific therapy techniques have been found to be effective with specific problems in only a very small number of clinical cases. In his view, patient and therapist attributes and their interaction exert the greatest influence in therapy outcomes. As he stated:

> In short, I remain persuaded that the technique that would be most effective for a particular patient–therapist combination would be one that best mobilizes the patient's hopes, enhances his or her self-efficacy, and so on. This depends, I believe, less on the patient's specific symptoms or difficulties or on the technique in itself than on attributes of the particular patient and therapist in interaction and the appropriateness of the technique to these. (p. 477)

This position has also received recent support from Lambert and Bergin (1983) who not only argue that patient–therapist variables are the most important factors in psychotherapy outcome but also make recommendations that we agree with and view as important in the direction of future research in this area. Summarizing their position, they stated:

> The question really is: Shall we invest our scientific attention in the investigation
> of specific procedures, or in understanding the meaning of influential relationship
> qualities that are at least partially dependent on the therapist's manner, or perhaps
> both? We do not feel that laboratory research which ignores therapist and patient
> factors should hold primary for future research in psychotherapy outcome. It has
> its place. But a more promising strategy for future research will be to study the
> interaction between the therapist–patient relationship and the technical proce-
> dures or techniques of therapy. (p. 225)

Until recently, the quality of the patient–therapist relationship was viewed mainly as a function of therapist interpersonal skills, such as accurate empathy, positive regard (nonpossessive warmth), and congruence or genuiness as proposed by Rogers (1959) and elaborated by others (Truax, Wargo, Frank, Imber, Battle, Hoehn-Saric, Nash, & Stone, 1966; Truax & Carkhuff, 1967). More recent reviewers, however, have suggested that although the patient–therapist relationship is critical, the research support regarding the importance of the client-centered therapist's attributes in psychotherapy is not as clearly positive as was once thought (cf. Gurman, 1977; Lambert, DeJulio, & Stein, 1978; Mitchell, Bozarth, & Krauft, 1977; Parloff, Waskow, & Wolfe, 1978).

It is becoming increasingly clear that the attributes of the patient, not the therapist, play the more important part in the quality of the therapeutic relationship and in the outcome of psychotherapy. Strupp (1980a,b,c,d) reported a series of case studies where two patients were seen by the same therapist in time-limited psychotherapy. In each case, one of the patients was seen as having a successful outcome while the other was considered to be a treatment failure. These case reports were part of a larger study that used extensive outcome measures and an analysis of patient–therapist interactions during the process of therapy. In each instance, the therapist was working with college males who were suffering from anxiety, depression, and social withdrawal. Although each therapist was seen as having good interpersonal skills, a different relationship developed with the two patients. In all four cases, the patients who had successful outcomes appeared more willing and able to have a meaningful relationship with the therapist, whereas the patients who did not do well in therapy did not relate well to the therapist and had a tendency to keep the interaction on a more superficial level.

In Strupp's analysis, the contributions of the therapists remained relatively constant throughout therapy and the difference in outcome could be attributed to patient factors such as the nature of the patient's personality makeup, including ego, organization, maturity, motivation, and ability to become productively involved in therapy. He adds:

> While these findings are congruent with clinical lore, they run counter to the view
> that "therapist provided conditions" are the necessary and sufficient conditions for

therapeutic change. Instead, psychotherapy of the variety under discussion can be
beneficial provided the patient is ready and able to avail himself of its essential
ingredients. If these preconditions are not met, the experience is bound to be dis-
appointing to the patient as well as the therapist. The fault for such outcomes may
lie not with psychotherapy as such, but rather with human failure to use it appropriately. (p. 602)

Strupp concludes further that:

given the "average expectable" atmosphere created by a person functioning in the
therapeutic role, that is, a person who is basically empathic and benign, the key
determinants of a particular therapeutic outcome are traceable to characteristics
of the patient that have been described: if the patient is a person who by virtue of
his past life experience is capable of human relatedness and therefore is amenable
to learning mediated within that context, the outcome, even though the individual
may have suffered traumas, reverses, and other viscissitudes, is likely to be positive. . . . If, on the other hand, his early life experiences have been so destructive
that human relatedness has failed to acquire a markedly positive valence, and elaborate neurotic and characterological malfunctions have created massive barriers
to intimacy (and therefore to "therapeutic learning") chances are that psychotherapy either results in failure or at best in very modest gains. (p. 716)

In addition to the preceding findings, a poor relationship between therapist
and patient has been shown to be associated with patient deterioration (Lambert, Bergin, & Collins, 1977). In connection with this, the finding that the
quality of the therapeutic relationship may depend most on the attributes of
the patient, particularly how the patient experiences the relationship offered by
the therapist, is quite sobering. Additional recent research evidence tends to
support the idea that the patient, not the therapist, is most responsible for the
quality of the therapeutic relationship. Marziali, Marmar, and Krupnick
(1981), for example, studied the relative contribution of the patient and therapist to the development of a therapeutic alliance and to psychotherapy outcome. These investigators found that the therapist's contribution to the alliance
(as measured by ratings on such items as: "The therapist is hopeful and encouraging, conveying the belief that the patient has made, is making, or can make
progress") did not discriminate between successful and unsuccessful patients.

Conversely, the ratings of patient's participation (based on such items as:
"The patient indicates that he or she experienced the therapist as understanding or accepting" or "The patient acts in a hostile, attacking and critical manner toward the therapist") in therapy discriminated very well between the
patients who profited from therapy and those who did not. The raters did not
judge the therapists of poor-outcome patients to be unempathic, critical, or
confused about focus. Instead, it appeared that poor-outcome patients brought
a negative disposition with them that persisted in therapy despite what seemed
to be appropriate therapist behavior.

This finding is in agreement with the results of the Vanderbilt psychotherapy project (Gomes-Schwartz, 1978) previously mentioned. In this study, 35 college males suffering from depression, anxiety, and social introversion were assigned to either a professional or nonprofessional therapist. A total of 35 patients were seen: 10 by analytic therapists, 10 by experiential therapists, and 15 by alternate therapists (college professors). Among other findings, results indicated that therapy outcome was most consistently predicted by the patient's willingness and ability to become actively involved in the therapy interaction. In addition, it was concluded that "the variables that best predicted change were not related to therapeutic techniques but to the positiveness of the patient's attitude toward his therapist and his commitment to work at changing" (p. 1032).

Along these same lines, research evidence indicates that patients who become involved in the therapy process from the onset of therapy—who acknowledge their responsibility for the quality of the patient–therapist interaction by taking responsibility for behavior change as well as in-therapy exploration of feelings and problems—are more likely to have positive outcomes (Kirtner & Cartwright, 1958; Rice & Wagstaff, 1967; Saltzman, Luetgert, Roth, Creaser, & Howard, 1976). On the other hand, patients who withdraw or distance themselves from the therapy interaction through defensive tactics such as intellectualization or who do not take responsibility for their problems but attribute them to external influences are more likely to receive no benefit from therapy (Heine & Trosman, 1960; Kirtner & Cartwright, 1958; Rice & Wagstaff, 1967).

Dynamic therapists (Malan, 1976a,b; Sifneos, 1979; Strupp, 1973) have also offered evidence in regards to the importance of the patient–therapist interaction, suggesting that patients must have the capacity and motivation to form an intense interpersonal relationship with the therapist. More specifically, Malan (1976a,b) has summarized a series of pioneering studies conducted at the Tavistok Clinic in London. In these naturalistic studies, series of 21 and 39 patients were treated with brief (10 to 40 sessions) psychoanalytically oriented psychotherapy. Among other purposes, the studies sought to discover whether brief treatment that applied the same types of interpretations as full-scale analysis could be as effective as traditional analysis with patients who varied considerably in degree of disturbance. Follow-up data were collected 5 to 6 years after treatment termination to determine whether changes induced by brief therapy would be enduring.

Although the results of these studies are undoubtedly limited by methodological problems (such as the use of session notes rather than recordings or transcripts of the interviews, and use of the ratings contaminated by raters' knowledge of treatment outcome), brief analytic treatment was shown to be effective with some patients. The characteristics of those who were helped most by this treatment could not be clearly specified, but the factors most highly related to positive outcome in both series of studies were (1) interpretive focusing on the transference/parent link, which involved interpretations linking early family experiences with the patient–therapist relationship, and (2) patient's motivation for change. In more general terms, positive outcome correlated with the process factor called "successful dynamic interaction," defined as a patient who wanted insight and accepted the interpretations of the therapist, especially as they related to the importance of early life relationships in shaping current behavior patterns. This type of "cooperative" patient could be identified early in the course of treatment.

In their study of transference interpretations in psychoanalysis and psychotherapy, Luborsky, Bachrach, Graff, Pulver, and Christoph (1979) discuss relationship factors in terms of the patient's experiencing a helping relationship. In their view, the experience of a helping relationship may be a function of the object-relations repertoire that the patient brings to treatment, the therapist's ability to impart that experience to the patient, or some combination of the two. In considering the impact of the therapeutic relationship on outcome, these authors hypothesized that the patient's ability to experience a positive along with a negative component of helping relationships may be one of the bases for the ability to derive benefit from interpretations. Following this line of reasoning, then, the therapeutic relationship as defined both by therapist and patient involvement has a major impact on the acceptance of in-therapy interpretations and on the outcome of therapy itself.

In our consideration of the research discussed above, two distinct conclusions can be drawn: (1) the therapeutic relationship between the patient and therapist plays a critical role in the process and outcome of psychotherapy and (2) the quality of the therapeutic relationship is influenced heavily by the contributions of the patient. Given these conclusions, it seems appropriate to turn our attention to those factors that seem to have the most bearing on the manner of the patient's participation in the therapeutic relationship.

Severity of Maladjustment

The patient's ability to benefit from the psychotherapeutic interaction has been shown to be related at least in part to the level and severity of his or her disturbance. Past research in this area generally shows that those patients who are the most integrated and least disturbed have a better chance at improving; stated another way, the more disintegrated and severely disturbed a patient is, the less likely are the chances of a positive therapeutic outcome.

Typical of studies in this area are the findings of Barron (1953a), who found that patients who were better integrated initially were better able to use psychotherapy. He concluded that patients who are not very sick in the first place are most likely to get well. Similar results were obtained by Katz *et al.* (1958) in a study involving 13 Veterans Administration clinics. They found that patients who were rated as less severely ill had a greater tendency to improve than those who were considered to be more severely disturbed.

Also, McNair *et al.* (1964) reported that patients who were more disturbed before therapy were also more disturbed after therapy. It was the contention of these authors that the initial disturbance scores constituted the best single predictor of later status. In a study involving hospitalized schizophrenic patients who were treated with client-centered therapy, Rogers, Gendlin, Kiesler, and Truax (1967) found a negative relationship between degree of disturbance and psychotherapy outcome. Specifically, they reported that patients who had high mental health ratings and a generally low level of manifest psychotic disturbance were most likely to have positive outcomes.

In their classic comparative study of behavior therapy and psychoanalytically oriented psychotherapy, Sloane, Staples, Cristol, Yorkston, and Whipple (1975) reported that the least disturbed patients, as measured by the MMPI, improved more with analytically oriented psychotherapy, whereas behavior therapy was reported as being equally successful regardless of whether the patient had a high or low degree of pathology. Results of this study also suggest that in some cases the type of therapy involved may be an important variable as well as the degree of maladjustment of the patient.

Two investigations by Luborsky and his associates also support the viewpoint that the least disturbed patients tend to have better outcomes. In their comprehensive review of this area, Luborsky *et al.* (1971) listed 55 separate studies under the heading of "Degree of Initial Disturbance versus Adequacy of Functioning." Of these 55 studies, 31 were reported as showing a positive relationship between adequacy of adjustment and outcome, one study was listed as showing a negative relationship, and 23 were categorized as inconclusive or with nonsignificant findings. From this evidence the authors concluded that "Initially sicker patients do not improve as much with psychotherapy as the initially healthier do" (p. 149). In a more recent report, Luborsky *et al.* (1979) discussed the responses of three patients to transference interpretations. For each of these patients there was a parallel between positive response to interpretations and overall outcome of therapy, which the authors believed was related to antecedent patient characteristics. For example, Health–Sickness Rating Scale scores showed that the patient who was most positive in response to interpretations was the least disturbed patient. It was also shown that patient readiness to experience a helping relationship was related to response to interpretations and outcome.

It should be pointed out that one of the problems involved in studying severity of patient maladjustment has to do with the large number of related but distinct variables that have been grouped under this heading. Some of these include diagnosis, premorbid state, type of onset, duration of illness, adjustment to current life circumstances, degree of social impairment, intensity of discomfort, symptom severity, initial level of disturbance, "most" severely disturbed patients, and so on. Severity of maladjustment can be identified through symptom rating scales, clinical impression, diagnosis, intensity of discomfort or symptoms, test results, and the like. Difficulties arise when researchers use different and, at times, incompatible definitions of severity of maladjustment in their investigations.

A variable that shows promise as a good prognostic indicator in this area is "symptom complexity." This variable can be defined in terms of ratings of premorbid adjustment, symptom duration, symptom severity, symptom pervasiveness, and so on. At present, this predictor seems to have prognostic value with a variety of patient groups, including hospitalized adults and adolescents, children, and adult outpatients. Whitehead (1976) utilized this approach in studying elderly psychiatric inpatients. He investigated improvement as related to patient sex, age, marital status, number and length of previous admissions, age of first admission, time since discharge, clinical state at admission, diagnosis, and psychological test scores. Although some success in predicting negative outcomes was noted, most of the prediction could be accounted for by degree of intellectual impairment. This study supports the notion that the more severely disturbed a patient is (e.g., chronic brain syndrome), the less favorable the outcome, and the less disturbed a patient is (e.g., affective disorder), the better the outcome. In general, however, the symptom complexity variable has not been utilized in current research; more definitive work is clearly needed in order to identify variables that interact with and modify the predictive validity of this important patient dimension.

Although most studies show that better outcomes are obtained in patients with the least amount of disturbance, a number of studies also offer evidence which shows no relationship between degree of maladjustment and improvement in psychotherapy. For example, Katz *et al.* (1958) found no relationship between degree of disturbance and outcome in a study involving 232 patients in Veterans Administration clinics. In this investigation, degree of disturbance was based on four types of diagnosis made by therapists: (1) psychoneurosis, (2) psychosomatic disorder, (3) character disorder, and (4) psychosis. Ratings of improvement as well as diagnosis were made by therapists. Klein (1960) found no evidence of differential change in 30 neurotic patients who were initially classified as having moderate or severe dysfunction. Cappon (1964) also reported no relationship between intensity of symptoms and outcome. Similar results are offered in studies by Cartwright and Roth (1957), Frank *et al.* (1957), and Seeman (1954).

Two studies using behavioral approaches have also reported no relationship between outcome and severity of maladjustment. Mathews, Johnston, Shaw, and Gelder (1974) attempted to evaluate several predictor variables in relation to outcome with flooding, desensitization, and a control group. Results revealed that severity of symptoms as well as low expectancy, low motivation, high anxiety, and neuroticism were not related to outcome. In a study utilizing modeling with snake phobics, Bandura, Jeffrey, and Wright (1974) reported that neither patients' initial attitudes toward snakes nor severity of phobic behavior correlated with degree of behavior change.

There is also a significant body of research supporting the view that the more severely disturbed patients show the greatest amount of change in psychotherapy. Nichols and Beck (1960) found that the more maladjusted the patients initial score on the CPI, the greater the change resulting from treatment. It should be pointed out, however, that therapists tended to rate as more improved those patients with well-adjusted initial scores, while the patients did not. Ewing (1964) noted that the largest change in a group of college counselees was shown to be in those who initially showed the greatest dissatisfaction in themselves. In a 5-year follow-up study that utilized a self-report scale tapping five areas of symptomatology as a measure of initial disturbance, Stone, Frank, Nash, and Imber (1961) found a significant relationship between greater initial disturbance and more positive outcomes. Similar results were obtained for group psychotherapy clients using four scales of the MMPI (Truax, Tunnell, Fine, & Wargo, 1966).

In their review of some of the literature in this area, Truax and Carkhuff (1967) attempted to explain the relationship between severity of maladjustment and therapy outcomes by hypothesizing that the patients with the greater felt disturbance (as indicated on self-report measures) but the least overt or behavioral disturbance are the ones who show the greatest therapeutic improvement. Unfortunately, this interesting suggestion has not been clearly documented. Other investigations have reported that patients who show the greatest amount of anxiety in their current situation at the initiation of therapy appear to have more favorable outcomes (Frank, 1974; Kernberg, Burstein, Coyne, Applebaum, Horwitz, & Voth, 1972; Luborsky *et al.,* 1971; Smith, Sjoholm, & Nielzen, 1975). The type and severity of anxiety, however, as well as the stimuli that influence it, must be considered in this context.

It seems clear that the bulk of the research evidence in this area supports the view that better therapeutic outcomes are obtained with patients who show the least amount of initial disturbance or maladjustment. There are, however, some important methodological considerations associated with the research findings supporting this conclusion, as well as with the findings of other research reports that do not support this viewpoint. The major problems in research dealing with severity of maladjustment and psychotherapy outcome seem most frequently to involve disagreement on definitions of severity (as mentioned previously), differences in outcomes measures or appraisals, variations in how test data are evaluated and interpreted, different types of therapy, and sample differences. Prager and Garfield (1972) have made two important observations concerning measurement problems that both clinicians and researchers should remember when evaluating patients on degree of disturbance. The first is that those studies finding that the more highly disturbed patients obtained more positive results than the less disturbed patients used the same instrument to measure both initial disturbance and outcome. Thus, the problem of regression toward the mean could be a significant factor in the findings. Second, the initial level of disturbance has been found to be inversely related to outcome primarily when global ratings of improvement have been used as the criteria of outcome at the termination of therapy. The significance here is that global ratings involve the rater's perception of change, as opposed to actual difference scores which are based on information regarding the client's symptoms and feelings at a particular time.

In connection with this, Mintz (1972) has pointed out that global ratings made at the end of therapy tend to be heavily influenced by the condition of the client at the time therapy is terminated, regardless of how the client functioned at the beginning of treatment. In this circumstance, the patient who begins therapy at a high level of functioning but makes only slight improvements is still rated as much improved in comparison to the individual who enters treatment at a low level of functioning and, while showing more improvement, does not function at the high level of the other patient at the end of treatment. These findings suggest that patients who begin therapy at high levels of functioning terminate at higher levels than those who enter treatment at low levels, although the former do not necessarily show the largest gains as a result of treatment.

In his review of the literature in this area, Garfield (1978) draws conclusions that agree with the above findings and which we see as instructive in a critical appraisal of psychotherapy outcome as it is associated with degree of patient maladjustment. In his view, actual amount of change in therapy is a poor outcome criterion because of methodological problems such as those involving individuals who are near the top of the adjustment scale and, thus, have a much smaller range of improvement open to them than do individuals who start at the lower end of the scale. Also, he suggests that standardized measuring instruments with equal intervals are needed to compare amounts of change at different parts of the outcome scale.

He further concludes:

> if the goal of therapy is some norm of personality functioning or adjustment, then
> it does seem tenable to say that those who begin therapy at higher levels generally

finish at higher levels than those who begin at lower levels. However, one cannot validly state that those who profit the most or show the most improvement are the ones who are the best adjusted at the beginning of therapy. They may be the most satisfied with therapy and their therapists may also find working with these clients most gratifying, but this is a different matter than amount of change. (p. 220)

It is recommended that future researchers be aware of the inherent methodological problems associated with the study of patient maladjustment and outcome of psychotherapy discussed above. It is particularly important to note that in drawing conclusions such as: "the least disturbed patients do better in therapy," it is essential that the outcome criteria be defined, especially as they relate to the measurement of actual amount of change in therapy or to some particular level of personality functioning attained at the end of treatment.

Also, at the present time the influence of the severity of patient maladjustment on the therapeutic relationship is somewhat unclear. Although better adjusted patients may bring more personal resources into the therapeutic situation, they do not necessarily show more therapeutic gains. Therefore, whether or not they actually form a better relationship compared to patients with more initial disturbance is a question that cannot be readily answered but will require careful, systematic research by future investigators.

Personality Characteristics

Another important set of variables that influence the manner of patient participation in the psychotherapeutic relationship has to do with the personality characteristics the patient brings into the therapeutic situation. These variables influence not only the way the patient perceives the quality of the therapeutic relationship but also how the patient participates in and utilizes this relationship to progress in therapy.

A number of patient personality characteristics have been considered in relation to therapy outcome, including hostility, suggestibility, ego strength, locus of control, introversion–extraversion, dogmatism, authoritarianism, and psychological mindedness. We will discuss those that seem to be the most promising as predictors of therapeutic outcome.

One such variable that has been a useful and consistent predictor of outcome is ego strength. The two most common measures of this variable are the Barron Ego Strength Scale and the Klopfer Rorschach Prognostic Rating Scale. These two scales show a high degree of intercorrelation and both assume that the better integrated the personality, the better the chances of improvement in psychotherapy.

The Barron Ego Strength Scale consists of 68 items from the MMPI that were found to correlate significantly with improvement in 33 neurotic patients treated in an outpatient clinic. Although Barron (1953b) provided evidence supporting the validity of this scale in predicting outcomes, subsequent research has failed to confirm these findings (cf. Endicott & Endicott, 1964; Fiske, Cartwright, & Kirtner, 1964; Gallagher, 1954; Getter & Sundland, 1962; Gottschalk, Fox, & Bates, 1973; Newmark, Finkelstein, & Frerking, 1974).

Another index of ego strength that has received more positive research support is the Klopfer Rorschach Prognostic Rating Scale (RPRS; Klopfer, 1951). This scale was derived from Rorschach determinants and was originally offered as a measure of potential ego strength that could be mobilized or made available through psychotherapy. A number of studies have shown a positive relationship between the RPRS and psychotherapy outcome, including an investigation by Mindess (1953) where outcomes of 80 schizophrenic, schizoid, neurotic, exhibitionist, and homosexual patients were significantly related to scores on the RPRS. Endicott and Endicott (1964) found positive correlations between the RPRS and outcome in 40 untreated and 21 treated patients where therapists ratings made up the outcome criteria. Positive results were also obtained in studies by Bloom (1956), Cartwright (1958), Kirkner, Wisham, and Giedt (1953), Newmark *et al.* (1974), Seidel (1960), and Sheehan, Frederick, Rosevear, and Spiegelman (1954). On the other hand, negative findings were obtained in investigations by Filmer-Bennett (1955), Fiske, Cartwright, and Kirtner (1964), and Schulman (1963).

Although the majority of the research reports on the RPRS support the view that it is a useful predictor of therapeutic outcome, Frank (1967) has asserted that the RPRS correctly predicts outcome in about two-thirds of the cases. This is not a particularly impressive figure. It suggests that although patients with good ego strength, as measured by the RPRS, have better outcomes than those with lower ego strength, these predictions may not exceed the base rate for therapeutic improvement. Again, an important consideration here, as discussed previously, is that patients who begin therapy at higher levels of functioning also end at higher levels. This does not mean, however, that those patients who terminate at higher levels of functioning actually show the largest therapeutic gains.

One of the more influential studies involving ego strength was the 18-year investigation carried out by the Menninger Foundation (Kernberg, Burstein, Coyne, Appelbaum, Horwitz, & Voth, 1972). In this study of 21 patients who received psychoanalysis and an additional 21 patients who received psychoanalytically oriented psychotherapy, initial ego strength was assessed by clinical judgments as opposed to specific rating scales. Ratings of ego strength, as defined in this investigation, were based on (1) integration of ego structures; (2) degree of deep, satisfying relations with others; and (3) symptom severity. Results of this study indicated that clinical appraisals of ego strength correlated positively with a measure of global improvement. Also, it was found that patients with some degree of ego strength had the ability to maintain interpersonal relationships and were able to cope with the stress involved in dynamic psychotherapy. Prognostically poor signs for improvement included low tolerance for both frustration and anxiety, especially when the patient was in more unstructured therapy. These findings have important significance in regard to the patient's ability to participate meaningfully in the therapeutic relationship. It appears that some degree of ego strength is required for the patient to be able to maintain and adequately utilize the intense interpersonal relationship with the therapist that is a necessary part of therapeutic improvement in many types of psychotherapy.

Another dimension of the patient's personality that has received considerable attention in recent research literature is locus of control. This variable taps the degree to which patients generally attribute consequences in their lives to events external to themselves or to their own efforts. Locus of control has been measured by the I-E Scale devised by Rotter (1966) and has generally been analyzed in interaction with therapist directiveness. The hypothesis that has guided most of the research in this area is that clients who see themselves as being more in control of and responsible for their fate are likely to prefer—and to show more improvement in—treatments that are minimally directive. On the other hand, patients who see themselves as the victims of fate, as being determined by influences outside themselves, will prefer and gain the most from directive therapies or those with the greatest structure.

The studies involving the I-E Scale have included as criterion both preference for therapist or therapy as well as actual therapy outcomes. Typical of research in this area was a study by Friedman and Dies (1974), who examined the differential responses of 36 test-anxious students who differed on the degree to which they had an internal versus external orientation to counseling and desensitization therapies. It was hypothesized that more "internal" individuals would respond better to counseling in which they felt more in control of the course of therapy, whereas external individuals would react more favorably to treatment that was directed and controlled more by the therapist. In accordance with this hypothesis, it was found that "external" subjects felt that they retained too much control of therapy; while "internals" generally indicated an optional amount of control in counseling. Also, it was reported that regardless of treatment conditions, internally oriented subjects would have preferred "more client control" in therapy than externally oriented subjects. Unfortunately, no outcome data were reported.

In a study involving distressed college students treated in group psychotherapy, Abramowitz, Abramowitz, Roback, and Jackson (1974) found that the more internally oriented clients were more responsive to nondirective group therapy, while externally oriented clients preferred the more directive group approach. Schwartz and Higgins (1979) studied the effects of a highly structured assertiveness training treatment on the behavior of a low-assertive college student population. Although patients who had an internal locus of control profited more from treatment than the control group patients, they improved less than the externally oriented patients. Internals also felt that treatment took away too much of their control of the environment and reported feeling more uncomfortable in treatment than did external patients. Other investigations that support the findings of these studies have been reported by Best (1975), Best and Steffy (1971), and Rosenbaum and Argon (1979).

A number of studies investigating locus of control as a predictor in psychotherapy have also reported nonsignificant or equivocal results. For example, Snowden (1978) tailored two covert sensitization procedures to 42 patients undergoing methadone treatment for heroin abuse. Patients with external locus-of-control scores received environmentally focused images, while those with more internal scores had sensitization focused on within-organism images. Although internals improved more from the internal rather than the external imagery procedures, externals did not profit significantly more from either treatment procedure. In their review of research on locus of control and alcoholism, Rohsenow and O'Leary (1978) report that most studies find alcoholics to be at the external end of the locus-of-control continuum, and an increase in internal locus of control is generally thought to be a desirable outcome of treatment. Of the actual studies reviewed, however, only two showed the expected change in locus of control, while three found no significant difference. It is clear from this review that the use of pretreatment scores to predict success and to assign alcoholic patients to preferred treatments has not yet been adequately investigated. A final study showing null results (Wallston, Wallston, Kaplan, & Maides, 1976) examined 34 overweight women who received either a self-directed program or a structured group treatment program. No differences in weight loss were found for participants based on locus-of-control ratings.

Judging from the research conducted thus far, it appears that the locus-of-control variable is modestly predictive of psychotherapy preferences when interaction effects are taken into account. However, the specific relationship between locus of control, therapy directiveness, and therapy outcome is unclear. At present, there is little evidence showing differential outcomes as a result of specific patient and treatment (directive vs. nondirective) matches. One factor here is that too many past investigations have utilized treatments that are relatively structured rather than being at clearly defined polar ends of the directiveness continuum. More research where treatment and outcome variables are adequately and clearly defined must be forthcoming before any conclusive statements can be made about the significance of locus of control for psychotherapy outcome.

A few other patient variables which have limited relevance in the course and outcome of psychotherapy can be mentioned briefly. The presence of anxiety in the patient has been reported in some studies to be positively related to therapeutic outcomes (Adams, 1961; Derr & Silver, 1962; Kernberg *et al.*, 1972; Luborsky *et al.*, 1971; Smith, Sjoholm, & Nielsen, 1975), although a number of studies have not supported these findings (Gottschalk *et al.*, 1967; Greenfield, 1958; Katz *et al.*, 1958; Morgenstern, Pearce, & Rees, 1965). It appears that some anxiety may be necessary to motivate a patient to seek treatment, but the relationship between level of anxiety and outcome is not clear. It should be noted that the type and severity of anxiety as well as the factors that influence it are important considerations when studying this variable.

Patient suggestibility has been shown to be related to length of stay in therapy in a study using the sway test (Imber *et al.*, 1956) as well as in a subsequent investigation by Frank *et al.*, (1957). However, this variable has not been implicated as a significant predictor of therapeutic outcomes (Frank *et al.*, 1957; Morgenstern *et al.*, 1965).

Although a number of patient personality characteristics have been suggested as being worthy of research investigation, it appears at this time that the most promising lines of research involve the study of internal–external locus of control and patient ego strength. Systematic research that considers both the defining criteria and measurement of those variables, as well as their specific predictive utility, would be very useful to clinicians and researchers alike.

PATIENT–THERAPIST MATCHING – SIMILARITY AND RELATED ISSUES

Although the issue of patient–therapist matching has been specifically discussed in those previous sections of this chapter that focused on the demographic variables of race and sex as well as on patient expectations, it seems appropriate at this point to consider some other types of research relating to this topic.

The area of patient–therapist matching has received a disproportionately small amount of attention in past psychotherapy research. Part of the problem appears to be confusion about which variables are important in this type of matching as well as resistances voiced by those who feel that "mechanistic" matching of patient and therapist would rob psychotherapy of its inherent uniqueness, spontaneity, and excitement.

Despite these problems, there has been some interesting and worthwhile research on this topic that should be considered. Carson and Heine (1962) found a curvilinear relationship between therapist–client similarity; they rated improvement by using the MMPI to compare therapists and clients. Lichtenstein (1966) and Carson and Llewellyn (1966), however, failed to replicate these findings. McLachlan (1972) found significant improvements, as reported in patients' ratings, when patients and therapists were matched for conceptual level, although staff ratings did not support this relationship.

Lanfield (1971) investigated the role of therapist–patient similarity in personal constructs as it influenced continuance and improvement in psychotherapy. Utilizing Kelly's personal construct theory and his Role Construct Repertory Test (REP) to assess construct similarity, Lanfield tested three types of hypotheses using eight therapists and college clinic patients of both sexes. The hypotheses involved the relationship between construct similarity of therapist and patient and premature termination, patient improvement, and therapist's ascribed pathology.

Results of this study indicated that therapy was enhanced if (1) participants perceived each other as meaningful figures, (2) participants saw each other's construct dimensions as meaningful, and (3) participants' constructs had similar content. Therapy outcomes were shown to be better when (1) participants organized their personal constructs differently, (2) participants showed converging construct systems over time, and (3) patients shifted their "present selves" toward therapists' "ideal selves." One of the positive points of this study was that the author made a distinction between content and structure of personal constructs, as well as emphasizing the importance of organizational differences in patient and therapist constructs. Further, it is his contention that differences in the organizational structures of client and therapist constructs are important in facilitating and maintaining interpersonal communication channels.

One of the more innovative and meaningful investigations of patient–therapist matching was carried out by Berzins (1977) in the Indiana Matching Project. This study, which was carried out over a 4-year time span, attempted to develop, implement, and evaluate a strategy for patient–therapist matching using short-term, crisis-oriented psychotherapy in a college clinic. Pretherapy assessment of 751 patients (291 male, 360 female) was based on measures of how patients fit four different role orientations: (1) avoidance of other's role (complaints of interpersonal anxiety and avoidance), (2) turning against the self role (depressive and somatic complaints), (3) dependency on other's role (approval and advice-seeking expectancies), and (4) the turning toward others and self role (audience and relationship-seeking expectancies). These roles were developed through rational and factor-analytic procedures; predictor scores were assessed using 50 Likert-type items. Predictor scores for therapists were based on Jackson's (1967) Personality Research Form and represented six personality dimensions (impulse expression, ambition, acceptance, dominance, caution, and abasement). Thus, therapists' predictor scores focused on their relatively stable personality characteristics, while patient predictor scores reflected more salient situational factors.

Results indicated that favorable patient–therapist pairings, resulting in patient improvement in three of the role orientations (avoidance of others, turning against the self, and dependency on others). In both patient sexes, such pairings were associated with dyadic complementarity along a superordinate dominance–submission dimension. Therefore, as Berzins states:

> Favorable pairings generally conjoined submissive, inhibited, passive patients with dominant, expressive, active, cue- and structure-emitting therapists, and sometimes vice versa, suggesting that improvement in brief psychotherapy was facilitated by pairing patients with therapists whose personalities embodied ingredients that complemented the patients' needs, deficits, expectations, or interpersonal stances. (p. 243)

Somewhat different results were obtained where the fourth role orientation (turning toward others and the self) was involved. Findings here showed that complementarity applied to dyads containing male patients, whereas "similarity" (e.g., verbal female patients improved more with spontaneous and forceful therapists) applied to dyads with female patients.

This study illustrates the importance of implementing matching procedures that are based on empirical findings where matching variables are appropriate to a particular clinical setting. Although cross-validation of this research in other clinic settings may be difficult to accomplish, the guidelines discussed in this study should be quite useful for future investigators.

One line of research that seems promising for future patient–therapist matching studies and has also been recommended by Berzins (1977) in his comprehensive review of research in this area has to do with the interpersonal circle proposed by Leary (1957). In an attempt to organize and extend some of Sullivan's ideas on interpersonal behavior in a diagnostic framework, Leary and his associates developed a comprehensive system for classifying persons in terms of their predominant interpersonal behaviors or "reflexes." In this system, interpersonal behaviors were categorized with respect to two orthogonal, bipolar dimensions, dominance–submission (a vertical dimension) and love–hate (a horizontal dimension).

These two dimensions were placed at right angles, and eight major subcategories were then organized into a circumplex (circular) arrangement. The eight subcategories used in the interpersonal circle were managerial-autocratic, responsible-hypernormal, cooperative-overconventional, docile-dependent, self-effacing-masochistic, rebellious-distrustful, aggressive-sadistic, and competitive-narcissistic. Two of the eight subcategories are contained in each of four quadrants. In the quadrant formed by the modal points dominance (12 o'clock) and hate (9 o'clock), for example, were placed aggressive-sadistic and competitive-narcissistic behaviors (the term after the hyphen is the extreme form of the behavior). Also, each octant was subdivided into two separate "reflexes." The aggressive–sadistic octant, for example, was divided up into the reflex "aggressive, firm actions" and the reflex "frank, forthright, critical actions." Each reflex, as Sullivan's theory of reciprocal emotion would predict, was presumed to pull its complement from the other person. For example, an aggressive reflex on the part of one person should provoke hostility in another.

As far as matching is concerned, both patient and therapist could be classified into the appropriate quadrants, octants, or reflexes through various measures. It was assumed that in dyadic interactions, individuals would reciprocate each other's affective orientation directly (i.e., love elicits love and hate elicits hate) and complement each other's dominance-submission behaviors (i.e., dominance elicits submission and vice versa).

Although Leary did not pursue patient–therapist matching research in a systematic manner, the interpersonal circle formulation has generated some meaningful investigations. For example, Heller, Myers, and Kline (1963) tested the hypothesis that dominance evokes submission (and vice versa) but that friendliness evokes friendliness and hostility evokes hostility. Four patients, who were actually actors, enacted the patient roles of the four main quadrants of the interpersonal circle: friendly-dominant, hostile-dominant, friendly-dependent, and hostile-dependent. The patients were seen by 34 graduate-student interviewers and interviewer behaviors were rated on the Leary Interpersonal Checklist by trained observers. Results indicated that both hypotheses were supported: patient friendliness evoked more interviewer friendliness than did patient hostility, and patient dependency evoked more interviewer dominance than did patient dominance.

In another study, Swensen (1967) investigated further the dominance-submission and love-hate dimensions of Leary's Interpersonal Circle. In this study, MMPI profiles from clinical psychology graduate students and the 74 patients they were treating were scored using a system suggested by Leary (1957) to determine dominance and love quadrant placements. Ratings of improvement were supplied by two raters who evaluated therapists' pre–post case summaries. It was found that improvement was greater when dyad members were complementary on the dominance dimension (therapist dominant, patient dominant, or vice versa) and similar on the love dimension.

The major advantages of Leary's Interpersonal Circle in patient–therapist matching research is that this approach involves a comprehensive sampling of interpersonal behaviors (as opposed to focusing on a particular attitude or personality variable) as well as theoretical propositions guiding research hypotheses.

One other area that we see as promising for future research on patient–therapist matching or similarity has to do with the idea of patient–therapist convergence in psychotherapy. This concept was originally formulated by

Pepinsky and Kurst (1964), who suggested that success in psychotherapy could be associated with convergence of patient's and therapist's belief and value systems. These authors also proposed that convergence might be predicted on the basis of initial patient and therapist similarity independent of the therapist's particular theoretical orientation.

The present status of convergence in counseling and psychotherapy was recently discussed in a comprehensive review by Beutler (1982). After reviewing the pertinent research studies in this area, he drew two major conclusions. First, convergence between a patient and his or her therapist (either in terms of personality or value constructs) is related to subsequent improvement in psychotherapy. This relationship seems not only characteristic of individual psychotherapy but has also been found in both group and family therapy. Second, it was concluded that initial patient–therapist dissimilarity predicts this convergence. In connection with this conclusion, the author notes that neither personality nor belief system similarity is more strongly related to convergence or improvement.

In discussing the complexity of the variables involved in the above conclusions, Beutler states:

> Nonetheless, one must conclude that the variables which contribute to the correspondence between outcome ratings and convergence in psychotherapy are not the same ones that contribute to the correspondence between patient and therapist initial discrepancy and convergence. In other words, while convergence between a patient and therapist may be associated both with initial dissimilarity and with improvement, these latter two variables are related to one another in a complex manner. (p. 89)

Beutler's recommendations for future research in the area of convergence involve (1) assessment of whether or not convergence interacts with initial dissimilarity to produce improvement, (2) the question of whether convergence is a *result* of improvement rather than a *cause* of improvement, and (3) the question of the extent to which a therapist's private and public beliefs are adopted as a model of "good adjustment" by patients. The study of patient–therapist convergence in psychotherapy seems particularly relevant in light of current issues relating to the impact of the therapist's belief system on the patient (e.g., the concept of interpersonal persuasion) as well as the question of what values, attitudes, or beliefs are most central to change in psychotherapy.

The area of patient–therapist matching in psychotherapy will necessarily require a substantial amount of research before any conclusive statements can be made about specific matching strategies that will produce significantly better therapy relationships and outcomes. In our view, matching research based on Leary's (1957) Interpersonal Circle as well as future research on patient–therapist convergence (Beutler, 1982) appears to be most promising, particularly in terms of scientific approachability as well as potential clinical utility.

SUMMARY

In summary, the research reviewed in this chapter suggests that patient variables are clearly important, if not the most important, factors affecting psychotherapy outcome. These variables have been shown to have an impact not only on the ultimate outcome of psychotherapy but also on the process of the therapeutic interaction. In the studies that were examined, the effects of patient variables have been measured via standard outcome assessment devices, attrition rates, preferences for and satisfaction with therapy, and so on. Conclusions about the role of patient factors have been shown to vary as a function of the dependent variable under consideration (e.g., outcome vs. preference). In general, the effects of patient variables are most obvious on immediate measures of outcome, such as premature termination of therapy. The importance of therapist characteristics is less obvious when rigorous measures of behavior change, aimed at the long-term effects of treatment, are applied.

A great deal of past research has focused on the effect of motivation, expectations, diagnosis, and demographic variables on psychotherapy outcome. A more contemporary and hopeful line of research has focused upon patient variables that directly affect the manner of participation in the therapeutic relationship. Future research will be most productive if its focus is on defining, measuring, and correlating these types of variables with the immediate and remote consequences of therapy.

We would also include among these variables process measures of patient participation based upon early reactions to the treatment relationship (e.g., acceptance of interpretations, negative attitudes toward the therapist, etc.). In addition, static traits that tell something of the patient's orientation to human relations (e.g., diagnosis, ego strength, historical data) are highly predictive of eventual outcome. Patients' past failures to use human support systems in the natural environment may also predict their failure to use the therapeutic relationship. Measuring these patient traits and taking appropriate action could enhance the value of treatment.

Finally we wish to emphasize the possibility that matching strategies may yet result in increased ability to use patient personality traits in therapy assignment decisions. Presently, this is more of a hope than a reality, although future researchers may be able to define the relevant patient personality dimensions that interact with therapist and treatment variables.

REFERENCES

Abramowitz, C. V., Abramowitz, S. I., Roback, H. B., & Jackson, C. Differential effectiveness of directive and nondirective group therapies as a function of client internal-external control. *Journal of Consulting and Clinical Psychology,* 1974, *42,* 849–853.

Acosta, F. Z., & Sheehan, J. G. Preferences toward Mexican-American and Anglo-American psychotherapists. *Journal of Consulting and Clinical Psychology,* 1976, *44,* 272–279.

Adams, S. *Interaction between individual interview therapy and treatment amenability in older youth authority wards* (Monograph No. 2). Sacramento, Calif.: Board of Corrections, State of California, 1961, pp. 27–44.

Affleck, D. C., & Garfield, S. L. Predictive judgments of therapists and duration of stay in psychotherapy. *Journal of Clinical Psychology,* 1961, *17,* 134–137.

Bailey, M. A., Warshaw, L., & Eichler, R. M. A study of factors related to length of stay in psychotherapy. *Journal of Clinical Psychology,* 1959, *15,* 442–444.

Baker, J., & Wagner, N. Social class and treatment in a child psychiatry clinic. *Archives of General Psychiatry,* 1966, *14,* 129–133.

Bandura, A., Jeffrey, R. W., & Wright, C. L. Efficacy of participant modeling as a function of response induction aids. *Journal of Abnormal Psychology,* 1974, *83,* 56–64.

Barron, F. Some test correlates of response to psychotherapy. *Journal of Consulting Psychotherapy,* 1953, *17,* 235–241. (a)

Barron, F. An ego-strength scale which predicts response to psychotherapy. *Journal of Consulting Psychology,* 1953, *17,* 327–333. (b)

Beck, J. C., Kantor, D., & Gelineau, V. A. Follow-up study of chronic psychiatric patients "treated" by college case-aide volunteers. *American Journal of Psychiatry,* 1963, *120,* 269–271.

Berzins, J. I. Therapist-patient matching. In A. S. Gurman & A. M. Razin (Eds.), *Effective psychotherapy: A handbook of research.* Elmsford, N.Y.: Pergamon, 1977.

Best, J. A. Tailoring smoking withdrawal procedures to personality and motivational differences. *Journal of Consulting and Clinical Psychology,* 1975, *43,* 1–8.

Best, J. A., & Steffy, R. A. Smoking modification procedures tailored to subject characteristics. *Behavior Therapy,* 1971, *2,* 177–191.

Beutler, L. E. Convergence in counseling and psychotherapy: A current look. *Clinical Psychology Review,* 1982, *1,* 79–101.

Bloom, B. L. Prognostic significance of the underproductive Rorschach. *Journal of Projective Techniques,* 1956, *20,* 366–371.

Borkovec, T. D., Grayson, J., B., & Cooper, K. M. Treatment of general tension: Subjective and psychological effects of progressive relaxation. *Journal of Consulting and Clinical Psychology,* 1978, *46,* 518–528.

Brady, J. P., Reznikoff, M., & Zeller, W. W. The relationship of expectation of improvement to actual improvement of hospitalized psychiatric patients. *Journal of Nervous Mental Disorders,* 1960, *130,* 41–44.

Brill, N. Q., & Storrow, H. A. Social class and psychiatric treatment. *Archives of General Psychiatry,* 1960, *3,* 340–344.

Cappon, D. Results of psychotherapy. *British Journal of Psychiatry,* 1964, *110,* 35–45.

Carson, R. C., & Heine, R. W. Similarity and success in therapeutic dyads. *Journal of Consulting Psychology,* 1962, *26,* 38–43.

Carson, R. C., & Llewellyn, C. E., Jr. Similarity in therapeutic dyads: A re-evaluation. *Journal of Consulting Psychology,* 1966, *30,* 458.

Cartwright, D. S. Success in psychotherapy as a function of certain actuarial variables. *Journal of Consulting Psychology,* 1955, *19,* 357–363.

Cartwright, D. S., & Roth, I. Success and satisfaction in psychotherapy. *Journal of Clinical Psychology,* 1957, *13,* 20–26.

Cartwright, R. D. Predicting response to client-centered therapy with the Rorschach PR scale. *Journal of Counseling Psychology,* 1958, *5,* 11–17.

Cartwright, R. D., & Lerner, B. Empathy, need to change, and improvement with psychotherapy. *Journal of Consulting Psychology*, 1963, *27*, 138–144.

Casner, D. Certain factors associated with success and failure in personal-adjustment counseling. *American Psychologist*, 1950, *5*, 348. (Abstract)

Cimbolic, P. Counselor race and experience effects on black clients. *Journal of Consulting and Clinical Psychology*, 1972, *39*, 328–332.

Clemes, S., & D'Andrea, V. J. Parents' anxiety as a function of expectation and degree of initial interview ambiguity. *Journal of Consulting Psychology*, 1965, *29*, 397–404.

Cole, N. J., Branch, C. H., & Allison, R. B. Some relationships between social class and the practice of dynamic psychotherapy. *American Journal of Psychiatry*, 1962, *118*, 1004–1012.

Costello, R. M., Baillargeon, J. G., Biever, P., & Bennett, K. Second-year alcoholism treatment outcome evaluation with a focus on Mexican-American patients. *American Journal of Drug and Alcohol Abuse*, 1979, *6*, 97–108.

Craig, T., & Huffine, C. Correlates of patient attendance in an inner-city mental health clinic. *The American Journal of Psychiatry*, 1976, *133*, 61–64.

Crowley, R. M. A low-cost psychoanalytic service: First year. *Psychiatric Quarterly*, 1950, *24*, 462–482.

Derr, J., & Silver, A. W. Predicting participation and behavior in group therapy from test protocols. *Journal of Clinical Psychology*, 1962, *18*, 322–325.

Duckro, P., Beal, D., & George, C. Research on the effects of disconfirmed client role expectations in psychotherapy: A critical review. *Psychological Bulletin*, 1979, *86*, 260–275.

Ellis, A. The effectiveness of psychotherapy with individuals who have severe homosexual problems. *Journal of Consulting Psychology*, 1956, *20*, 191–195.

Endicott, N. A., & Endicott, J. Prediction of improvement in treated and untreated patients using the Rorschach prognostic rating scale. *Journal of Consulting Psychology*, 1964, *28*, 342–348.

Ends, E. J., & Page, C. W. Functional relationships among measures of anxiety, ego strength and adjustment. *Journal of Clinical Psychology*, 1957, *13*, 148–150.

Ewing, T. N. Changes during counseling appropriate to the client's initial problem. *Journal of Counseling Psychology*, 1964, *11*, 146–150.

Ewing, T. N. Racial similarity of client and counselor and client satisfaction with counseling. *Journal of Consulting Psychology*, 1974, *21*, 446–469.

Feldman, M. P., & MacCulloch, M. The application of anticipatory avoidance learning to the treatment of homosexuality. *Behavior Research and Therapy*, 1965, *2*, 165–172.

Filmer-Bennett, G. The Rorschach as a means of predicting treatment outcome. *Journal of Consulting Psychology*, 1955, *19*, 331–334.

Fiske, D. W., Cartwright, D. S., & Kirtner, W. L. Are psychotherapeutic changes predictable? *Journal of Abnormal and Social Psychology*, 1964, *69*, 418–426.

Frank, G. H. A review of research with measures of ego strength derived from the MMPI and the Rorschach. *Journal of General Psychology*, 1967, *77*, 183–206.

Frank, J. D. The dynamics of the psychotherapeutic relationship. *Psychiatry*, 1959, *22*, 17–39.

Frank, J. D. Relief of distress and attitudinal change. *Science and Psychoanalysis*, 1961, *4*, 107–124.

Frank, J. D. Therapeutic components of psychotherapy. A 25-year progress report of research. *The Journal of Nervous and Mental Disease*, 1974, *159*, 325–342.

Frank, J. D. Reply to Telch. *Journal of Consulting and Clinical Psychology*, 1981, *49*, 476–477.

Frank, J. D., Gliedman, L. H., Imber, S. D., Nash, E. H., Jr., & Stone, A. R. Why patients leave psychotherapy. *Archives of Neurology and Psychiatry*, 1957, *77*, 283–299.

Frank, J. D., Gliedman, L. H., Imber, S. D., Stone, A. R., & Nash, E. H. Jr. Patients' expectancies and relearning as factors determining improvement in psychotherapy. *American Journal of Psychiatry*, 1959, *115*, 961–968.

Friedman, H. J. Patient expectancy and symptom reduction. *Archives of General Psychiatry*, 1963, *8*, 61–67.

Friedman, M. L., & Dies, R. R. Reactions of internal and external test anxious students to counseling and behavior therapies. *Journal of Consulting and Clinical Psychology*, 1974, *42*, 921.

Gallagher, E. B., Levinson, D. J., & Erlich, I. Some sociolopsychological characteristics of patients and their relevance for psychiatric treatment. In M. Greenblott, D. J. Levinson, & R. Williams (Eds.), *The patient and the mental hospital*. New York: Free Press, 1957.

Gallagher, E. B., Sharaf, M. R., & Levinson, D. J. The influence of patient and therapist in determining the use of psychotherapy in a hospital setting. *Psychiatry*, 1965, *28*, 297–310.

Gallagher, J. J. Test indicators for therapy prognosis. *Journal of Consulting Psychology*, 1954, *18*, 409–413.

Garfield, S. L. A note on patients' reasons for terminating therapy. *Psychological Reports*, 1963, *13*, 38.

Garfield, S. L. Research on client variables in psychotherapy. In S. L. Garfield & A. E. Bergin (Eds.), *Handbook of psychotherapy and behavior change* (2nd ed.). New York: Wiley, 1978.

Garfield, S. L., & Affleck, D. C. An appraisal of duration of stay in outpatient psychotherapy. *Journal of Nervous and Mental Disease*, 1959, *129*, 492–498.

Garfield, S. L., & Affleck, D. C. Therapists' judgments concerning patients considered for psychotherapy. *Journal of Consulting Psychology*, 1961, *25*, 505–509.

Garrison, J. E. Written vs. verbal preparation of patients for group psychotherapy. *Psychotherapy: Theory, Research, and Practice*, 1978, *15*, 130–134.

Gaylin, N. Psychotherapy and psychological health: A Rorschach function and structure analysis. *Journal of Consulting Psychology*, 1966, *30*, 494–500.

Gendlin, E. T. Initiating psychotherapy with "unmotivated" patients. *Psychiatric Quarterly*, 1961, *35*, 134–139.

Getter, H., & Sundland, D. M. The Barron ego strength scale and psychotherapeutic outcome. *Journal of Consulting Psychology*, 1962, *26*, 195.

Goldstein, A. P. Therapist and client expectation of personality change in psychotherapy. *Journal of Counseling Psychology*, 1960, *3*, 180–184. (a)

Goldstein, A. P. Patient's expectancies and non-specific therapy as a basis for (un)-spontaneous remission. *Journal of Clinical Psychology*, 1960, *16*, 399–403. (b)

Goldstein, A. P. *Therapist-patient expectancies in psychotherapy*. New York: Pergamon, 1962.

Goldstein, A. P. *Structured learning therapy: Toward a psychotherapy for the poor*. New York: Academic Press, 1973.

Goldstein, A. P., & Shipman, W. G. Patient expectancies, symptom reduction and aspects of initial psychotherapeutic interviews. *Journal of Clinical Psychology*, 1961, *17*, 129–133.

Gomes-Schwartz, B. Effective ingredients in psychotherapy: Prediction of outcome from process variables. *Journal of Consulting and Clinical Psychology*, 1978, *46*, 1023–1035.

Gottschalk, L. A., Mayerson, P., & Gottlieb, A. A. Prediction and evaluation of outcome in an emergency brief psychotherapy clinic. *Journal of Nervous and Mental Disease*, 1967, *144*, 77–96.

Gottschalk, L. A., Fox, R. A., & Bates, D. E. A study of prediction and outcome of a mental health crisis clinic. *American Journal of Psychiatry*, 1973, *130*, 1107–1111.

Greenfield, N. S. Personality patterns of patients before and after application for psychotherapy. *Journal of Consulting Psychology*, 1958, *22*, 280.

Greer, F. L. Prognostic expectations and outcome of brief therapy. *Psychological Reports,* 1980, *46,* 973–974.

Grotjahn, M. Learning from dropout patients: A clinical view of patients who discontinued group psychotherapy. *International Journal of Group Psychotherapy,* 1972, *22,* 306–319.

Gurman, A. S. The patient's perception of the therapeutic relationship. In A. S. Gurman & A. M. Razin (Eds.), *Effective psychotherapy.* New York: Pergamon, 1977.

Hall, S. M., Bass, A., & Monroe, J. Continued contact and monitoring as follow-up strategies: A long-term study of obesity treatment. *Addictive Behavior,* 1978, *3,* 139–147.

Halperin, K. M., & Snyder, C. R. Effects of enhanced psychological test feedback on treatment outcome: Therapeutic implications of the Barnum effect. *Journal of Consulting and Clinical Psychology,* 1979, *47,* 140–146.

Hamburg, D. A., Bibring, G. L., Fisher, C., Stanton, A. H., Wallerstein, R. S., Weinstock, H. I., & Haggard, E. Report of ad hoc committee on central fact gathering data of the American Psychoanalytic Association. *Journal of the American Psychoanalytic Association,* 1967, *15,* 841–861.

Heilbrun, A. B., Jr. Prediction of continuing client contact in a brief treatment campus mental health agency. *Psychological Report,* 1978, *42,* 481–482.

Heine, R. W. (Ed.). *The student physician as psychotherapist.* Chicago: University of Chicago Press, 1962.

Heine, R. W., & Trosman, H. Initial expectations of the doctor-patient interaction as a factor in the continuance in psychotherapy. *Psychiatry,* 1960, *23,* 275–278.

Heitler, J. B. Preparation of lower class patients for expressive group psychotherapy. *Journal of Consulting and Clinical Psychology,* 1973, *41,* 251–260.

Heitler, J. B. Preparatory techniques in initiating expressive psychotherapy with lower-class, unsophisticated patients. *Psychological Bulletin,* 1976, *83,* 339–352.

Heller, K., Myers, R. A., & Kline, L. V. Interviewer behavior as a function of standardized client roles. *Journal of Consulting Psychology,* 1963, *27,* 117–122.

Hoehn-Saric, R., Frank, J. D., Imber, S. D., Nash, E. H., Stone, A. R., & Battle, C. C. Systematic preparation of patients for psychotherapy. Effects on therapy behavior and outcome. *Journal of Psychiatric Research,* 1964, *2,* 267–281.

Hollingshead, A. B., & Redlich, F. C. *Social class and mental illness: A community study.* New York: Wiley, 1958.

Imber, S. D., Frank, J. D., Gliedman, L. H., Nash, E. H., Jr., & Stone, A. R. Suggestibility, social class, and the acceptance of psychotherapy. *Journal of Clinical Psychology,* 1956, *12,* 341–344.

Imber, S. D., Nash, E. H., & Stone, A. R. Social class and duration of psychotherapy. *Journal of Clinical Psychology,* 1955, *11,* 281–284.

Imber, S. D., Pande, S. K., Frank, J. D., Hoehn-Saric, R., Stone, A. R., & Wargo, D. G. Time-focused role induction. *Journal of Nervous and Mental Disease,* 1970, *150,* 27–30.

Jackson, D. N. *Personality research form manual.* Goshen, N.Y.: Research Psychologists Press, 1967.

Jeffrey, R. W., Vender, M., & Wing, R. R. Weight loss and behavior change one year after behavioral treatment for obesity, *Journal of Consulting and Clinical Psychology,* 1978, *46,* 368–369.

Jones, E. E. Effects of race on psychotherapy process and outcome: An exploratory investigation. *Psychotherapy: Theory, Research, and Practice,* 1978, *15,* 226–236.

Jones, E. E., & Zoppel, C. L. Impact of client and therapist gender on psychotherapy process and outcome. *Journal of Consulting and Clinical Psychology,* 1982, *50,* 259–272.

Kadushin, C. *Why people go to psychiatrists.* New York: Atherton, 1969.

Kamin, C., & Caughlan, J. Subjective experiences of outpatient psychotherapy. *American Journal of Psychotherapy*, 1963, *17*, 660–668.

Kaplan, A. G. Toward an analysis of sex-role related issues in the therapeutic relationship. *Psychiatry*, 1979, *42*, 112–119.

Karasu, T. B., Stein, S. P., & Charles, E. S. Age factors in patient-therapist relationships. *The Journal of Nervous and Mental Disease*, 1979, *167*, 100–104.

Katahn, M., Strenger, S., & Cherry, N. Group counseling and behavior therapy with test-anxious college students. *Journal of Consulting Psychology*, 1966, *30*, 544–549.

Katz, J., & Solomon, R. Z. The patient and his experiences in an out-patient clinic. *Archives of Neurology and Psychiatry*, 1958, *80*, 86–92.

Katz, M. M., Lorr, M., & Rubenstein, E. A. Remainder patients attributes and their relation to subsequent improvement in psychotherapy. *Journal of Consulting Psychology*, 1958, *22*, 411–413.

Kazdin, A. E., & Wilcoxon, L. A. Systematic desensitization and nonspecific treatment effects: A methodological evaluation. *Psychological Bulletin*, 1976, *83*, 719–758.

Kernberg, O. F., Burstein, E. D., Coyne, L., Appelbaum, A., Horwitz, L., & Voth, H. Psychotherapy and psychoanalysis: Final report of the Menninger Foundation's Psychotherapy Research Project. *Bulletin of the Menninger Clinic*, 1972, *36* (1–2).

Kirkner, F. J., Wisham, W. W., & Giedt, F. H. A report on the validity of the Rorschach prognostic rating scale. *Journal of Protective Techniques*, 1953, *17*, 465–470.

Kirshner, L. A. Effects of gender on psychotherapy. *Comprehensive Psychiatry*, 1978, *19*, 79–82.

Kirtner, W. L., & Cartwright, D. S. Success and failure in client-centered therapy as a function of initial in-therapy behavior. *Journal of Consulting Psychology*, 1958, *22*, 329–333.

Klein, H. R. A study of changes occurring in patients during and after psychoanalytic treatment. In P. H. Hoch & J. Zibin (Eds.), *Current approaches to psychoanalysis*. New York: Grune & Stratton, 1960.

Klopfer, B. Introduction: The development of a prognostic rating scale. *Journal of Projective Techniques*, 1951, *15*, 421.

Knapp, P. H., Levin, S., McCarter, R. H., Wermer, H., & Zetzel, E. Suitability for psychoanalysis: A review of 100 supervised analytic cases. *Psychoanalytic Quarterly*, 1960, *29*, 459–477.

Koran, L. M., & Costell, R. M. Early termination from group psychotherapy. *International Journal of Group Psychotherapy*, 1973, *23*, 346–359.

Krebs, R. L. Some effects of a white institution on black psychiatric outpatients. *American Journal of Orthopsychiatry*, 1971, *41*, 589–597.

Lambert, M. J. *The effects of psychotherapy* (Vol. 1). Montreal: Eden, 1979.

Lambert, M. J., & Bergin, A. E. Therapist characteristics and their contribution to psychotherapy outcome. In C. E. Walker (Ed.), *Handbook of clinical psychology: Theory, research and practice*. Homewood, Ill.: Dorsey, 1983.

Lambert, M. J., Bergin, A. E., & Collins, J. L. Therapist induced deterioration in psychotherapy. In A. S. Gurman & A. S. Razin (Eds.), *Effective psychotherapy: The therapist's contribution*. New York: Pergamon, 1977.

Lambert, M. J., DeJulio, S. S., & Stein, D. M. Therapist interpersonal skills: Process, outcome, methodological considerations, and recommendations for future research. *Psychological Bulletin*, 1978, *85*, 467–489.

Lambert, M. J., & Utic, J. The relationships of client characteristics and outcome. In M. J. Lambert (Ed.). *The effects of psychotherapy* (Vol. 2). New York: Human Sciences Press, 1982.

Lambert, R. G., & Lambert, M. J. The use of a general role induction interview to increase therapy participation in Polynesian immigrants. *Journal of Community Psychology,* in press.

Lanfield, A. W. *Personal construct systems in psychotherapy.* Chicago: Rand McNally, 1971.

Lazarus, A. A. The results of behavior therapy in one hundred and twenty-six cases of severe neurosis. *Behaviour Research and Therapy,* 1963, *1,* 69–79.

Leary, T. *Interpersonal diagnosis of personality.* New York: Ronald, 1957.

Lichtenstein, E. Personality similarity and therapeutic success: A failure to replicate. *Journal of Consulting Psychology,* 1966, *30,* 282.

Lick, J., & Bootzin, R. Expectancy factors in the treatment of fear: Methodological and theoretical issues. *Psychological Bulletin,* 1975, *82,* 917–931.

Liljestrand, P., Gerling, E., & Saliba, P. A. The Effects of social sex-role stereotypes and sexual orientation on psychotherapeutical outcomes. *Journal of Homosexuality,* 1978, *3,* 361–372.

Lipkin, S. Client's feelings and attitudes in relation to the outcome of client-centered therapy. *Psychological Monographs,* 1954, *68* (Whole No. 372.).

Lorion, R. D. Socio-economic status and traditional treatment approaches reconsidered. *Psychological Bulletin,* 1973, *79,* 263–270.

Lothstein, L. M. The group psychotherapy dropout phenomenon revisited. *American Journal of Psychotherapy,* 1978, *135,* 1492–1495.

Lubin, B., Hornstra, R. K., Lewis, R. V., & Bechtel, B. S. Correlates of initial treatment assignment in a community mental health center. *Archives of General Psychiatry,* 1973, *29,* 497–504.

Luborsky, L., Chandler, M., Auerbach, A. H., Cohen, J., & Bachrach, H. M. Factors influencing the outcome of psychotherapy: A review of quantitative research. *Psychological Bulletin,* 1971, *75,* 145–185.

Luborsky, L., Bachrach, H., Graff, H., Pulver, S., & Christoph, P. Preconditions and consequences of transference interpretations: A clinical-quantitative investigation. *Journal of Nervous and Mental Disorders,* 1979, *167,* 391–401.

Luborsky, L., Mintz, J., & Christoph, P. Are psychotherapeutic changes predictable? Comparison of a Chicago counseling center project with a Penn psychotherapy project. *Journal of Consulting and Clinical Psychology,* 1979, *47,* 469–473.

Luborsky, L., Mintz, J., Auerbach, A., Christoph, P., Bachrach, H., Todd, T., Johnson, M., Cohen, M., & O'Brien, C. P. Predicting the outcome of psychotherapy. *Archives of General Psychiatry,* 1980, *37,* 471–481.

Malan, D. H. *The frontier of brief psychotherapy.* New York: Plenum Press, 1976. (a)

Malan, D. H. *Toward the validation of dynamic psychotherapy: A replication.* New York: Plenum Press, 1976. (b)

Martin, P. J., Sterne, A. L., & Hunter, M. L. Share and share alike: Mutuality of expectations and satisfaction with therapy. *Journal of Clinical Psychology,* 1976, *32,* 677–683.

Marziali, E., Marmar, C., & Krupnick, J. Therapeutic alliance scales: Development and relationship to psychotherapy outcome. *American Journal of Psychiatry,* 1981, *138,* 361–364.

Mathews, A. M., Johnston, D. W., Shaw, P. M., & Gelder, M. G. Process variables and the prediction of outcome in behavior therapy. *The British Journal of Psychiatry,* 1974, *125,* 256–264.

McLachlan, J. F. C. Benefit from group therapy as function of patient-therapist match on conceptual level. *Psychotherapy: Theory, Research & Practice,* 1972, *9,* 317–323.

McNair, D. M., Lorr, M., Young, H. H., Roth, I., & Boyd, R. W. A three-year follow-up of psychotherapy patients. *Journal of Clinical Psychology,* 1964, *20,* 258–264.

Meltzoff, J., & Kornreich, M. *Research in psychotherapy.* New York: Atherton, 1970.

Miles, H. H. W., Barrabee, E. L., & Finesinger, J. E. Evaluation of psychotherapy. *Psychosomatic Medicine,* 1951, *13,* 83–105.

Milton, F., & Hafner, J. The outcome of behavior therapy for agoraphobia in relation to marital adjustment. *Archives of General Psychiatry,* 1979, *36,* 806–811.

Mindess, H. Predicting patients' responses to psychotherapy: A preliminary study designed to investigate the validity of the Rorschach prognostic rating scale. *Journal of Projective Techniques,* 1953, *17,* 327–334.

Mintz, J. What is "success" in psychotherapy? *Journal of Abnormal Psychology,* 1972, *80,* 11–19.

Mintz, J., Luborsky, L., & Auerbach, A. H. Dimensions of psychotherapy: A factor analytic study of ratings of psychotherapy sessions. *Journal of Consulting and Clinical Psychology,* 1971, *36,* 106–120.

Mitchell, K. M., Bozarth, J. D., & Krauft, C. C. A reappraisal of the therapeutic effectiveness of accurate empathy, nonpossessive warmth, and genuineness. In A. S. Gurman & A. M. Razin (Eds.), *Effective psychotherapy: A handbook of research.* New York: Pergamon, 1977.

Morgan, W. G. Nonnecessary conditions or useful procedures in desensitization: A reply to Wilkins. *Psychological Bulletin,* 1973, *79,* 373–375.

Morgenstern, F. S., Pearce, J. F., & Rees, W. L. Predicting the outcome of behavior therapy by psychological tests. *Behaviour Research and Therapy,* 1965, *2,* 191–200.

Myers, J. K., & Schaffer, C. Social stratification and psychiatric practice: A study of an outpatient clinic. *American Sociological Review,* 1954, *19,* 307–310.

Newmark, C. S. Finkelstein, M., & Frerking, R. A. Comparison of the predictive validity of two measures of psychotherapy prognosis. *Journal of Personality Assessment,* 1974, *38,* 144–148.

Nichols, R. C., & Beck, K. W. Factors in psychotherapy change. *Journal of Consulting Psychology,* 1960, *24,* 388–399.

Overall, B., & Aronson, H. Expectations of psychotherapy in lower socioeconomic class patients. *American Journal of Orthopsychiatry,* 1962, *32,* 271–272.

Parloff, M. B., Waskow, I. E., & Wolfe, B. E. Research on therapist variables in relation to process and outcome. In S. L. Garfield & A. E. Bergin (Eds.), *Handbook of psychotherapy and behavior change* (2nd ed.). New York: Wiley, 1978.

Pepinsky, H. B., & Kurst, T. O. Convergence: A phenomenon in counseling and psychotherapy. *American Psychologist,* 1964, *19,* 333–338.

Perotti, L. P., & Hopewall, C. A. *Expectancy effects in psychotherapy and systematic desensitization: A review.* Paper presented at the seventh annual meeting of the Society for Psychotherapy Research, San Diego, Calif., June 18, 1976.

Perry, C., Gelfand, R., & Marcovitch, P. The relevance of hypnotic susceptibility in the clinical context. *Journal of Abnormal Psychology,* 1979, *88,* 592–603.

Pettit, I., Pettit, T., & Welkowitz, J. Relationship between values, social class and duration of psychotherapy. *Journal of Consulting and Clinical Psychology,* 1974, *42,* 482–490.

Piper, W. E., & Wogan, M. Placebo effects in psychotherapy: An extension of earlier findings. *Journal of Consulting and Clinical Psychology,* 1970, *34,* 447.

Pomerleau, O., Adkins, D., & Pertschuk, M. Predictors of outcome and recidivism in smoking cessation treatment. *Addictive Behaviors,* 1978, *3,* 65–70.

Prager, R. A., & Garfield, S. L. Client initial disturbance and outcome in psychotherapy. *Journal of Consulting and Clinical Psychology,* 1972, *38,* 112–117.

Rachman, S. Current status of behavior therapy. *Archives of General Psychiatry,* 1965, *13,* 418–423.

Rice, L. N., & Wagstaff, A. K. Client voice quality and expressive style as indices of productive psychotherapy. *Journal of Consulting Psychology,* 1967, *31,* 557–563.

Rogers, C. R. A theory of therapy, personality, and interpersonal relationships as developed in the client-centered framework in psychology: A study of science. In S. Koch (Ed.), *Formulations of the person and the social context.* New York: McGraw-Hill, 1959.

Rogers, C. R., Gendlin, E. T., Kiesler, D. J., & Truax, C. B. (Eds.). *The therapeutic relationship and its impact.* Madison, Wis.: University of Wisconsin Press, 1967.

Rohsenow, D. J., & O'Leary, M. R. Locus of control research on alcoholic populations: A review in development, scales, and treatment. *International Journal of Addiction,* 1978, *13,* 55–78.

Rosenbaum, J., Friedlander, J., & Kaplan, S. Evaluation of results of psychotherapy. *Psychosomatic Medicine,* 1956, *18,* 113–132.

Rosenbaum, M., & Argon, S. Locus of control and success in self-initiated attempts to stop smoking. *Journal of Clinical Psychology,* 1979, *35,* 870–872.

Rosenthal, D., & Frank, J. D. The fate of psychiatric clinic outpatients assigned to psychotherapy. *Journal of Nervous Mental Disorders,* 1958, *127,* 330–343.

Ross, A. O., & Lacey, H. M. Characteristics of terminators and remainers in child guidance treatment. *Journal of Consulting Psychology,* 1961, *25,* 420–424.

Ross, M., & Mendelsohn, F. Homosexuality in college. *Archives of Neurological Psychiatry,* 1958, *80,* 253–263.

Rotter, J. B. Generalized expectancies for internal versus external control of reinforcement. *Psychological Monographs,* 1966, *80* (1, Whole No. 609).

Ryan, W. (Ed.) *Distress in the city.* Cleveland: The Press of Case Western Reserve University, 1969.

Saltzman, C., Luetgert, M. J., Roth, C. H., Creaser, J., & Howard, L. Formation of a therapeutic relationship: Experiences during the initial phase of psychotherapy as predictors of treatment duration and outcome. *Journal of Consulting and Clinical Psychology,* 1976, *44,* 546–555.

Salzman, C., Shader, R. I., Scott, D. A., & Binstock, W. Interviewer anger and patient dropout in walk-in clinic. *Comprehensive Psychiatry,* 1970, *11,* 267–273.

Schaffer, L., & Myers, J. K. Psychotherapy and social stratification: An empirical study of practice in a psychiatric outpatient clinic. *Psychiatry,* 1954, *17,* 83–93.

Schmidt, E., Castell, D., & Brown, P. A retrospective study of 42 cases of behavior therapy. *Behaviour Research and Therapy,* 1965, *3,* 9–19.

Schroeder, P. Client acceptance of responsibility and difficulty of therapy. *Journal of Consulting Psychology,* 1960, *24,* 467–471.

Schulman, R. E. Use of the Rorschach prognostic rating scale in predicting movement in counseling. *Journal of Counseling Psychology,* 1963, *10,* 198–199.

Schwartz, R. D., & Higgins, R. L. Differential outcome from automated assertion training as a function of locus of control. *Journal of Consulting and Clinical Psychology,* 1979, *47,* 686–694.

Seeman, J. Counselor judgments of therapeutic process and outcome. In C. R. Roger & R. F. Dymond (Eds.), *Psychotherapy and personality change.* Chicago: University of Chicago Press, 1954.

Seidel, C. The relationship between Klopfer's Rorschach prognostic rating scale and premature termination from psychotherapy. *Journal of Consulting Psychology,* 1960, *24,* 46–49.

Sheehan, J. G., Frederick, C. J., Rosevear, W. H., & Spiegelman, M. A validity study of the Rorschach prognostic scale. *Journal of Projective Techniques,* 1954, *18,* 233–239.

Siegel, N., & Fink, M. Motivation for psychotherapy. *Comprehensive Psychiatry,* 1962, *3,* 170–173.

Sifneos, P. E. *Short-term psychotherapy and emotional crisis*. Cambridge, Mass.: Harvard University Press, 1972.

Sifneos, P. E. *Short-term dynamic psychotherapy*. New York: Plenum Press, 1979.

Sloane, R. B., Staples, F. R., Cristol, A. H., Yorkston, N. J., & Whipple, K. *Psychotherapy versus behavior therapy*. Cambridge, Mass.: Harvard University Press, 1975.

Smith, G. J. W., Sjoholm, L., & Nielzen, S. Individual factors affecting the improvement of anxiety during a therapeutic period of $1\frac{1}{2}$ to 2 years. *Acta Psychiatrica Scandinavica*, 1975, *52*, 7–22.

Snowden, L. R. Personality tailored covert sensitization of heroin abuse. *Addictive Behavior*, 1978, *3*, 43–49.

Stone, A. R., Frank, J. D., Nash, E. H., & Imber, S. D. An intensive five-year follow-up study of treated psychiatric outpatients. *Journal of Nervous and Mental Disease*, 1961, *133*, 410–422.

Strickland, B. R., & Crowne, D. P. Need for approval and the premature termination of psychotherapy. *Journal of Consulting Psychology*, 1963, *27*, 95–101.

Strupp, H. H. Toward a reformulation of the psychotherapeutic influence. *International Journal of Psychiatry*, 1973, *11*, 263–327.

Strupp, H. H. Success and failure in time-limited psychotherapy. *Archives of General Psychiatry*, 1980, *37*, 595–603.(a)

Strupp, H. H. Success and failure in time-limited psychotherapy. *Archives of General Psychiatry*, 1980, *37*, 708–716.(b)

Strupp, H. H. Success and failure in time-limited psychotherapy. *Archives of General Psychiatry*, 1980, *37*, 831–841.(c)

Strupp, H. H. Success and failure in time-limited psychotherapy. *Archives of General Psychiatry*, 1980, *37*, 947–954.(d)

Strupp, H. H., & Bloxom, A. L. Preparing lower-class patients for group psychotherapy: Development and evaluation of a role induction film. *Journal of Consulting and Clinical Psychology*, 1973, *41*, 373–384.

Strupp, H. H., Wallach, M. S., Wogan, M., & Jenkins, J. W. Psychotherapists' assessment of former patients. *Journal of Nervous Mental Disorders*, 1963, *137*, 222–230.

Sue, S., McKinney, H., Allen, D., & Hall, J. Delivery of community mental health services to black and white clients. *Journal of Consulting and Clinical Psychology*, 1974, *42*, 794–801.

Sullivan, P. L., Miller, C., & Smelzer, W. Factors in length of stay and progress in psychotherapy. *Journal of Consulting Psychology*, 1958, *22*, 1–9.

Swensen, C. H. Psychotherapy as a special case of dyadic interaction: Some suggestions for theory and research. *Psychotherapy: Theory, Research, and Practice*, 1967, *4*, 7–13.

Szapocznik, T., Scoppetta, M. A., & King, O. E. Theory and practice in matching treatment to the special characteristics and problems of Cuban immigrants. *Journal of Community Psychology*, 1978, *6*, 112–122.

Telch, M. J. The present status of outcome studies: A reply to Frank. *Journal of Consulting and Clinical Psychology*, 1981, *49*, 472–475.

Tollinton, H. J. Initial expectations and outcome. *British Journal of Medical Psychology*, 1973, *46*, 251–257.

Tolman, R. S., & Meyer, M. M. Who returns to the clinic for more therapy? *Mental Hygiene*, 1957, *41*, 497–506.

Truax, C. B., & Carkhuff, R. R. *Toward effective counseling and psychotherapy: Training and practice*. Chicago: Aldine, 1967.

Truax, C. B., Tunnell, B. T., Jr., Fine, H. L., & Wargo, D. G. *The prediction of client outcome during group psychotherapy from measures of initial status.* Unpublished manuscript, Arkansas Rehabilitation Research and Training Center, University of Arkansas, 1966.

Truax, C. B., Wargo, D. G., Frank, J. D., Imber, S. D., Battle, C. C., Hoehn-Saric, R., Nash, E. H., & Stone, A. R. Therapist's contribution to accurate empathy, nonpossessive warmth and genuineness in psychotherapy. *Journal of Clinical Psychology,* 1966, *22,* 331–334.

Uhlenhuth, E., & Duncan, D. Subjective change in psychoneurotic outpatients with medical student therapists. Some determinants of change. *Archives of General Psychiatry,* 1968, *18,* 532–540.

Volsky, T., Magoon, T. M., Norman, W. T., & Hoyt, D. P. *The outcome of counseling and psychotherapy: Theory and research.* Minneapolis: University of Minnesota Press, 1965.

Vriend, J., & Dyer, W. W. Counseling the reluctant client. *Journal of Counseling Psychology,* 1973, *20,* 240–246.

Wallston, B. S., Wallston, K. H., Kaplan, G. D., & Maides, S. A. Development and validation of the health locus of control scale (HLC). *Journal of Consulting Clinical Psychology,* 1976, *44,* 580–585.

Warren, N. C., & Rice, L. Structure and stability of psychotherapy for low-prognosis clients. *Journal of Consulting and Clinical Psychology,* 1972, *39,* 173–181.

Warren, W. A study of adolescent psychiatric in-patients and the outcome six or more years later: II. The follow-up study. *Journal of Child Psychological Psychiatry,* 1965, *6,* 141–160.

Whitehead, A. The prediction of outcome in elderly psychiatric patients. *Psychological Medicine,* 1976, *6,* 469–479.

Wilkins, W. Expectancy of therapeutic gain: An empirical and conceptual critique. *Journal of Consulting and Clinical Psychology,* 1973, *40,* 69–77.

Willer, B., & Miller, G. H. On the relationship of client satisfaction to client characteristics and outcome of treatment. *Journal of Clinical Psychology,* 1978, *34,* 157–160.

Wolberg, L. R. *The technique of psychotherapy* (3rd ed.). New York: Grune & Stratton, 1977.

Yalom, I. D. A study of group therapy dropouts. *Archives of General Psychiatry,* 1966, *14,* 393–414.

Yamamoto, J., & Goin, M. K. Social class factors relevant for psychiatric treatment. *Journal of Nervous and Mental Disease,* 1966, *142,* 332–339.

Yamamoto, J., James, Q. C., & Palley, N. Cultural problems in psychiatric therapy. *Archives of General Psychiatry,* 1968, *19,* 45–49.

11

Therapist Variables

CURTIS L. BARRETT and JESSE H. WRIGHT

INTRODUCTION

In this chapter we consider therapist variables in psychotherapy and psychotherapy research. Our endeavor is not at all a new one, since the topic has been of interest from the very beginning of systematic thought about the complex, dynamic set of processes that define psychotherapy. It is a difficult endeavor, however, because of the massive literature, including expert opinion and systematic research, that has addressed the topic. Fortunately, in the last decade or so two major critical surveys (Meltzhoff & Kornreich, 1970; Parloff, Waskow, & Wolfe, 1978) as well as numerous specialized reviews (e.g., Dent, 1978; Kilman, Scovern, & Moreault, 1979) have systematized the literature either generally, to a specified point in time, or with regard to a particular issue or perspective. Our strategy will be to summarize briefly the conclusions that previous reviewers reached and then, in the process of updating the literature, to comment on selected issues that we view as enduring or promising. One departure from the approach that other reviewers have taken will be our use of the term *psychotherapist* to include persons, regardless of professional discipline, who synthesize pharmacological and psychological treatment of emotional or behavioral disorders. Also, we will use the terms *psychotherapy* and *psychotherapist* in a generic sense to include a wide range of intervention techniques or strategies and persons who practice them. Parloff *et al.* (1978) critically surveyed the literature on therapist variables through about 1977 and systematically considered most of what had been covered in the earlier major review by Meltzhoff and Kornreich (1970). We consider both of these reviews

---

CURTIS L. BARRETT and JESSE H. WRIGHT ● University of Louisville School of Medicine, Louisville, Kentucky 40292.

required reading for the serious researcher, therapist, or planner, since they often critique studies from somewhat different methodological or philosophical perspectives. However, here we will draw mainly on Parloff *et al.* (1978), since it is the most recent and, in our opinion, more readily accessible survey.

It is important to note that Parloff *et al.* (1978), like Meltzhoff and Korneich (1970), examined therapist variables in relation to psychotherapy process and outcome and not in isolation. Unfortunately, the range of variables to be related to measures of psychotherapy process and outcome is almost endless. Limitations on the study of process and outcome variables, in turn, have defined what could be learned about therapist variables. Quite often, researchers who have considered therapist variables have continued to choose what Parloff *et al.* (1980, p. 273) have termed "simplistic, global concepts." As a result, they say that this field has been brought to a state of "terminal vagueness." Seldom has it been possible to separate effect of therapist from other uncontrolled variables. When that has been possible, the following conclusions seem to be warranted:

1. Therapists' emotional problems may interfere with effective treatment.
2. These attractive therapist variables have unclear, weak, or unconfirmed effects:
    a. Personality of therapist
    b. Personal therapy of therapist
    c. Sex of therapist
    d. Therapist experience
3. Therapist qualities that were formerly hypothesized to be "necessary and sufficient" (e.g., accurate empathy, warmth, and genuineness) now appear to be related to other variables in complex ways and do not stand alone.

When therapist variables are combined with patient variables research supports the following conclusions:

1. Dropout rate (but not therapy effectiveness) appears to be related to the congruence of therapist and patient expectations.
2. The effect of psychosocial treatment may be enhanced if the patient is adequately prepared for it.
3. "Matching" of patients and therapists on personality dimensions and on cognitive dimensions has not been shown to be useful, but some evidence suggests that this could be fruitful.
4. It may be useful within a given treatment setting to use factor analytic techniques to match patients and therapists in order to utilize treatment staff more efficiently.
5. The following have not been adequately tested:

    a. Interactive roles of therapist and patient on variables of race, social class, and sex

    b. Effects of therapist–patient value similarity

6. The personality variable that has been most extensively used to predict therapy outcome (the A-B Scale) has been brought into question by an increasing number of studies.

These are, of course, our versions of the Parloff *et al.* (1978) conclusions. We have not attempted to duplicate here the extensive and painstaking review of studies, including methodology, that went into those conclusions. Instead, we acknowledge their invaluable contribution and move on to consideration of selected studies that, for the most part, have been published since the Parloff *et al.* (1978) review was completed. These studies concern the following therapist variables: therapist personality, race, sex/sexism, values, religious values, personal therapy, experience, and attitude toward selected patient groups. That done, we will consider some of the therapist variables as they have been presented in studies that utilized pharmacotherapy and psychotherapy together.

PERSONALITY OF THE THERAPIST

It makes intuitive sense that the personality of the therapist should have a major effect on psychotheraphy process and/or outcome. Yet this variable was not found among those that Parloff *et al.* (1978) listed as having known effects. The conventional wisdom regarding the role of therapeutic technique and the personality of the person using the technique has been expressed by Jung (1934).

> It is in fact immaterial what sort of technique (the therapist) used, for the point is not the technique but the person who uses the technique. . . .the personality and attitude of the doctor are of supreme importance—whether he appreciates this fact or not.

It is interesting that the most durable of the attempts to study therapist personality as a variable began with an observation of therapist differential effectiveness with schizophrenic patients. Whitehorn and Betz (1954), in their now famous paper, designated therapists who were effective with schizophrenics Type A and those who were not effective with schizophrenics Type B. In time, Type A's were characterized as persons who were active in therapy, used personal participation, and showed the ability to establish a trusted relationship. Further, on the Strong Campbell Vocational Interest Blank (SVIB), Type A's showed interests in occupations such as lawyer or certified accountant. Type B's, in addition to being less successful in treating schizophrenics, were characterized as passive and prone to using interpretations and an instructional style of therapy. On the SVIB they were not interested in occupations that attracted Type A's and, instead, showed career interests as printers or as mathematics or science teachers. Later, a specific 23-item scale was derived; research has suggested correlated personality patterns in the Personality Research Form. This research suggested an empirical method by which therapists could respond to the Fromm-Reichman (1950, p. 40) charge that therapists should discover the types of persons that they are capable of treating and concentrate on them.

We have begun our discussion of the therapist's personality as a variable in psychotherapy with consideration of the A-B therapist type mostly out of respect for its durability. Parloff *et al.* (1978) appeared to have had much the same reaction to it, but concluded:

> a wide range of research over the past decade on the A-B variable reveals very little support for the hypothesis that the A-B Scale has reliability in predicting long term outcome, and there is inconsistent and weak support for the possibility that the A-B Scale might predict short term outcome. Nevertheless, research in this area continues. (p. 268)

Parloff *et al.* then referred to imaginative but unpublished work by Dent. This work is now published (Dent, 1978) and, in our judgment, ranks as the best single source of information on the A-B Scale and other attempts to explore the personality of effective therapists. Through interviews with persons who worked on the original Whitehorn-Betz studies, Dent has managed to put life into the idea of relating the personality of the therapist to therapy outcome. This helps to counter the often sterile, obligatory statements that one reads in introductory sections of journal articles.

Dent (1978) concludes that there is little to support the idea that, in the general sense, there is a personality pattern that defines "the effective therapist." Instead, evidence suggests that different disorders require different treatments and that therapists vary in their ability to conduct those treatments. This, or course, makes sense, but it is necessary to look further. For example, do therapists actually differ in personality as measured by the A-B Scale?

Geller and Berzins (1976) sought to determine the A-B therapist characteristics of a sample of nationally prominent psychotherapists. Their scale was such that the individuals could be classified as A's; A-B's, at the midrange of the scale; or B's. Naturally, definition of "nationally prominent psychotherapist" was difficult, but most of the therapists' names were easily recognizable. Most remarkable is the fact that so many of the therapists permitted themselves to be publicly "classified." When this was done, some of the pioneers in psychotherapy with schizophrenic/schizoid patients (e.g., S. Arieti, D. Jackson) did appear as A's. However, some specialists in the treatment of schizophrenics were classified as B's (e.g., John Rosen), and some as A-B's (e.g., O.

A. Will). Much the same thing happened at the B end of the scale. Expected names (e.g., C. Rogers, J. Barron) were present and accounted for as B's; others were not. The A-B (midrange) classified therapists included a number who claim to be eclectic (e.g., V. Raimy, F. Thorne).

It is indeed interesting that the venerable and battered A-B therapist variable fared as well as it did in this study. Some would question whether the results would have been the same if other scales had been chosen (Seidman, Golding, Hogan, & LeBow, 1974). Certainly, others might have classified the Geller and Berzins study as "waste," along with A-B Scale studies that classified trainees or undergraduates as "therapists" (e.g., Barnes & Berzins, 1978; Kennedy & Chartier, 1976). Cox (1978), for example, asked: "Where are the A and B therapists?" (1970–1975) and answered: "A and B therapist types don't exist and, for that matter . . . they never have existed" (p. 119).

One innovative way of determining whether A therapists exist was reported by King and Blaney (1977). They reasoned that if a therapist (in this case, psychologist) was effective with a group of clients, his or her colleagues would know it. Thus, one could determine what sorts of patients colleagues would refer to them. This is a naturalistic approach that reflects the conventional wisdom, and it stays pretty far from the more common laboratory model. However, results showed that, indeed, schizophrenic patients were more likely to be referred to Type A therapists, and Type B's were referred more neurotics. The finding held whether the A-B categorization was made using the median split or the outer quartile separations of the distributions. These results are strongly reminiscent of the Ricks (1974) report of "Supershrink." It was common knowledge in a clinic that one of two highly trained and experienced therapists worked very differently than another with preschizophrenic adolescent males. In time, those differences showed up in differential rates of hospitalization and adult diagnosis. King and Blaney (1977) collected no such data, of course, and actually we do not know whether differential treatment or results actually existed once the clients were referred.

James and Foreman (1973) reported that A-B personality variables might affect a person's success in utilizing Mowrer's conditioning apparatus with enuretic children. It was found that B therapists, actually mothers enlisted as "technicians," were more effective than A therapists (also mothers) with the conditioning technique. This study has been strongly criticized by Eysenck for calling the mothers "therapists" (Eysenck, James, & Foreman, 1975). However, James and Foreman, in response, have noted that the technician-mothers were given full responsibility for the treatment program, including purchase of the apparatus, repairing it, and so on. Differential success, favoring B's, would not be at all unreasonable on the basis of the Barnes and Berzins (1978) suggestion that Type B's are more dominant, risk-taking, and self-assured, perceiving situations as challenges. These may have been just the behaviors to help one through the problem of treating enuresis. Also, the fact that apparatus was involved may have attracted the "engineering" interests of B's (Delk & Ryan, 1975).

The complexity of the therapist personality variable has led some investigators to novel solutions to control for it. Lang (1968), for example, developed a Device for Automated Desensitization, known—for psychodynamically unclear reasons—as DAD. DAD presented hierarchies for fear-evoking stimuli, gave relaxation instructions with remarkable fidelity, and obtained good results. Indeed, DAD could be regarded as the prototype of the psychotherapist who demonstrates "integrity" (Yeaton & Sechrest, 1981). *Integrity* refers to adherence to a protocol and delivery of treatment as it is intended to be delivered.

Using a procedure similar to DAD, Morris and Suckerman (1974) found that a "warm," automated-desensitization "therapist" was more effective than a "cold" automated-desensitization procedure in reducing snake phobia. *Warm* and *cold* referred to the recorded voices that were used. Rosen, Glasgow, and Barrera (1976) and Kirsch (1978) pushed the logic a bit further in eliminating the "therapist" altogether. Rosen *et al.* (1976) treated highly anxious self-referred snake phobics using (1) therapist-administered Systematic Desensitization Therapy (SDT); (2) self-administered SDT, with weekly telephone therapist support; (3) totally self-administered SDT; or (4) self-administered double-blind placebo control. A "no-treatment" condition was also provided. Results indicated comparable dropout rates (low) and treatment effects (moderate) across conditions. Of particular note is the fact that Rosen *et al.,* like others before them (e.g., Barrett 1969; Paul, 1966) utilized a therapy manual "written so as to closely parallel all procedures used by therapists in the clinic" (p. 210). Kirsch's (1978) approach was simply to teach a therapeutic skill for the patient to use; he did not address the problem of how the "client–therapist" happened to use the therapy. That, of course, has long been the problem with self-administered therapies (e.g., weight-control diets, self-administered insulin, self-administered Antabuse.)

A number of researchers (e.g., Barrett, 1969; Miller, Barrett, Hampe, & Noble, 1972; Shaw, 1977) have attempted to hold the therapist personality variable constant by having him or her conduct more than one treatment. Shaw (1977), for example, treated 32 depressed male and female clients, aged 18 to 26, using cognitive therapy, behavior therapy, and "nondirective" therapy. The therapist (Shaw) followed "detailed manuals" and "attempted to behave consistently in each treatment group" (p. 546). A fellow graduate student monitored sessions to assure compliance with treatment protocols. Shaw's results, incidentally, favored cognitive therapy, a field in which he has come to be regarded as an expert.

Attempts to control the therapist variable (e.g., with manual-guided therapies) overlap with attempts to standardize therapies. These attempts are aided by demonstrations that raters can recognize different therapies when applied by the same therapist (Barrett, 1969; Miller, Barrett, Hampe, & Noble, 1972) or different therapists (Luborsky, Woody, McLellan, O'Brien, & Rosenzweig, 1982). Further, therapies, once recognized, can be rated effectively for degree of compliance to the prescribed therapeutic procedures without considering the therapist's personality *per se* (Barrett, 1969). Thus, the near future for psychotherapy research may see even more emphasis on technique as opposed to emphasis on the technician or therapist.

As noted above, "therapist personality" as a therapist variable has not yielded a productive line of research. In part, this must be attributed to the current status of personality theory and personality measurement. Just as behavior therapy was unable to anchor in monolithic learning theory (Kiesler, 1966), psychotherapy research has never had monolithic personality theory to use in studying the effect of therapist personality.

The past decade has seen a resurgence of questioning as to whether the basic units of personality study (dispositions, traits) are sufficiently stable to permit prediction of behavior across situations. Mischel (1973), for example, has suggested that personality be reconceptualized within the cognitive social learning framework. In a later paper, he (Mischel, 1977) suggested that person variables (read "psychotherapist personality variables") may become potent for prediction when the same person is responding to the same situation or when situational variables are weak. As one might expect, this "situationist" position has been critiqued vigorously, and alternatives have been proposed (e.g., Bowers, 1973). Mischel (1979), having apparently attained his announced goal of protecting individuals (patients *and* psychotherapists, we suppose) from a form of clinical hostility (using a few behavioral signs such as the SVIB to categorize persons), has tempered his view somewhat. He now suggests that we go beyond the person–situation debate to a study of at least the cognitive processes of all members in an interacting system. This, of course, is tall cotton for a field that found the Whitehorn-Betz variables hard to assimilate. However, it also raises hope that the conventional wisdom, based as it is in multidimensional, historically founded analyses, can be right even though we do not yet know how to show it. That is, therapist's "personality" may, after all, be shown to make a difference.

At the practical level, the psychotherapy researcher who wants to identify therapist personality variables now faces some distinct problems. For example, Nicholls, Licht, and Pearl (1982) suggest that we may not be able to accept personality questionnaires as ways to measure therapist's personality (e.g., even a dimension such as "masculinity"). Similarly, Hogan, Desoto, and Solano (1977) warn that personality assessment, partly because it is easy, leads to "mindless" research. But they also note that the view "that human personality is too complex to be described in terms of a few concepts or dimensions may be the counsel of wisdom or the counsel of despair" (1977, p. 262). The counsel of despair has not served other fields, and it will not help the psychotherapy researcher. Caution, or alternative approaches such as those described above, may.

THERAPIST'S RACE

For the foreseeable future, the condition described by Jones and Seagull (1977), that black clients will be seen in therapy by white therapists, will continue. This reflects nothing more nor less than the fact that whites far outnumber blacks in the helping professions. Jones and Seagull (1977) make a number of suggestions about what white therapists can or should do to minimize problems when conducting therapy with blacks. This centers around understanding one's own prejudices, examining one's motivations for working with minority clients, and protecting against the stereotyping of others. Education to dispel stereotypes and to learn the minority perspective is also recommended. These suggestions serve the useful purpose of heightening awareness, but unfortunately Jones and Seagull (1977) cite opinion and *possibly* relevant research, not data. The psychotherapy researcher or planner is given little help by them.

Jones (1978) approached the issue of race in the psychotherapy process from the perspective of literature, suggesting that outcome may be enhanced if clients/patients and therapists are "matched" on certain psychological and social dimensions. In the study, therapists (two black and three white) were all clinical psychologists, presumably at the doctoral level, with an average of 3.6 years of experience. The author judged them to be similar in theoretical orientation and use of techniques. Fourteen patients, evenly divided by race, served as subjects. They were matched for type of psychopathology (all were neurotics) and all were women. This last fact reflects the scarcity of black males in outpatient clinics, rather than the investigator's choice. Although there was no time limit for the study, data were collected after 10 once-weekly sessions (unless therapy had terminated earlier).

Therapy outcome was not found to be related to race of client or therapist. This finding must be tempered somewhat by noting that the criteria for improvement relied on therapists' judgment to a large degree. Process, on the other hand, was affected by race, but in complex ways.

> If the client was white, it mattered little if the therapist was black or white. . . .
> Among black clients, regardless of the race of the therapist, there evolved an
> important aspect of therapy that centered around the discussion of race-related

> concerns; associated with this was the salient role the client's race assumed in the
> manner which the therapist viewed his client and thought about the case. (p. 234)

It was also noted that erotic transference was less likely when therapist and client were racially dissimilar. However, matched black dyads were observed to develop a stronger erotic transference. It was speculated that this happened because black clients felt that they were being fully understood and developed a strong sense of trust in their therapist.

As mentioned above, Jones's (1978) findings are complex and must not be generalized to patient or client groups inappropriately. However, they also suggest that quite potent factors may operate on the *process* of therapy without showing on *measured* therapy outcome. Whether these effects would have been described as "outcome" within a different research design remains unanswered.

Turner and Armstrong (1981) examined the questionnaire responses of 37 black and 41 white therapists from a sample of 170 black and 250 white psychotherapists who were asked to return questionnaires. The psychotherapists averaged 8.2 years of postdoctoral experience and carried an average of 28 patients, 7 of whom were cross-racial. The black psychologists had, on the average, less training and fewer years of clinical practice but saw more cross-racial patients.

Since neither process nor outcome measures were used, the Turner and Armstrong (1981) study is valuable only (1) in that it partially confirms Jones's (1978) results and (2) in that it points out differences between responses of black and white therapists in cross-racial therapy. In the latter case, black therapists reported more awareness of racial similarities and differences but were relatively unbothered by doing cross-racial therapy; white therapists were less aware of racial concerns but were more uneasy about the process.

For the time being, it would seem prudent for psychotherapy researchers to balance or to control racial characteristics of clients or therapists. However, there is still no strong evidence that the race of the therapist would be a major predictor of therapy process or outcome.

SEX OF THERAPIST/SEXISM

In our clinical experience, the wisdom of assigning a given patient to a therapist of the same or different sex has been debated often. The issue has been extended somewhat when, on our ward, we sometimes ask that an inpatient mother bring her child to the ward as part of her treatment. Should her therapist be a female? Should the therapist be female and a mother? Or should parenthood rather than sex of therapist be the defining variable in selecting a therapist? Does it make any appreciable difference?

The women's movement has vigorously raised the question of whether male therapists have the ability to understand and to deal with the emotional problems of their female patients. Orlinsky and Howard (1976) have pointed out that logically, alleged behavior of male therapists (i.e., subtle sexism) may or may not affect the treatment process. Using data collected for another purpose, they sought to resolve this question at least partly on an empirical basis.

Subjects were 118 women, 78 of whom received therapy from a male therapist and 40 from a female. All 27 therapists were dynamically oriented but came from different professions: psychiatric social work, psychiatry, and clinical psychology. After each therapy session, patients filled out a "Therapy Session Report" indicating

> what they had talked about, what they had hoped to gain, what their concerns had been, what feelings they had experienced, how they had acted toward the therapists, how they had reacted to themselves, what they felt they had gained, and how they had evaluated the session. Patients were also asked to describe how their therapists had acted towards them, and how their therapists seemed to feel. (p. 82)

Of 46 dimensions of therapy experience, Orlinsky and Howard found 15 to show statistically significant differences as a function of sex of therapist. For example, women with male therapists talked more about the oppostie sex, expressed a greater desire to achieve insight through therapy, and saw their therapists as more demanding and detached. The women who had female therapists had, as a group, higher levels of satisfaction with regard to the encouragement that they received in therapy. It was also clear that impact of therapist gender was not the same across life status categories (e.g., young single women, young divorced mothers). Young single women with male therapists

> reported more intense desire for active input from the therapist; greater concern with identity, fear and anger; felt more inhibited, less self-possessed, and less open; and viewed the therapist as feeling more detached and demanding. (p. 84)

Orlinsky and Howard (1976) also found that the reaction of women to therapists varied with the sex of the therapist and the patient's diagnosis. Single women with depressive reactions felt more support and satisfaction when they were in treatment with female therapists.

The Orlinsky and Howard (1976) study ranks as an important one for psychotherapy researchers who consider using therapists who differ in sex and with patients of varying diagnoses. Since Klein (1976) has given some guidelines for outcome measures that would be compatible with feminist concepts of mental health, combining these criteria with Orlinsky and Howard's methodology could help to determine more appropriately the effect of sex of therapist, within a diagnostic group, on outcome of therapy for women.

Whereas Orlinsky and Howard (1976) observed patient response as a function of therapist sex, Stearns, Penner, and Kimmel (1980) investigated sexism as expressed by psychotherapists in response to clients. The 86 psychotherapists viewed videotaped, staged case conferences depicting a patient's situation. Data indicated that overall, therapists did not stereotype according to sex. Instead, they responded to the case presentation (e.g., presenting problem, personal history). However, female therapists tended to see female patients as more depressed than male therapists saw them.

Jones and Zoppel (1982) reviewed the limitations on most studies of the effect of client and therapist gender on psychotherapy process and outcome. While noting that the results of client–therapist matching may reflect an implicit rule that the clients should be matched to a counselor of the same gender, regardless of the presenting problem (Shullman & Betz, 1979), no data lead to such decisions. They also point out (1) that therapist gender may be confounded with experience and status, (2) that analogue studies do not reflect clinical reality, and (3) that research has usually examined the effect of therapist gender on women patients. They proposed that more could be learned by investigating all possible therapist–gender matches.

In two studies, Jones and Zoppel (1982) attempted to investigate gender and psychotherapy, first from the therapist's perspective and then from that of the client. In the first study, 160 recently terminated clients who had been treated for neurotic, personality, depressive, and adjustment disorders were selected as subjects. Forty subjects were included in each of four cells: male therapist–male clients; female therapist–female clients; and so on. A total 140 therapists (69 women and 70 men) contributed to the sample. Few therapists treated more than one client for the study. Therapists represented the three major mental health disciplines, but there were marked differences in the gender distribution of the groups: psychology, 64% of therapists, 50% women; social work, 20% of the sample, 80% women; psychiatry, 16% of the sample, 4% women. Therapists were fairly experienced (average of 4.4 years) and therapy was "dynamic, insight-oriented." Clients received an average of 35 hours of treatment (range 8 to 100 plus). Therapists used a set of rating scales and completed a questionnaire to report therapy outcome; they used an adjective checklist to describe patient personality characteristics. Therapists also completed a questionnaire describing their reactions to the client. Finally, therapists used Gough's Adjective Check List to describe their clients in therapy.

Although this was not a focus of the study, Jones and Zoppel report the usual finding that male and female clients carried different diagnoses in the patient sample (e.g., men, more personality disorders; women, more hysterical syndromes). As therapists, women viewed patients of *either* gender as having greater problems in sexual adjustment and as having more difficulties with their spouses and children. Looking at outcome of therapy, all therapists saw their patients as improved, but female therapists saw more improvement in five areas: symptoms, happiness, ability to enjoy life, ability to get along with spouse and children, and ability to handle personal problems. Also, *women* therapists rated their *women* patients as significantly more improved than their men patients in symptoms; ability to enjoy life; and relationship with spouse, children, parents, siblings and other close relatives. *Male* therapists rated their *male* patients as more improved on the "happiness" scales than they rated their women patients. Compared to male therapists, women therapists rated their patients (regardless of gender) as having greater success in therapy, more basic personality change, being more satisfied with treatment, and having developed a better working relationship. Therapists of both sexes rated same-gender patients better in prognosis and more enjoyable in therapy.

Male and female therapists tended to describe male clients similarly or in a way that balanced positive and less desirable characteristics. However, they described female clients differently. That is, female therapists were more sympathetic and captured personality competencies and strengths in their ratings; they saw female clients as "shy" and "emotional." Male therapists used terms such as *wary* and *temperamental* and, in the author's opinion, assumed a more critical or judgmental stance (e.g., using adjectives such as *affected* or *awkward*).

This study seems to establish rather clearly that the male and female therapists represented by this sample differ in the way they perceive therapy outcome and their clients. Further, some interaction of patient and therapist gender affects perception of the client. This suggests that therapist gender should be looked at more closely than we would have expected on the basis of the Parloff *et al.* (1978) conclusions.

In their second study, Jones and Zoppel (1982) selected 99 former therapy clients, 40 of whom had been evaluated by therapists in Study 1. Except for the cell "female therapist–male client," each of four "gender–therapist/gender–client" cells contained 25 clients. Former patients were given a structured interview; they also completed the same scales assessing therapy outcome as did therapists. Patients agreed with therapists, overall, that therapy had been effective. Women patients reported that before and after therapy they, more than men, had problems getting along with spouse and children and that they experienced anxiety. Regardless of gender, clients seen by women therapists reported having more energy to do things. Also, clients in therapy with women were more likely to find a match between what happened in therapy and what they expected to happen. Finally, same-gender therapist dyads reported therapies that were longer in duration. (Whether this is interpreted as positive or negative may depend on whether one is a third-party carrier or the director of a fee-for-service mental health clinic!)

Patients did not perceive their psychotherapy in uniform ways. A factor labeled "therapeutic alliance" was purported to

> tap aspects of the therapist's personal stance or attitude toward the client as well as intervention technique. It gauges therapist interest in, acceptance of, and respect for the client as well as the extent to which his or her manner was warm and attentive. (Jones & Zoppel, 1982, p. 267)

This complex factor also tapped the "trust" dimension. Men and women clients who were treated by women therapists reported higher scores on therapeutic alliance than did those who were treated by men. A second factor, "formality/detachment," was not found to discriminate between male and female therapists/clients. There was some tendency for women clients with women therapists to experience therapy with greater intensity. Also, clients in same-gender pairings viewed their therapists as neutral or nondirective as compared with those in cross-gender pairings. Finally, Jones and Zoppel (1982) found that, regardless of the gender of their therapist, women clients were more likely than men "to experience being deprecated by their therapists" (p. 269).

Without doubt, Jones and Zoppel have made a significant contribution to our understanding of the effects that therapist gender may have on therapy process and outcome, at least with some sorts of clients and in a "dynamic, insight-oriented" therapy. Whether these findings will hold with other diagnostic groups or therapies remains open question.

Jones and Zoppel also found that male and female therapists may see outcome differently and may use rating scales differently. This suggests that outcome measures may need to be standardized on samples of male and female therapists in order to preclude artifactual findings (e.g., of differential improvement for male and female therapists' groups). Suffice it to say, therapist gender, confounded as it is with professional discipline, is a variable that deserves close attention in the psychotherapy research protocols of the 1980s.

PERSONAL THERAPY OF THE THERAPIST

Garfield and Kurtz (1976) and Buckley, Karasu, and Charles (1981) have addressed the issue of personal therapy for psychotherapists. Buckley *et al.* (1981) surveyed psychotherapists who had had personal therapy and concluded that the majority felt that they had benefited from the experience. There was no evidence that specific skills in psychotherapy or effectiveness of psychotherapy were related to the therapists' having had personal therapy. The authors suggest that any such effects contribute to "non-specific factors."

Peebles (1980) attempted to determine whether the number of hours that psychology graduate students received in psychotherapy would be related to ability to show warmth and empathy with their clients. Results did not indicate that this was so. However, this study is quite limited by the definition of *therapist* and the specification of personal therapy (e.g., by reports that students had had therapy).

To summarize, it is clear that personal therapy remains a popular enterprise for psychotherapists. However, there has been little progress in recent years in determining whether this activity makes a difference or what the difference might be.

THERAPIST VALUES

The importance attached to values in psychotherapy, at least in the opinion of experts, may be gauged by the fact that the entire Winter 1980 issue of the journal *Psychotherapy: Theory, Research and Practice* was devoted to the topic. This issue makes for interesting reading and certainly has considerable heuristic worth. However, it does not provide much in the way of researchable questions in the context of today's available methodology. We must take a wait-and-see attitude, for now, about what will develop that can be applied to the issue of therapist values.

The obvious fact that psychotherapists may have religious beliefs or values has been generally neglected in psychotherapy research. Recently, Bergin (1980a,b) opened discussion of this therapist variable, and we see it as one that may stimulate a considerable body of research in coming years. As could be expected, there was vigorous response to Bergin's suggestion. Ellis (1980), in a carefully reasoned statement, pointed out contrasting hypotheses about the psychotherapy process and its goals that would follow from what he labeled theistic, clinical humanistic, and clinical–humanistic–atheistic positions. Similarly, Walls (1980) suggested that accepting Bergin's position that psychotherapists can be guided by divine authority is a potentially dangerous one. He suggests that psychologists, and we suppose some other psychotherapists, have a responsibility to submit all values to rational scrutiny.

As yet there is no empirical basis that psychotherapists or psychotherapy researchers may use for deciding about the effect of therapists' religious values on psychotherapy. Once more, we have a variable that the conventional wisdom and expert opinion say is important. Also, we know that, in our clinical setting, patients frequently ask about the religious background of our therapists. What we do not know is whether it matters.

THERAPIST EXPERIENCE

At first glance, therapist experience appears to be a clear-cut variable that would be certain to affect psychotherapy in some way. It has not turned out that way. However, the interim since the Parloff *et al.* (1978) review has brought one mild concession in the controversy over the effect of therapist experience in the treatment of schizophrenia (Karon & VandenBos, 1970, 1972, 1975; Tuma & May, 1975). Tuma and May argued that psychotherapy had added little to the effect of medication in treatment of schizophrenics, but Karon and VandenBos countered with the view that appropriate psychotherapy by experienced persons had not been provided to their patients. That is, Tuma and May had used psychiatric residents as therapists, and these persons knew little or nothing about formal psychotherapy or psychotherapy with schizophrenics. The Michigan State Psychotherapy Research Project (MSPRP; Karon & VandenBos, 1970) demonstrated fairly well that experienced therapists obtained better results with schizophrenics than had Tuma and May's residents. Unfortunately, funding was withdrawn from the MSPRP before it was completed, and the results are not as convincing as they might have been. Recently Tuma, May, Yale, and Forsythe (1978) reanalyzed their original data to address the psychotherapist experience issue. They found that their therapists differed among themselves in clinical experience in treating psychiatric patients and in "general ability" as reported by their supervisor. These differences were not related to outcome. However, Tuma *et al.* (1978, p. 1124) now acknowledge that while their results are possibly applicable to run-of-the-mill psychotherapy as practiced in combination with medication, they do not apply "to the [small and select number of] highly experienced persons who specialize in the treatment of schizophrenia." Perhaps this opens the way again to systematic research on psychotherapy with schizophrenic patients. If so, it is heartening to clinicians who have found schizophrenics difficult to treat even when the most advanced psychopharmacological methods were at hand.

Experience in treating patients, we are reminded, does not imply success in treating patients, schizophrenics or others. The therapists in the Ricks (1974) study were both highly experienced, but "Supershrink" was more effective.

The fact that training and experience can alter the behavior of interviewers has been demonstrated by Pope, Nudler, Van Korff, and McGee (1974) and Pope, Nudler, Norden, and McGee (1976). Over a 3-year training period, novice interviewers increased facilitation skills, decreased in anxiety, and increased in evoking self-disclosure. The degree to which these therapist skills relate to outcome remains to be demonstrated. Also, Loo (1979) found that specific skill training did predict help-oriented behavior in interviewers, whereas more academic or classroom learning did not. Here again, it is obvious that therapist training or experience, to have any likelihood of showing an effect, must be closely related to a definable task. As methodologies such as manual-guided therapies become more widely used, skills that therapists must demonstrate compliance with may become better defined. At that point the therapist experience variable could become crucial.

SPECIAL PROBLEMS OF THERAPIST ATTITUDE

Langer and Abelson (1974) have shown that the generally recognized effects of the labeling process can apply to clinicians as well as to the public at large. This has increased significance, since psychotherapists are beginning to deal with persons about whom they, as members of the larger society, may have biases. Graham (1980, p. 797) has reported that "when offender status is introduced as a [patient] variable, it has a negative effect on the acceptance phase of the therapy." She goes on to note that perception of the offender's *personality* is not affected much. Therapists, in this study, were 100 persons (at the master's level or above) who were employed in community mental health centers in a rural northeastern state. Graham speculates that this therapist variable contributes to the very high early dropout rate of offenders from therapy. She then suggests that therapist training or selection procedures could be useful in solving this problem. From the planner's perspective (e.g., Parloff, 1979), such findings suggest that it may do little good to staff community mental health centers with psychotherapists in an effort to rehabilitate offenders unless those therapists can accept those offenders for treatment. To do otherwise is to assure that there will be an underutilized staff.

Mintz, Steiner, and Jarvik (1981) point to a similar problem in the treatment of the growing number of depressed, elderly persons. These persons often are excluded from research programs on depression (e.g., Lewinsohn, Steinmetz, & Antonnuccio, 1982), even though estimates indicate that about 10% of the elderly population suffer from depression. This includes the massive study sponsored by the NIMH, since persons accepted for treatment cannot be more than 60 years of age. It is well known that elderly persons often suffer from medical ailments that contraindicate treatments using antidepressant medication. Thus, excluding the elderly from the NIMH study is no small matter.

Mintz *et al.* indicate that a number of familiar psychotherapy research problems (e.g., the myth of patient homogeneity; Kielser, 1966) and some new ones will be involved in study or treatment of the depressed elderly. Two are therapist variables: (1) therapist experience and attitudes, and (2) therapist's age. Just as psychotherapists who work effectively with schizophrenics and see value in treating such persons are rare, so too are therapists who are experienced and hopeful in the treatment of the depressed elderly. One need only visit the nearest nursing home to become convinced of that. Stereotypes of rigidity and untreatability are common, as are tendencies to attribute all these patients' problems to organic brain syndrome. It is doubtful that any psychotherapy would succeed in the face of such attitudes.

The problem of the therapist's age in relation to the client has not, to our knowledge, received attention from researchers dealing with either the young

(Barrett, Miller, & Hampe, 1978) or the elderly. Probably, most therapists for the elderly will be younger than their clients. Further, as Mintz *et al.* (1981) point out, it is likely that younger therapists will be trained in therapeutic approaches that do not at all match the patient's expectancies. This could easily add to the elderly person's sense of hopelessness, helplessness, and worthlessness. It must be hoped that the Mintz *et al.* (1981) message will not be lost and that research with the depressed elderly will soon move out of the pilot stage.

THERAPIST AND THERAPY DIFFERENCES YIELDING SIMILAR OUTCOME

A common thread in discussion of psychotherapy research for decades has been that common factors (or basic ingredients), some of which have to do specifically with the therapist, account for changes that occur in behavior or "personality" change. Among those who have spoken eloquently and frequently to this point are Frank (1979), Garfield (1973), Mahrer (1978), and Strupp (1973a,b, 1978, 1979, 1981). Recently, this position was reiterated in slightly different form in a call by Rappaport and Rappaport (1981) for an integration of "scientific and traditional" healing. Certainly, firsthand observation of African "native healers," such as the senior author has had, when coupled with findings from psychotherapy research, makes the argument an appealing one.

Relevant to this issue is one of the most intriguing of the findings in psychotherapy research: namely, that even when very clear-cut differences between therapies and therapists have been found, outcome at times has been similar and positive (e.g., Barrett, 1969; Gomes-Schwartz, 1978; Staples, Sloane, Whipple, Cristol, & Yorkston, 1976). In the time period covered by this chapter, the Gomes-Schwartz report on the "Vanderbilt study" is, of course, the most dramatic example. This effort must be regarded as one of the most sophisticated and methodologically sound to have been undertaken in some time. Yet the results were much the same as those that have been found in much less impressive research. Simply put, there seem to be several effective ways "to skin a cat," and we are currently identifying some of them rather well.

Two fairly recent examples of determining differences between therapies/ therapists without relating these differences to outcome are described in this connection because of their sound methodology. The first, Brunink and Schroeder (1979), investigated the in-therapy verbal behavior of "expert" psychoanalytically oriented, gestalt, and behavior therapists. Results indicated that gestalt therapists differed from the other two groups in that they "provided more direct guidance, less verbal facilitation, less focus on the client, more self-disclosure, greater initiative, and less emotional support" (p. 572). Psychoanalytically oriented and behavior therapy–oriented therapists were very much alike except that behavior therapists "provide more direct guidance and greater emotional support" than did psychoanalytically oriented therapists (p. 572).

Greenwald, Kornblith, Hersen, Bellack, and Himmelhoch (1981, p. 757) investigated two "fundamentally different approaches to treatment of depression"; namely, social skills training and dynamically oriented psychotherapy. Again, in-therapy behavior of the therapists, in this case psychologists who are Ph.D.'s but not identified as "experts," differed. For example, the social skills therapists were more directive and took more initiative. Interestingly, in this case the patients/clients were depressed women, and Beck, Rush, and Shaw (1979) include "social skills" type behaviors in their prescriptions for treatment of depression.

It would be easy to dismiss these studies as simply throwbacks to the studies of process for process's sake that characterized the very earliest of the Rogerian investigations. That is not the case. There is a clear trend now toward identifying what actually happens when therapists of different persuasions approach a treatment problem. Possibly "common ingredients" will, in fact, be found. But it may just as well be that very different procedures, in the hands of very different persons, will lead to common positive, negative, or neutral results. We know of no good reason to lament such findings.

THERAPIST VARIABLES IN PHARMACOTHERAPY AND COMBINED PHARMACOTHERAPY AND PSYCHOTHERAPY RESEARCH

Most pharmacotherapy trials have not recognized or assessed therapist variables. This has occurred despite warnings early in the history of psychopharmacology that the doctor–patient relationship was of considerable importance in determining drug effect (Beuscher, Carmichael, Hasenbush, Mackenzie, & Semrad, 1972; Feldman, 1956; Sabshin & Ramot, 1956; Sheard, 1963). Pharmachotherapy studies that have tackled the issue of therapist variables have generally found that therapists' attitudes and behaviors influence outcome. These studies are reviewed and discussed here for the purpose of illustrating how drug research may offer valuable opportunities for study of therapist variables.

Uncontrolled Studies

The introduction of tranquilizers and antidepressants in the 1950s was greeted by widely divergent attitudes toward the place of chemotherapy in the treatment of mental disorders. A series of reports on uncontrolled observations of physician attitudes toward drug use soon followed. Early experiences with chlorpromazine at the Michael Reese Hospital led Sabshin and Ramot (1956) to suggest that attitudes toward drug therapy were influencing outcome. This observation prompted an uncontrolled study of 338 patients from many different diagnostic groups who were treated with a wide variety of drugs and other therapies. (Eisen, Sabshin, & Heath, 1959). Doctors rated themselves on a global four-point scale that described their attitudes toward drug therapy. Outcome was not appreciably different in doctors with opposing views on the value of drugs. However, because so many of the other uncontrolled variables could have affected outcome, it is not surprising that a relationship between attitude toward drugs and outcome was not substantiated. The importance of negative attitude toward drugs was emphasized by Stone and co-workers (Stone, Green, Gleser, Whitman, & Foster, 1975) in a study of combined analytic therapy and minor tranquilizers. They attributed an 82% dropout rate to the analytically oriented therapists' negative beliefs about drugs.

In another uncontrolled study, Feldman found success in therapy with chlorpromazine to be associated with a "wholehearted" acceptance of the drug by the physician (Feldman, 1956). He reported on discharge rates of schizophrenics treated by a large group of therapists who had different levels of acceptance of drug therapy. A positive attitude toward drugs was also found to be associated with better response to antidepressant therapy in an early study at the Institute of Living (Sheard, 1963). Doctors, depending on their attitude toward drugs, were observed to give different communications to patients. Examples of cues given by doctors included: "This medication is going to make you feel better," as compared with "By the way, Mrs. G., I thought we might as well try a little medication."

These early reports pointed out that the introduction of powerful new drugs in the treatment of emotionally disturbed persons did not eliminate the physician–patient relationship as one of the most important elements in the therapeutic process. Although it was clear from the onset of pharmacotherapy research that a psychotherapeutic relationship existed between the patient and the doctor administering the drug, only a few efforts have been made to study this process under controlled research conditions.

Partially Controlled Studies

Haefner, Sachs, and Mason (1980) studied the attitude toward chemotherapy in physicians who administered five different phenothiazine-derivative drugs to acute schizophrenics in a Veterans Administration cooperative study. The drugs were given under double-blind conditions, but outcome ratings were confounded by including the treating physician as part of the rating team. All the drug conditions were combined to perform the data analysis. It was concluded that positive attitudes toward drug use were associated with a more favorable outcome on the Lorr Scale than were negative attitudes.

Veterans Administration cooperative studies provided data for two additional reports that considered the dimension of the physician's attitude toward the patient (Honingfeld, 1962, 1963). In these studies, the investigator's coworkers used the Chemotherapy Attitude Scale (CAS) and the Custodial Mental Illness Scale (CMIS) to categorize attitudes of physicians participating in studies on schizophrenia and depression. Unfortunately, the data analysis used a correlation matrix between therapists' answers to individual scale items from patient outcome data. There were so many possible combinations that the several significant associations (e.g., a positive relationship between sympathetic attitudes toward patients and improvement in withdrawn behavior; a negative relationship between belief in drugs and improvement in psychotic thought disorder) are of dubious validity. As with the previous Veterans Administration study, several drugs were used, and all of the drug groups were considered equivalent for statistical analysis—a strategy that is highly questionable, given the current knowledge of differential effects of related drugs. Another drawback is that the CMIS focused primarily on attitudes toward patients as a general group. An example of a question from this inventory is "One of the main causes of mental illness is a lack of moral strength." Doctors' attitudes toward chemotherapy or toward the patient could be expected to vary from case to case, but this was not studied.

Data from a unique study of treatment of anxious patients by two general physicians in North Carolina suggested the possibility that the patient's response to drugs may be an important variable affecting the doctor's attitude (Rickels, Boren, & Stuart, 1964). The doctors perceived active drug patients (e.g., those taking meprobamate or phenobarbital) as making them feel much more comfortable than patients who were receiving placebo. Patients who received active drug also were much more likely to improve than those who received placebos. In another study, Rickels and co-workers (1979) reported that the doctor's liking for the patient was associated with good outcome in anxious patients treated with minor tranquilizers. However, no information was given on methods of assessing therapists' attitudes or outcome measures.

As previously noted, Karon and VandenBos (1970) found that experienced psychotherapists did better than inexperienced ones in treating schizophrenics regardless of medication use, therapeutic rationale, or technique. However, use of medication was critical for inexperienced therapists. This widely quoted study has been criticized on a number of methodological issues (Group for the Advancement of Psychotherapy, 1975). For example, attitudes and techniques of the doctor who managed the chemotherapy were not assessed. But the study does raise questions about how experience level might interact with or even override drug effect. A missing part of the analysis, the communication between the chemotherapist and the patient, deserves comment. Although many clinics use "primary" psychotherapists and pharmacotherapists as a team, the interaction between technique and variables involved in the different procedures has not yet been addressed.

The personality of the psychotherapist, but not the pharmacotherapist, was also the focus of a report by Shader, Grinspoon, and Harmatz, (1971), who reanalyzed data from two earlier studies of psychotherapy plus drug or placebo in the treatment of schizophrenia. They were troubled by their prior consideration of psychotherapy as a homogeneous variable. In an attempt to rectify this problem, they later classified therapists as Type A or Type B based on their responses to the SVIB. In the first study, chronic schizophrenics received psychotherapy from senior, analytically oriented therapists; they were also given thioridazine, a butyrophenone, or placebo. The small number of subjects in each treatment cell caused problems with statistical analysis, but there was a trend for the best results to occur in the combination of Type A therapists plus active drug and the worst results with Type B therapists plus placebo. The investigators reported anecdotally that regardless of drug or therapist type, the patients who improved most had therapists who encouraged the discussion of anger. Their second study utilized first-year psychiatric residents who treated acute schizophrenics with psychotherapy plus thioridazine or placebo. Type A therapists had better results with active drug patients, but there were no differences in outcome of Type A compared with Type B therapists in patients treated with placebo. It was suggested that the match between therapist and patient was a crucial variable. However, it is unclear from the data how matching was or could be assessed.

Controlled Studies

Uhlenhuth, Canter, Neustadt, and Payson (1959) used the SVIB to choose two psychiatrists to participate in a double-blind study of phenobarbital, meprobamate, and placebo in the treatment of anxious neurotics. Outcome was assessed by a 45-item Symptom Checklist and a global rating by the patient and by the doctor. Since doctors' guesses about which patients were on active drug were no better than chance, the authors concluded that the double-blind was upheld. Therapy visits were limited to 10 to 20 minutes and were designed to be highly structured around collecting research data. They found that Dr. A's patients responded equally well to drug and placebo, but Dr. B's patients did better on active drug than on placebo. Dr. B was perceived by his patients as being more positive, helpful, and dependable. He had predicted before the study that there would be a difference in results between drug and placebo patients, while Dr. A believed that there would be no difference. Although the doctors apparently could not detect which patients were on active drug, it is quite possible that patients did know. Doubts have been raised elsewhere about whether active psychotropic drugs can be studied in true double-blind fashion (Friedman, 1975). The results in this study could also be attributed to the fact that Dr. B's patients satisfied the demand characteristics of his anticipation of a better response to active drug (Orne, 1962). Although the mechanism of the effect was not clear, this study demonstrated that individual differences in doctors' attitudes could be associated with variation in treatment response, even if sessions were limited to brief, highly structured interviews.

Individual differences in pharmacotherapists were studied extensively in a later investigation by Uhlenhuth, Rickels, Fishers, Park, Lipman, and Mock (1966). Psychiatric residents at three university clinics were trained to behave in a T (therapeutic) or an E (experimental) role in their treatment of anxious neurotics. The residents were supervised extensively through a one-way mirror and were judged before beginning the study to be acting in the T role (competent, enthusiastic, believed that drug was effective treatment) or the E role (detached, uncertain, believed that drug was of uncertain value). The patients were treated with meprobamate or placebo and were seen for an initial 30-minute visit followed by 15-minute medication checks. All therapists were specifically instructed *not* to do psychotherapy. Outcome was evaluated with the Symptom Check List (SCL) and global ratings by the doctor and patients. Despite the aggressive attempt to control therapist variables by training for specific behaviors, it was concluded that all doctors did not take equally well to their role assignments. The finding that the T and E roles were associated with different outcomes at only one of the three clinics was attributed to a greater ease of identification with the T role by doctors at this clinic. However, patients perceived all doctors in a positive way and did not appear to be able to discriminate between the T and E roles. This effort at studying therapist variables seems to confirm the importance of personality characteristics of the doctor. Ironically, instructing doctors what to do and attempting to train them to perform roles was confounded by the process that they were attempting to study—the personality of the therapist. It would appear unlikely that therapists can ever be trained to perform identically in attitude and behavior, even in a very limited and circumscribed therapy. This may have implications for manual-guided therapies even if they do not involve use of medication.

Several large, controlled pharmacotherapy–psychotherapy interaction studies are notable for their lack of attention to therapist variables. After the incisive review by Uhlenhuth, Lipman, and Covi (1969) that pointed out the need for rigorously controlled research on pharmacotherapy–psychotherapy interaction, four research groups initiated important studies on combinations of drug and psychotherapy in the treatment of depression (Covi, Lipman, Derogatis, Smith, & Pattison 1976; Friedman, 1975; Klerman, DiMascio, Weissman, Prusoff, & Paykel, 1974; Weismann, Prusoff, DiMascio, Neu, Goklaney, & Klerman, 1979). Recent reviews have critiqued these studies on such methodological problems as categorization of depression, attrition rates, control of drug therapy, and control of psychotherapy. Hollon and Beck (1978) and Wright and Looney (1979) have questioned the overall lack of evidence for either positive or negative interaction between therapies. Nonetheless, these studies are frequently cited as evidence for drug effectiveness as compared with psychotherapy alone or combined treatment. Therapist variables—such as attitudes, skill, adherence to the treatment model, and empathy—were not systematically assessed in any of the studies for either the primary psychotherapist or physicians who provided medication management. Data analysis was based on the assumption that psychotherapy and pharmacotherapy were discrete homogeneous processes, while it is quite possible that considerable "psychotherapy" was performed in the drug-alone groups, and that unmeasured therapist variables in psychotherapy and chemotherapy groups contributed heavily to outcome.

Discussion

The studies reviewed here shed little light on the nature of therapist variables that presumably interact with pharmacotherapy or combined pharmacotherapy and psychotherapy. The limitations of these studies are primarily in the area of experimental design, but a constricted theoretical view of therapist variables is also at issue.

Problems in experimental design have received thorough attention in reviews of pharmacotherapy–psychotherapy interaction research (Group for the Advancement of Psychiatry, 1975; Hollon & Beck, 1978), so these points will not be addressed at length here. However, it should be pointed out that the familiar problems of categorization of illness, randomization, management of attrition, assessment of patient variables, independent assessment of outcome, and control of drug and psychotherapy conditions will need to be managed effectively before any conclusions can be reached about the role of therapist variables in pharmacotherapy or combined therapy. Several recent advances should be helpful in this regard. First, publication of specific treatment manuals and ratings of actual performance in videotaped interviews in cognitive therapy and interpersonal therapy have raised hopes for increasing the data base on how therapists actually perform in sessions (Beck *et al.,* 1979; DeRubeis, Hollon, Evans, & Bemis, 1982; Klerman *et al.,* 1974). These techniques are now being used to assess criterion-level performance in several studies including one in the authors' clinic. Second, new findings from drug research have offered additional possibilities for improving research designs. For example, pharmacokinetic studies have revealed widely variant individual differences in the metabolism of drugs (Risch, Huey, & Janowsky, 1979a). Since many psychotropic medications have long half-lives and are strongly bound to tissue, short "washout" periods may not ensure drug-free states before beginning research studies. The development of blood-level assays has allowed for evaluation of drug-free status (at least in plasma) and measurement of adequacy of drug therapy during treatment (Risch *et al.*, 1979a; Risch *et al.*, 1979b).

Even with improved control of the experimental condition, assessment of therapist variables in drug research meets formidable obstacles. Among these are several concepts that have appeared to bind investigators to rather concrete ways of viewing the therapists' contribution to outcome. An artificial dichotomy between pharmacotherapy and psychotherapy has been almost uniformly accepted. This has occurred despite the common knowledge among practicing therapists that sharp boundaries between these treatments do not exist. Even the most entrenched biological psychiatrist will admit that supportive psychotherapy is needed in chemotherapy management. Data from research by Uhlenhuth *et al.* (1966) suggest that psychiatrists utilize psychotherapy techniques in pharmacotherapy, even if they are instructed not to do so. This is not surprising, since few physicians are ready to discard the doctor–patient relationship as a central focus of their daily activities. It even is possible that an experienced pharmacotherapist may be able to utilize effective psychotherapeutic techniques in a very short time period. Whether this is true remains to be proven, but it is one possible explanation for the findings, in some studies, that pharmacotherapy plus medication checks is as effective as extensive psychotherapy plus pharmacotherapy (Friedman, 1975; Klerman *et al.*, 1974).

Another concept that has limited research on therapist variables in drug research is the linear model that underlies the randomized double-blind design. Attitudes or other personality variables are assumed to be static entities that, if measured at one point in time, hold fast as a constant variable throughout the study. This assumption runs contrary to common sense as well as recent developments in personality research mentioned earlier. Even for a variable as simple as positive or negative attitude of the doctor toward the drug, shifts could occur from patient to patient and time to time, depending upon what the therapist feels and observes. The common practice has been to assume uniformity of these variables throughout the study and to comment on statistical "interaction," such as addition, potentiation, reciprocation, or inhibition by measuring differential outcome in treatment cells (Karon & VandenBos, 1970;

Klerman *et al.,* 1974). Thus, little is known about how drug, patient, and therapist variables actually interact in the process of therapy.

A logical first step to counter these conceptual problems might be to perform a series of naturalistic studies that attempt a description of what therapists actually do in pharmacotherapy management. What cues do pharmacotherapists give to patients? What patient variables change therapist behavior? How much time is used? How empathic are pharmacotherapists? What psychotherapeutic beliefs or strategies are utilized? These are examples of questions for which there are currently no clear answers.

Although this chapter deals primarily with therapist variables in psychotherapy research, a review of drug research has revealed that therapist variables are also important in pharmacotherapy. An increased awareness of this influence could make a valuable contribution to future research for several reasons: (1) Comparisons of psychotherapy to pharmacotherapy could consider the contribution of therapist variables in "drug-alone" treatment groups. (2) Study of therapist variables could help identify active ingredients of treatment and lead to an increased efficiency and effectiveness of medication-check therapy. (3) Research on interaction of therapist variables in teams of "primary" psychotherapists and pharmacotherapists could help improve this common method of therapy. (4) Compared with intensive psychotherapy, pharmacotherapy could offer a less complicated experimental model for study of therapist variables. If these issues are addressed, perhaps a response could then be made to Houston's statement of over 40 years ago: "A great deal is said about changes to be made in a patient's outlook and viewpoint, not much about what the emotional attitude of the doctor should be" (Houston, 1938).

SUMMARY

In this chapter we have chosen to look primarily at recent studies of therapist variables in psychotherapy and to address selected issues. The roughly 5-year period since the last major critical review of the literature on this topic has produced no significant breakthroughs that we could identify. Instead, we perceive a trend toward careful, methodologically sound exploitation of previous findings that should eventually provide a firmer base for the field of psychotherapy. Some issues of importance, in the United States at least, will probably receive more attention in the near future. These include previously emphasized therapist variables such as race, sex/sexism, and therapist styles or personality, but with altered directions. In addition, therapist variables that could preclude or inhibit attention to pressing problems (e.g., depression in the elderly) will probably gain in importance. Finally, it appears that the therapist variable is becoming an increasingly attractive one for study, as there is movement toward integrating the therapeutic wisdom of a wide variety of approaches: behavior therapy, psychodynamically oriented psychotherapy, psychopharmacology, and even traditional healing. We come away from this review encouraged, not because we have discovered new frontiers but because we sense a growing maturity about the way therapist variables are being treated in psychotherapy research.

REFERENCES

Antonnuccio, D. O., Lewinsohn, P. M., & Steinmetz, J. L. Identification of therapist differences in a group treatment for depression. *Journal of Consulting and Clinical Psychology,* 1982, *50,* 433–435.

Barnes, D. F., & Berzins, J. I. A and B undergraduate interviewers of schizophrenic and neurotic inpatients: A test of the interaction hypothesis. *Journal of Consulting and Clinical Psychology,* 1978, *46,* 1368–1373.

Barrett, C. L. Systematic desensitization versus implosive therapy. *Journal of Abnormal Psychology,* 1969, *75,* 587–592.

Barrett, C. L., Miller, L. C., & Hampe, I. E. Research on psychotherapy with children. In S. Garfield & A. E. Bergin (Eds.), *Handbook of psychotherapy and behavior change.* New York: Wiley, 1978.

Beck, A. T., Rush, A. J., & Shaw, B. F. *Cognitive therapy of depression.* New York: Guilford, 1979.

Bergin, A. E. Psychotherapy and religious values. *Journal of Consulting and Clinical Psychology,* 1980, *48,* 95–105.(a)

Bergin, A. E. Religious and humanistic values: A reply to Ellis and Walls. *Journal of Consulting and Clinical Psychology,* 1980, *48,* 642–645.(b)

Beuscher, W. F., Carmichael, W., Hasenbush, L. L., Mackenzie, J., & Semrad, E. The psychotherapeutic experience with chronic schizophrenics. In L. Grinspoon, J. Ewalt, & R. Shrader (Eds.), *Schizophrenia: Pharmacotherapy and psychotherapy.* Baltimore: Williams & Wilkins, 1972.

Bowers, K. S. Situationism in psychology: An analysis and a critique. *Psychological Review,* 1973, *80,* 307–336.

Brunnink, S. A., & Schroeder, H. E. Verbal therapeutic behavior of expert psychoanalytically oriented, gestalt, and behavior therapists. *Journal of Consulting and Clinical Psychology,* 1979, *47,* 567–574.

Buckley, P., Karasu, T., & Charles, E. Psychotherapists view their personal therapy. *Psychotherapy: Theory, Research and Practice,* 1981, *18,* 299–305.

Covi, L., Lipman, R. S., Derogatis, L. R., Smith, J. E., & Pattison, J. H. Drugs and group psychotherapy in neurotic depression. *American Journal of Psychiatry,* 1976, *133,* 502–508.

Cox, W. M. Where are the A and B therapists, 1970–1975? *Psychotherapy: Theory, Research and Practice,* 1978, *15,* 108–121.

Delk, J. L., & Ryan, T. T. Sex role stereotyping and A-B therapist status: Who is more chauvinistic? *Journal of Consulting and Clinical Psychology,* 1975, *43,* 589.

Dent, J. K. *Exploring the psychosocial therapies through the personalities of effective therapists.* National Institute of Mental Health: DHEW Publication No. (ADM) 77–527. Washington D.C.: U.S. Government Printing Office, 1978.

DeRubeis, R. J., Hollon, S. D., Evans, M. D., & Bemis, K. M. Can psychotherapies for depression be discriminated? A systematic investigation of cognitive therapy and interpersonal therapy. *Journal of Consulting and Clinical Psychology,* 1982, *50,* 744–756.

Eisen, S. B., Sabshin, M., & Heath, H. A comparison of the effects of investigators' and therapists' attitudes in the evaluation of tranquilizers prescribed to hospital patients. *Journal of Nervous & Mental Diseases,* 1959, *128,* 256–261.

Ellis, A. Psychotherapy and atheistic values: A response to A. E. Bergin's "Psychotherapy and religious values." *Journal of Consulting and Clinical Psychology,* 1980, *48,* 635–639.

Eysenck, H. J., James, L. E., & Foreman, M. E. Some comments on the relationship between A-B status of behavior therapists & success of treatment. *Journal of Consulting and Clinical Psychology,* 1975, *43,* 86–88.

Feldman, P. E. The personal element in psychiatric research. *American Journal Psychiatry,* 1956, *113,* 52–54.

Frank, J. D. The present status of outcome studies. *Journal of Consulting and Clinical Psychology,* 1979, *47,* 310–316.

Friedman, A. S. Interaction of drug therapy with marital therapy in depressive patients. *Archives of General Psychiatry,* 1975, *32,* 619-737.

Fromm-Reichman, F. *Principles of intensive psychotherapy.* Chicago: University of Chicago Press, 1950.

Garfield, S. L. Basic ingredients or common factors in psychotherapy? *Journal of Consulting and Clinical Psychology,* 1973, *41,* 9-12.

Garfield, S. L., & Kurtz, R. Personal therapy for the psychotherapist: some findings & issues. *Psychotherapy: Theory, Research and Practice,* 1976, *13,* 188–192.

Geller, J. D., & Berzins, J. L. A-B distinction in a sample of prominent psychotherapists. *Journal of Consulting Clinical Psychology,* 1976, *44,* 77–82.

Gomes-Schwartz, B. Effective ingredients in psychotherapy: prediction of outcome from process variables. *Journal of Consulting and Clinical Psychology,* 1978, *46,* 1023–1035.

Graham, S. A. Psychotherapists attitudes toward offender clients. *Journal of Consulting and Clinical Psychology,* 1980, *48,* 796–797.

Greenwald, D. P., Kornblith, S. J., Hersen, M., Bellack, A. S., & Himmelhoch, J. M. Differences between social skills therapists & psychotherapists in treating depression. *Journal of Consulting and Clinical Psychology,* 1981, *49,* 619–626.

Group for the Advancement of Psychiatry, Committee on Research. *Pharmacotherapy and psychotherapy: Paradoxes problems and progress.* New York: Brunner/Mazel, 1975.

Haefner, D. P., Sacks, J. M., & Mason, A. S. Physicians' attitudes toward chemotherapy as a factor in psychiatric patients' responses to medication. *Journal of Nervous and Mental Disease,* 1980, *131,* 64–69.

Hogan, R., DeSoto, C. B., & Solano, C. Traits, tests & personality research. *American Psychologist,* 1977, *32,* 255–264.

Hollon, S. D., & Beck, A. T. Psychotherapy and drug therapy: comparison and combinations. In S. Garfield & A. E. Bergin (Eds.), *Handbook of psychotherapy and behavior change* (2nd ed.). New York: Wiley 1978.

Honingfeld, G. Relationships among physicians' attitudes and response to drugs. *Psychological Reports,* 1962, *11,* 683–690.

Honingfeld, G. Physician and patient attitudes as factors influencing the placebo response in depression. *Diseases of the Nervous System,* 1963, *16,* 343–347.

Houston, W. R. The doctor himself as a therapeutic agent. *Annals of Internal Medicine,* 1938, *11,* 1416–1425.

James, L. E., & Foreman, M. E. A-B status of behavior therapy technician as related to success of Mowrer's conditioning treatment for enuresis. *Journal of Consulting and Clinical Psychology,* 1973, *41,* 224–229.

Jones, A., & Seagull, A. A. Dimensions of the relationship between black clients and white therapists: A theoretical overview. *American Psychologist,* 1977, *32,* 850–855.

Jones, E. E., & Zoppel, C. L. Impact of client and therapist gender on psychotherapy process and outcome. *Journal of Consulting and Clinical Psychology,* 1982, *50,* 259–272.

Jones, E. E. The effects of race on psychotherapy process and outcome: An exploratory investigation, *Psychotherapy: Theory, Research and Practice,* 1978, *15,* 226–236.

Jung, C. G. The state of psychotherapy today. *Collected Works.* (Vol. 10). *Civilization in transition.* Princeton, N. J.: Princeton University Press, 1964. (Originally published in 1934.)

Karon, B. P., & VandenBos, G. R. Experience, medication and the effectiveness of psychotherapy with schizophrenics: A note on Drs. May and Tuma's conclusions. *British Journal of Psychiatry,* 1970, *116,* 427–428.

Karon, B. P., & VandenBos, G. R. The consequences of psychotherapy for schizophrenic patients. *Psychotherapy: Theory, Research and Practice,* 1972, *9,* 111–119.

Karon, B. P., & VandenBos, G. R. Issues in current research on psychotherapy vs. medication in treatment of schizophrenia. *Psychotherapy: Theory, Research and Practice,* 1975, *12,* 143–148.

Kennedy, J. L., & Chartier, G. M. Effects of client and therapist A–B status on the process of psychotherapy. *Psychotherapy: Theory, Research and Practice,* 1976, *13,* 412–417.

Kielser, D. J. Some myths of psychotherapy research and the search for a paradigm. *Psychological Bulletin,* 1966, *65,* 110–136.

Kilman, P. R., Scovern, A. W., & Moreault, D. Factors in the patients–therapist interaction and outcome: A review of the literature. *Comprehensive Psychiatry,* 1979, *20,* 132–146.

King, D. G., & Blaney, P. H. Effectiveness of A and B therapists with schizophrenic and neurotic: A referral study. *Journal of Consulting and Clinical Psychology,* 1977, *45,* 407–411.

Kirsch, I. Teaching clients to be their own therapists: A case study illustration. *Psychotherapy: Theory, Research and Practice,* 1978, *15,* 302–305.

Klein, M. H. Feminist concepts of therapy outcome. *Psychotherapy: Theory, Research and Practice,* 1976, *13,* 89–95.

Klerman, G. L., Dimascio, A., Weissman, M., Prusoff, B., & Paykel, E. S. Treatment of depression by drugs and psychotherapy. *American Journal of Psychiatry,* 1974, *131,* 186–191.

Lang, P. J. Fear reduction and fear behavior: Problems in treating a construct. In J. M. Shlien (Ed.), *Research in psychotherapy* (Vol. 3). Washington, D.C.: APA, 1968.

Langer, E. J., & Abelson, R. P. A patient by any other name: Clinician group difference in labeling bias. *Journal of Consulting and Clinical Psychology,* 1974, *42,* 4–9.

Loo, C. Measures of predicting therapeutic talent and pre-and post-training differences. *Psychotherapy: Theory, Research and Practice,* 1979, *16,* 460–466.

Luborsky, L., Woody, G. E., McLellan, A. T., O'Brien, C. P., & Rosenzweig, J. Can independent judges recognize different psychotherapies? An experience with manual-guided therapies. *Journal of Consulting and Clinical Psychology,* 1982, *50,* 49–62.

Mahrer, A. The therapist-patient relationship: Conceptual analysis and a proposal for a paradigm-shift. *Psychotherapy: Theory, Research and Practice,* 1978, *15,* 201–215.

Meltzhoff, J., & Kornreich, M. *Research in psychotherapy.* New York: Atherton, 1970.

Miller, L. C., Barrett, C. L., Hampe, E., & Noble, H. Comparison of reciprocal inhibition, psychotherapy and waiting list control for phobic children. *Journal of Abnormal Psychology,* 1972, *79,* 269–279.

Mintz, J., Steuer, J., & Jarvik, L. Psychotherapy with depressed elderly patients: Research considerations. *Journal of Consulting and Clinical Psychology*, 1981, *49*, 542–548.

Mischel, W. Toward a cognitive social learning reconceptualization of personality. *Psychological Review*, 1973, *80*, 252–283.

Mischel, W. On the future of personality measurement. *American Psychologist*, 1977, *32*, 246–254.

Mischel, W. On the interface of cognition and personality: beyond the person–situation debate. *American Psychologist*, 1979, *34*, 740–754.

Morris, R. J., & Suckerman, K. R. Therapist warmth as a factor in automated desensitization. *Journal of Consulting and Clinical Psychology*, 1974, *42*, 244–250.

Nicholls, J. G., Licht, B. G., & Pearl, R. A. Some dangers of using personality questionnaires to study personality. *Psychological Bulletin*, 1982, *92*, 572–580.

Orlinsky, D. E., & Howard, K. I. The effects of sex of therapist on the therapeutic experiences of women. *Psychotherapy: Theory, Research and Practice*, 1976, *13*, 82–88.

Orne, M. T. On the social psychology of the psychological experiment: With particular reference to demand characteristics and their implications. *American Psychologist*, 1962, *117*, 776–83.

Parloff, M. B. Can psychotherapy research guide the policy maker? *American Psychologist*, 1979, *34*, 296–306.

Parloff, M. B., Waskow, I. E., & Wolfe, B. E. Research on therapist variables in relation to process and outcome. In S. Garfield & A. E. Bergin (Eds.), *Handbook of psychotherapy and behavior change*. New York: Wiley, 1978.

Paul, G. L. *Insight versus desensitization in psychotherapy*. Stanford, Calif.: Stanford University Press, 1966.

Peebles, M. J. Personal therapy and ability to display empathy, warmth and genuiness in psychotherapy. *Psychotherapy: Theory, Research and Practice*, 1980, *17*, 258–262.

Pope, B., Nudler, S., Norden, J. S., & McGee, J. P. Changes in nonprofessional (novice) interviewers over a 3-year training period. *Journal of Consulting and Clinical Psychology*, 1976, *44*, 819–825.

Pope, B., Nudler, S., Van Korff, M. R., & McGee, J. B. The experienced professional interviewer versus the complete novice. *Journal of Consulting and Clinical Psychology*, 1974, *42*, 680–690.

Rappaport, H., & Rappaport, M. The intergration of scientific and traditional healing. *American Psychologist*, 1981, *36*, 774–781.

Rickels, K., Baumm, N. C., & Taylor, W. Humanism in clinical research. *Psychosomatics*, 1964, *5*, 315–316.

Rickels, K., Boren, R., & Stuart, H. M. Controlled psychopharmacological research in general practice. *Journal of New Drugs*, 1964, 138–147.

Ricks, D. F. Supershrink: methods of a therapist judged successful on the basis of adult outcomes of adolescent patients. In D. F. Ricks, A. Thomas, & M. Roff (Eds.), *Life history research in psychopathology*. Minneapolis: University of Minnesota Press, 1974.

Risch, S. C., Huey, L. Y., & Janowsky, D. S. Plasma levels of tricyclic antidepressants and clinical efficacy: Review of the literature—part 1. *Journal Clinical Psychiatry*, 1979, *40*, 4–16.(a)

Risch, S. C., Huey, L. Y., & Janowsky, D. S. Plasma levels of tricyclic antidepressants and clinical efficacy: Review of the literature—part 2. *Journal Clinical Psychiatry*, 1979, *40*, 58–69.(b)

Rosen, G. M., Glasgow, R. E., & Barrera, M. A controlled study to assess the clinical efficacy of totally self-administered systematic desensitization. *Journal of Consulting and Clinical Psychology*, 1976, *44*, 208–217.

Sabshin, M., & Ramot, J. Pharmacotherapeutic evaluation and the psychiatric setting. *AMA Archives of Neurology and Psychiatry,* 1956, *75,* 362–370.

Seidman, E., Golding, S. L., Hogan, T. P., & LeBow, M. D. A multidimensional interpretation & comparison of three A-B scales. *Journal of Consulting and Clinical Psychology,* 1974, *42,* 10–20.

Shader, R. I., Grinspoon, L., & Harmatz, J. S. The therapist variable. *American Journal of Psychiatry,* 1971, *127,* 49–52.

Shaw, B. F. Comparison of cognitive therapy and behavior therapy in the treatment of depression. *Journal of Consulting and Clinical Psychology,* 1977, *45,* 543–551.

Sheard, M. H. The influence of doctor's attitude on the patient's response to antidepressant medication. *Journal of Nervous and Mental Disease,* 1963, *136,* 555–560.

Shullman, S. L., & Betz, N. E. An investigation of the effects of client sex and presenting problem in referral from intake. *Journal of Counseling Psychology,* 1979, *26,* 140–145.

Staples, F. R., Sloane, R. B., Whipple, K., Cristol, A. H., & Yorkston, N. Process and outcome in psychotherapy and behavior therapy. *Journal of Consulting and Clinical Psychology,* 1976, *44,* 340–350.

Stearns, B. C., Penner, L. A., & Kimmel, E. Sexism among psychotherapists: A case not yet proven. *Journal of Consulting and Clinical Psychology,* 1980, *48,* 548–550.

Stone, W. N., Green, B. L., Gleser, G. C., Whitman, R. M., & Foster, B. B. Impact of psychosocial factors on the conduct of combined drug and psychotherapy research. *British Journal Psychiatry,* 1975, *127,* 432–439.

Strupp, H. H. On the basic ingredients of psychotherapy. *Journal of Consulting and Clinical Psychology,* 1973, *41,* 1–8.(a)

Strupp, H. H. The interpersonal relationship as a vehicle for therapeutic learning. *Journal of Consulting and Clinical Psychology,* 1973, *41,* 13–15.(b)

Strupp, H. H. The therapist's theoretical orientation: An overated variable. *Psychotherapy: Theory, Research and Practice,* 1978, *15,* 314–317.

Strupp, H. H. A psychodynamicist looks at modern behavior therapy. *Psychotherapy: Theory, Research and Practice,* 1979, *16,* 124–131.

Strupp, H. H. Clinical research, practice & the crisis of confidence. *Journal of Consulting and Clinical Psychology,* 1981, *49,* 216–219.

Tuma, A. H., & May, P. R. A. Psychotherapy, drugs and therapist experience in the treatment of schizophrenia. *Psychotherapy: Theory, Research and Practice,* 1975, *12,* 138–142.

Tuma, A. H., May, P. R. A., Yale, C., & Forsythe, A. B. Therapist experience, general clinical ability, and treatment outcome in schizophrenia. *Journal of Consulting and Clinical Psychology,* 1978, *46,* 1120–1126.

Turner, S., & Armstrong, S. Cross-racial psychotherapy: What the experts say. *Psychotherapy: Theory, Research and Practice,* 1981, *18,* 375–378.

Uhlenhuth, E. H., Canter, A., Neustadt, J. O., & Payson, H. E. The symptomatic relief of anxiety with meprobamate, phenobarbital and placebo. *American Journal of Psychiatry,* 1959, *115,* 905–910.

Uhlenhuth, E. H., Rickels, K., Fisher, S., Park, L. C., Lipman, R. S., & Mock, J. Drug, doctor's verbal attitude and clinic setting in the symptomatic response to pharmacotherapy. *Psychopharmacology,* 1966, *9,* 392–418.

Uhlenhuth, E. H., Lipman, R. S., & Covi, L. Combined pharmacotherapy and psychotherapy. *Journal of Nervous and Mental Disease,* 1969, *148,* 52–64.

Walls, G. B. Values and psychotherapy: A comment on "Psychotherapy and religious values. *Journal of Consulting and Clinical Psychology,* 1980, *48,* 640–641.

Weismann, M. M., Prusoff, B. A., Dimascio, A., Neu, C., Goklaney, M., & Klerman, G. L. The efficacy of drugs and psychotherapy in the treatment of acute depressive episodes. *American Journal Psychiatry,* 1979, *136,* 555–558.

Whitehorn, J. C., & Betz, B. J. A study of psychotherapeutic relationships between physician and schizophrenic patients. *American Journal of Psychiatry,* 1954, *111,* 321–331.

Wogan, M. Effect of therapist–patient personality variables on therapeutic outcome. *Journal of Consulting and Clinical Psychology,* 1970, *35,* 356–361.

Wright, J. H., & Looney, J. J. *Psychotherapy in depression: A general systems analysis.* Presented at Annual Meeting of the Academy of Psychosomatic Medicine, San Francisco, Calif., October 30, 1979.

Yeaton, W. H., & Sechrest, L. Critical dimensions in the choice and maintenance of successful treatments: Strength, integrity, and effectiveness. *Journal of Consulting and Clinical Psychology,* 1981, *49,* 156–167.

12

Ethical Issues

GEARY S. ALFORD and WILLIAM G. JOHNSON

INTRODUCTION

By its very nature, psychotherapeutic research involves some of the most complex and difficult ethical issues and problems confronted within the diverse fields of human investigation. Unlike objective, experimentally derived concepts of natural law verifiable by reference to an enduring, universal, external reality, ethical concepts emerge from subjective, emotional, and even esthetic judgments. Moreover, values and beliefs from which ethical propositions emerge are not static but constantly evolving. What is considered ethical behavior today may be considered unethical within the span of just a few years. To recognize the ultimate subjectivity of ethical propositions is not to render the philosophy of ethics moot. Ethics are more like social than natural laws. While interpretations and applications may change as a consequence of the vagaries of human emotion and sensibilities, certain fundamental notions tend to endure and are accorded almost universal endorsement. From such core concepts, as from simple propositions in the U.S. Constitution or historical and legal precedents, systematic codes of conduct can be derived and guidelines for specific, discrete cases articulated.

The notion that a single, individual human life has value, and as such is to be accorded protection, preservation, respect, and dignity, is such a core concept. Such a prescriptive axiological proposition is not verifiable by reference to empirical examination of natural phenomena or natural law. Rather, its validity arises from and rests on other sets of propositions, primarily legal codes and religious and humanistic texts (in which the right of life and civil

GEARY S. ALFORD and WILLIAM G. JOHNSON ● Department of Psychiatry and Human Behavior, University of Mississippi Medical Center, Jackson, Mississippi 39216.

liberty have been, as a reflection of the values of general societies, explicity and implicity incorporated). Legal proscriptions against murder clearly imply a value against the illegal taking of life; hence, they also imply that societies in which such laws exist value—or at least intellectually claim to value—individual human life. While individual ethical values and sensibilities may not be instated by legislation, laws and regulations governing behavior can be. The history of human experimentation, particularly within the past 35 years, is marked by both an effort to enhance the *spirit* of ethical guidelines and, perhaps more importantly, by a more clearly and finely drawn *letter,* in the form of increasingly specific regulations governing ethical conduct with human subjects.

Experimentation with human subjects is not a new phenomenon. Alexander the Great tested various battle formations and tactics with his Macedonian troops; Socrates investigated different teaching methods with young students; and, of course, the history of medicine is one of human experimentation. Explicit and systematic ethical prescriptions for human investigation that go beyond criminal law, however, are relatively new. Until the middle of the 20th century, most societies relied on professional researchers (usually among the most educated and respected members of a community) to adhere not only to criminal and civil laws but to conduct themselves according to the highest ideals of human behavior. These *ideals* were rarely explicitly specified but were assumed to be commonly understood.

Today we may look back and shudder at such episodes as the Tuskegee syphilis study (Department of Health, Education and Welfare, 1973), but it was the discovery at the end of World War II of the monstrous Nazi medical experiments that shocked the world and demonstrated the extent to which human experimentation could be perverted. While these "medical experiments" were gruesome, horrifying, and even psychopathic in their nature, a perhaps less horrifying but equally frightening discovery was subsequently made. Many of the individuals who conducted these experiments were not Nazi fanatics but prior to the war had been solid, respectable physicians and scientists. It would be too easy to attribute their behavior to fanatic ideological excesses of Nazism or to individual psychopathology. Although this was no doubt true in some cases, in others transcripts from the Nuremberg trials suggest that at least a few had somehow convinced themselves that their actions and "studies" were justifiable. That previously sane, rational, "average" human beings would engage in such cruel and inhuman behavior is not only perplexing but frightening (Mitscherlich & Mielke, 1949). A similar, though correspondingly less intense reaction appeared in response to the results of Stanley Milgram's (1963) studies in obedience. What these revelations so clearly demonstrated was that sane, rational researchers, even clinical scientists otherwise dedicated to human welfare and the relief of individual suffering, could under certain circumstances violate even basic precepts of appropriate, desirable, ethical human interaction. What are these certain circumstances? In the case of Nazi "experiments," the scientists were operating within a global pathophilosophic political and social system. But was that the crucial condition? Unfortunately the answer seems to be no. Milgram's (1963) subject–experimenters were "normal," "average" volunteers simply participating in a "learning experiment." Willingness to inflict pain and even risk of death to another human being was evoked solely within the context of a "scientific experiment." That the "pain" was faked and the risk of harm or death in fact nonexistent is irrelevant. That the "true" subjects were the "experimental assistants," who were volunteers from the local community, does not mitigate the implications for ostensibly more mature, trained senior scientific investigators. In his review of the social nature of psychological research, Friedman (1967) examined the sources and influences of bias on the experimenter and the experimenter's behavior as well as on the subjects and independent events-variables. We as experimenters are not immune to the effects of bias, whether there is intrinsic commitment to a favored theory or belief or to extrinsic, contextual demand characteristics. Specific codes of scientific conduct are necessary to fulfill the justifiable demands of an informed and sensitized public that the humanism of scientists be enhanced. Furthermore, it is precisely because of our fundamental commitment to human welfare (to enhance, preserve, and protect life), basic human rights and dignities that we as human investigators need specific ethical guidelines against which we can check our research protocols. This is similar to the way even the most experienced pilot, wishing to maximally ensure the preservation of his or her own life, carefully and systematically goes through a standard preflight checklist. While humanitarian sensitivity, professional training, sound judgment, and experience may be necessary, they are not sufficient.

The Nuremberg Code was one of the first explicit sets of standards for human investigation to be widely circulated, broadly endorsed, and generally accepted and incorporated into diverse formulation of ethical research conduct. This document set forth 10 basic principles fundamental to *ethical* experimentation with human subjects. These 10 points have been combined and summarized into 5 critical guidelines:

1. Informed consent. A research subject must be given all the information about his or her participation in an experiment, which must be truly voluntary.
2. Freedom to withdraw. A research subject must be free to withdraw from an experiment at any time without negative consequences.
3. Minimize risks. All unnecessary risk should be eliminated from the experiment.

4. Benefit of research to the subject or to society should clearly outweigh risks to the subject.
5. Competence of experimenter. Experimental procedures should be performed or carried out only by individuals qualified to do so.

These basic principles have been incorporated into various professional ethical codes and have been adopted as legal requirements by the Department of Health, Education, and Welfare (1971) and, subsequently, by the Department of Health and Human Services (Federal Register, 1981) for research supported by federal grants or conducted within institutions receiving federal support. In addition, investigators are required to submit protocols for proposed research to human investigation committees for review and approval in their respective institutions. The purpose of these committees is to provide an independent, objective review of proposed research to further ensure adherence to these guidelines.

While these guidelines initially appear simple, clear, and straightforward, in actual application and practice various ambiguities and interpretations emerge. For example, even an ostensibly simple and clear question as to who should be considered an experimental subject can turn out to be neither simple nor clear. For instance, does any variation in a therapeutic procedure with a given patient constitute an experiment? Does it become an experiment only if the therapist keeps data or plans to publish the results? It is an experiment if there is no known standard, generally accepted, efficacious treatment regimen for a given condition? Does not any clinical treatment *per se* consititute an experiment, even if the therapist has only clinical and not research or possible publication ends in mind? Such issues become even more complex where research is not designed to benefit the patient or subject directly. Should passersby on a street, surreptitiously observed donating money to a male versus female "invalid," be considered experimental subjects from whom prior, informed consent is a necessity?

Such issues are partially resolved by considering the ethical principles together rather than separately, one at a time. For example, prior informed consent might be a desirable ideal for anyone participating as a subject in any research activity. Of course, to require such would preclude the pursuit of many research activities of benefit to society and for which the risk or inconvenience to the subject is almost nonexistent. The fundamental purpose of ethical guidelines is the protection of the subject—his or her health, feelings, dignity, and general welfare. Thus, risk, possibility of physical or psychological injury, or potential adverse effect on the individual's general well-being consititutes the crucial element against which all other considerations must be weighed and, indeed, from which other points in the guidelines (e.g., competence of experimenter) are derived. Although experimental conditions involving minimal risks do not completely obviate the importance of other ethical constraints (e.g., prior informed consent), the lower the risk factors, the relatively greater weight can be given to consideration of the scientific or social benefit of the proposed study. Passersby asked by a young woman versus an older one—or male versus female solicitor—to donate to a charitable organization need not be previously informed of the nature of the study and asked to agree to participate and sign a consent form. Such a study would not significantly inconvenience the subject. There is no *a priori* reason to suspect any form of physical or psychological injury. Such research introduces no new or unusual experience for subjects. Subjects are, in essence, free to "withdraw" by refusing to donate or even by ignoring the solicitor. On the other hand, such a study should be expected to contribute to our knowledge base about social prejudices (e.g., gender, age) and to provide information useful in increasing charitable contributions from which society as a whole might benefit.

The foregoing example demonstrates the kind of multifaceted issues inherent in even simple, relatively noninvasive human investigation. As exemplified above, the two crucial issues at the core of all other ethical considerations are (1) the relative invasiveness of the procedure—that is, to what extent might the experiment induce change or harm in the participant and (2) if such a possibility exists, the informed, voluntary participation of the subject.

The nature of psychotherapeutic research presents particularly complex and, at times, subtle ethical problems for investigators. Rather than simply assessing an acute, discrete behavioral reaction in a specific environment (with a noninvasive procedure as in the social psychological research example described above), psychotherapeutic research is usually designed to induce significant and enduring change in the subject or patient. Indeed, the scientific societal value of clinical research is judged primarily by the potency or effectiveness in producing change (efficacy); the amount, significance, and type of behavior change (clinical importance); and the reliability or duration of the change (long-term outcome). In this sense, psychotherapy research, in fact psychotherapy itself, is very much an invasive procedure. Simply observing and measuring certain abnormal behaviors may be done in a manner that avoids influencing the subject. Psychotherapeutic research, however, specifically entails an intention to generate change in the patient–subject. Therein lies the inherent possibility of harm or risk. Harmful or iatrogenic effects arising from an intervention procedure might include (1) exacerbating rather than ameliorating the patient's problem; (2) instating a new, additional problem; and (3) delaying or diverting the patient from receiving an alternative, more effective procedure. Further, certain therapeutic interventions may involve equipment risks (e.g., electrical shock, etc.) or tasks that might endanger a patient's health or safety.

INFORMED CONSENT

The notion of informed consent is the bedrock upon which the relationship between the therapist (investigator) and subject rests. It is based on the idea of free will and voluntary participation. That is, the basic ethical issue is for subjects to agree to participate in a project and with full knowledge of any documented or anticipated side effects that may occur during the implementation of the procedures. A most important issue involves the question of whether individuals are competent to give such consent. For example, by reason of age, mental retardation, or psychoticism, some individuals may not be in a position to give their consent. They may not be able effectively to evaluate issues involved in therapeutic interventions or treatments or to ascertain whether their rights are being violated. This does not mean that because an individual is in an institution or under age that he or she is automatically not competent to give informed consent. Investigators, however, must be extremely cautious in order to avoid the potential of coercion whether from project staff or others who have contact with the potential subject, including employees or hospital administrators.

Other dilemmas present themselves after a subject is enrolled in a study. Even though the subject may have given informed consent regarding the untoward consequences of his or her participation, he or she *may,* when such side effects or aversive features of the intervention manifest themselves, be inclined to withdraw but decide not to do so because of the earlier "commitment" to the investigators. In contrast to nonclinical research, not only must the investigator inform the prospective subject of the nature, risks, and so on of the intended experiment but must also discuss with the patient appropriate alternative treatments. The fact that for some problems no particular intervention procedure has been documented as reliable and effective—or that controversy surrounds various current alternatives—does not obviate the necessity of discussing these diverse treatment options with the experimental subject.

Similarly, the clinician must be sensitive to his or her special relationships with prospective subjects. Because of the "doctor–patient" relationship, the patient may be unusually compliant or susceptible to influence and thus may "volunteer" under pressure. This same subtle, even unintended pressure may influence choice of treatment goals within or outside the context of research. The Association for Advancement of Behavior Therapy (AABT) developed and adopted an ethical practice statement that translated conventional ethical principles into a set of guidelines for practitioners to follow (Association for Advancement of Behavior Therapy, 1977). These guidelines go beyond the constraints enumerated in the Nuremberg Code in that they specifically address ethical issues within the realm of clinical or professional service delivery in contrast to human experimentation *per se.*

In addition to adopting the Nuremberg positions with regard to voluntary consent to treatment, freedom to withdraw, and competency of therapist to conduct the procedures, the AABT guidelines also specifically address crucial issues of therapeutic goals and efficacy of diverse treatment alternatives. Since psychotherapeutic research almost invariably involves a patient–client population (or in the case of analogue studies, nonpatients who nevertheless are placed in patient roles to undergo some "therapeutic" behavior change procedure), clinical research must adhere to ethical constraints of both human experimentation and clinical service provisions. Issues arising from both sets of guidelines must be considered during research planning, addressed in discussion with prospective patient–subjects, and at least be summarized in a written consent form.

In addition to questions in regard to who may provide consent and under what circumstances, there are questions as to what and how much information about the proposed research protocol must be given to prospective subjects.

INFORMED CONSENT AND DECEPTION

There may be some instances when informing subjects of all aspects of the intervention and its side effects may compromise the experiment. For example, informing subjects that they may be placed on an active treatment regimen or one that is an "attention–placebo" can result in their search for the type of intervention they in fact are receiving. This may lead to spurious compliance, hence biasing the results. Additionally, informing subjects that they may be assigned to a control group may jeopardize potential results. Generally, whether or not to engage in deception is based on a cost-benefit analysis. Accordingly, in order to maximize the value of the experiment while minimizing its cost and risk, the experimenter must first note that there is no essential risk to the subject for participating in the experiment. If this is so, then deception may be implemented where the experimenter fails to mention a key aspect of the experiment (which would jeopardize the results if the subject were aware).

Deception essentially may take two forms. In one instance the experimenter will deliberately fail to inform the subject about an aspect of the study. In the second case the subject will be misled. In light of the above, it should be noted that individuals will often behave in "socially desirable ways" if, in fact, they are cognizant that they are under study. In cases of deception, the following guidelines apply: (1) there should not be any documented and/or potential risk of the intervention; (2) the benefits of deception should far outweigh any cost to the subject's dignity as a result of the deception process; (3) an analysis in the form of a pilot study should be conducted to actually determine whether or not subjects would, in fact, behave differently if they were not deceived; and (4) mechanisms for debriefing should be maintained, carried out, and documented.

WITHHOLDING TREATMENT

An additional difficulty is often encountered in clinical psychotherapeutic trials. This is so when one compares several forms of treatment that have been empirically documented to be effective with a new treatment lacking such support. Does the promise of the new strategy, no matter how great, warrant withholding the established procedure?

In our second instance the ethical dilemma is similar. Here, a treatment that has already proven to be of some utility is to be contrasted with one that is expected to achieve much more beneficial results. In both cases there is obviously some conflict between the dictates of the experimental design and the ethical constraints favoring application of the known treatment. Of course, there is the option to use the known treatment later, should the experimental treatment result in failure.

Then there are ethical dilemmas that are often encountered during the course of the study. In this particular case, subjects may be assigned to a treatment condition that is obviously not effective for their problem. For example, in a recent research project involving bulimia, subjects were randomly assigned to exposure with response prevention or eating habit control. During the course of the study, it was obvious to the investigators that some subjects would have responded much better to the alternative approach. In one case the subject became disenchanted with her treatment. The investigators then had to make a choice between maintaining this individual in the study or removing her and offering another clinical intervention. However, the saving grace in this study was that both treatments were implemented, but in randomized order.

INFORMED CONSENT – EMPIRICAL APPROACH

There is little actual information on the decision-making behavior of subjects consenting to participate in an experiment. Recently, Michaels and Oetting (1979) examined whether subjects would participate in various studies if, in fact, they were informed. Male and female introductory psychology students indicated, from a list of physically and/or psychologically stressful procedures, which they would be willing to undergo (given two levels of personal or social benefits or no particular benefits). In most cases, benefit had no effect on the amount of stress each subject would be willing to tolerate. Additionally, males showed more tolerance of psychologically stressful procedures than females.

Such studies may reveal factors that influence the individual's willingness to participate. They may also provide insight into the process of deciding to serve as a research subject. In particular, this work may help us to identify those risk factors and other potentially aversive components involved in various types of human experimentation that lead individuals to decline to participate. Once such factors are identified (and if present in a proposed project), it becomes incumbent upon the investigator to clearly inform potential subjects of them. Arguments for the necessity of using naive subjects for a particular project become moot if that study entails factors that previously have led most of the queried recruits to decline participation.

REVIEW OF CURRENT LITERATURE

Although a review of the current relevant clinical research literature is conducted with a view to helping the investigator to formulate new and mean-ingful questions and to design an appropriate and sound methodology, there are important ethical aspects to this exercise as well. A careful review can reveal complications or even contraindications related to specific treatment or patient variables known to the ethical investigator. The investigator must be knowledgeable of current efficacy and outcome data related to diverse treat-ment options so that prospective patient–subjects are informed of alternative treatments, their relative effectiveness, time commitment, possible difficulties or discomfort, and other factors (e.g., costs that might influence a patient's decision to participate as a subject). The investigator needs to assess the pro-posed experimental procedure carefully to determine its consistency with estab-lishing basic scientific principles of behavior and its relationship to current therapeutic procedures. The greater the inconsistency with basic psychological principles, reliance on otherwise unsupported theory, and/or divergence from currently accepted clinical practices, the greater the ethical burden on inves-tigators to examine their experimental procedures for potential harm. The eth-ical experimenter must carefully and clearly inform potential subjects of the untested and experimental nature of the treatments while also discussing alter-native procedures that the patient might wish to elect. Finally, literature reviews may help investigators design methodologies that can answer experi-mental questions within the context of providing competent clinical service while minimizing inconvenience, delay, discomfort, or risk of harm to patient–subjects.

DESIGNING THE METHODOLOGY

Investigators need to recognize that, when dealing with clinical populations, the clinical care and treatment of the patient must remain the central focus of intervention efforts. Thus, any methodological component or variation that might not only constitute a direct risk but may otherwise harm the patient by delaying treatment or diminishing treatment effectiveness must be carefully considered. Unfortunately, almost every paradigm entails experimental treatment conditions that have the effect of (1) delaying treatment (e.g., waiting-list control and placebo–expectancy conditions); (2) delaying parts or components of treatment (e.g., multiple-baseline designs; see Hersen & Barlow, 1976); (3) at best temporarily diminishing treatment effectiveness (e.g., in component analyses); or (4) allow undesirable, symptomatic behavior to recur during non-treatment phases (e.g., in reversal or return to baseline condition phases). Even the use of a "standard" or "traditional" treatment against which the experimental condition is to be compared does not avoid this dilemma if, as has often been the case in recent psychotherapeutic research, the "conventional therapy" is one that has been found to be of questionable efficacy. The investigator must explore the possible adverse effects of such treatment delays or interruptions and carefully weigh these against potential benefits to patients.

Because clinical populations serve as subjects, therapeutic investigations require direct supervision by clinicians competent in both the therapeutic procedure under investigation and in working with the specific clinical problem or population being studied. The clinical investigator need not necessarily rely on the patient to request termination of his or her participation. To the contrary, the experimenter (clinician) should carefully monitor the patient's general clinical condition and remove him or her if evidence indicates an adverse reaction to experimental treatment. This procedure should also be carried out if the patient's clinical status changes so that it contraindicates continuation in such an experimental project. Ethical clinical investigators (indeed ethical clinicians) should be prepared to abandon a treatment procedure not only if it results in undesirable or adverse effects but also if it fails to generate therapeutic improvement within a reasonable time frame, particularly with respect to alternative interventions whose effectiveness has been documented. In designing a study, the investigator should determine whether experimental personnel are appropriately trained, knowledgeable, and skilled to carry out their respective clinical research duties.

CONSULTATION AND INSTITUTIONAL REVIEW

Although discussing a proposed research project with professional colleagues is always a useful and desirable part of planning, formal consultation should always be obtained (1) whenever significant potential risks are involved in a procedure, (2) where a procedure is or could reasonably be expected to be controversial, (3) where high-risk patient populations or problems are involved, or (4) where principal investigators are not qualified or not fully trained, experienced, or skilled in one or more aspects of the assessment or therapeutic procedure. When the principal investigators have finished drafting the protocol for the proposed research, it will, depending upon the policies of the institution in which the research is to be conducted, require submission to institutional review boards. Also to be considered are state and federal regulations that apply with regard to specific grant support. These regulations apply even if no part of the study itself is being supported by federal funds.

The Institutional Review Board (IRB) is charged with the responsibility and authority to review, approve, disapprove, or require changes in research activities involving human subjects. The institution provides a meeting space for the IRB and sufficient staff to support IRB's review and record-keeping duties. It is also designed to encourage and promote communication among various administrative heads, clinical directors, research investigators, clinical care staff, and other institutional officials involved with human subjects.

Research investigators are respomsible for obtaining informed consent and ensuring that no human subjects are involved in research prior to the attainment of such consent unless otherwise authorized by the IRB. Legally effective informed consent shall:

1. Be obtained from the subject or the subject's legally authorized representative.
2. Be stated in language understandable to the subject or the authorized representative.
3. Be obtained under circumstances that offer the subject or authorized representative sufficient opportunity to consider whether or not he/she should participate.
4. Not include exculpatory language in which the subject and the representative are made to waive or appear to waive any legal rights, releases, or appears to release the research investigator, the sponsor, the institution, or its agency from liability or negligence.

Informed consent includes the following:

1. A clear statement that the subject's participation involves research, the explanation of the purposes of the research, the expectation regarding subject participation, a description of the procedures and interventions, and an identification of any procedures which are deemed experimental. Also, a description of any foreseeable risk and/or discomfort must be provided, as well as a description of any benefits which may accrue to the subjects as a result of their participation. A statement regarding

the confidentiality of records must also be incorporated. When there is more than minimal risk, an explanation is required as to whether any compensation for injury will be available. Additionally, a statement should be given as to an explanation of whom to contact for questions regarding the research. Also, a clear statement is needed that participation is voluntary, that refusing to participate will involve no penalty or loss of benefits, and that the subject may discontinue participation at any time without penalty or loss of benefits.

2. Research investigators are required to document informed consent by use of a written form, and this document should be retained for approximately three years. They also are required to submit a progress report on a yearly basis to their IRB.

3. All research conducted with human participants must be reviewed except for certain exemptions. Exempted research includes certain categories of no risk or very low risk involving human subjects. Exempted research includes studies conducted in established or currently accepted educational settings, involving normal educational practices (e.g., as research on regular or special educational instructional strategy), or the comparison among instruction techniques or classroom management method. Additionally, research involving use of an educational task (cognitive, diagnostic, aptitude, etc.), if information taken from these sources is recorded in such a manner that subjects cannot be identified, is exempt. Also exempt is research involving surveys or interview procedures, except where all the following exist: (a) the subjects can be identified, (b) responses could place the subject at risk for criminal or civil liability, and (c) the research deals with sensitive aspects of the subject's behavior such as illegal conduct, drug use, sexual behavior, etc. Research involving the observation of public behavior is generally exempt in the aforementioned cases. Also exempt is research involving the collection or study of existing data, documents, records, pathological specimens, if these are publicly available or if the information is recorded by the investigator in such a manner that subjects cannot be identified. The definition of minimal risk means that it is not greater, considering probability and magnitude, than that ordinarily encountered in daily life or during the performance of routine physical or psychological examinations or tests.

PUBLICATION OF CLINICAL RESEARCH FINDINGS

Clinical researchers have traditionally endorsed the policy that research findings be reported conservatively, avoiding excessive, sensational, or overgeneralized claims. With few exceptions, at least in the major professional journals, authors (with the aid of able editors and reviewers) have adhered to this position in practice. This tradition is an important one and should be followed when interpreting results and formulating discussion or summary sections of manuscripts. Of course, serious investigators are naturally enthusiastic when significant, potent, and even astonishing positive results are obtained. It is all too easy for such enthusiasm to exert a mildly intoxicating effect on judgment. Unfortunately, such results might be translated into unwarranted conclusions which, if they slip by editors, can mislead practicing clinicians as well as other researchers and, perhaps, even institutional policymakers to the ultimate detriment of patients and society. The history of psychotherapeutic research is a history of overly optimistic conclusions. Indeed, each new therapeutic school or approach has hailed its first glowing successes only to subsequently limit and qualify its claims in light of cooler, if not more sober, independent evaluations (Tourney, 1967).

SUMMARY

Over the last half century, the role of ethical issues in human investigation has evolved from one of essentially elective sensitivity and adherence to prevailing religious and humanistic ideals to increasingly explicit sets of guidelines, proscriptions, and even prescriptions for ethical conduct. Currently, much of what was traditionally considered the *duty* of ethical investigators is now *required* by law. To what extent statutory regulations governing the conduct of human investigation will continue to grow in specificity and scope is unclear. Likewise, it is difficult to determine if and when expanded ethical controls reach a point where individuals and society no longer derive additional benefit from tighter more restrictive controls and, in fact, risk losing what would have been valuable findings.

Similarly, research topics entailing sociopolitically controversial issues call for particular care in terms of how scientific questions are framed, projects conducted, and results discussed. It is unfortunate, however, whenever prevailing political winds exert a chilling effect on legitimate scientific inquiry or on publication or free, open speech (see Herrnstein, 1982). Clearly, ethical guidelines and constraints always need to address the conduct and actual operations of research. Speculations as to motives of the investigator or how his or her results might be misused, while interesting, are most certainly tertiary questions that should have little or no place in evaluating projects with respect to their adherence to ethical requirements. One cannot praise the burning of *Mein Kampf* or *The Gulag Archipelago* and condemn the burning of *Das Kapital* or *Tropic of Cancer*.

Ethical standards may be proscriptive with respect to *how* research is conducted, but they should never prescribe *what* is discovered. The purpose of ethical standards is not to constrain truth but only to ensure that the pursuit of truth is conducted with respect for the dignity, health, and freedom of all involved.

REFERENCES

American Psychological Association. *Ethical principles in the conduct of research with human participants*. Washington, D.C.: Author, 1977.

Association for Advancement of Behavior Therapy. Ethical issues for human service. *Behavior Therapy,* 1977, *8,* 763–764.

Department of Health, Education and Welfare. Publications No. (NIH) 72–102. Washington, D.C.: U.S. Government Printing Office, Dec. 1, 1971.

Department of Health, Education and Welfare. *Final report of the Tuskegee syphilis study ad hoc advisory panel*. Washington, D.C.: U.S. Government Printing Office, 1973.

Federal Register, Final regulations amending basic HHS policy for the protection of human research subjects. *Federal Register,* 1981, *46,* 8366–8392.

Friedman, N. *The social nature of psychological research: The psychological experiment as social interaction*. New York: Basic Books, 1967.

Herrnstein, R. J. I.Q. Testing and the media. *The Atlantic Monthly,* August 1982.

Hersen, M., & Barlow, D. H. *Single case experimental designs: Strategies for studying behavior change*. New York: Pergamon, 1976.

Michaels, T. F., & Oetting, E. R. The informed consent dilemma: Empirical approach. *Journal of Social Psychology,* 1979, *109,* 223–230.

Milgram, A. Behavioral study of obedience. *Journal of Abnormal and Social Psychology,* 1963, *67,* 371–378.

Mitscherlich, A., & Mielke, F. *Doctors of infamy*. New York: Schuman, 1949.

Tourney, C. A history of therapeutic fashions in psychiatry, 1800–1966. *American Journal of psychiatry,* 1967, *124,* 784–796.

13

Technical Diversity

NORMAN I. HARWAY

INTRODUCTION

The diversity of psychotherapy—both in theoretical formulation and in practice—and disagreements as to the very nature of the phenomena that psychotherapy is designed to influence are so clearly evident in the literature that recognition of this state of affairs may well represent the strongest point of agreement among therapy watchers. Research in and on psychotherapy has expanded exponentially from the early systematic studies four decades ago. A bibliography of research on individual psychotherapy with adults (Strupp, 1964) cited approximately 1,000 studies. Some $3\frac{1}{2}$ years later, an updated compendium (Strupp & Bergin, 1968) contained almost 2,800 items. Now, 15 years later, compiling such a listing would be an almost unimaginably heroic undertaking. Almost a decade ago, Bordin (1974) noted that a survey of this massive research must result "in confusion because of the apparent lack of connection between the efforts of different investigators" (p. 4). The comment is still applicable today, though a degree of substantive and methodological continuity can be found in the work of investigators who share a common theoretical commitment.

Most clinicians, at least among psychologists, report that they do not subscribe to a single specific theory of psychotherapy; rather, they describe their approach as eclectic. Surveys of clinical psychologists reveal that over half "disavow any systematic, organized, articulated theory held in common with others" (Sundland, 1977, p. 190). It is doubtful that there is a single expla-

---

NORMAN I. HARWAY ● Department of Psychology, University of Pittsburgh, Pittsburgh, Pennsylvania 15260.

nation for this state of affairs, but it does indicate that for many the more or less cohesive alternative theories of psychotherapy and the research deriving from them do not independently provide an adequate basis for dealing with the complexities therapists face in clinical practice.

One can distinguish two broadly contrasting attitudes regarding the nature of a therapeutic activity: one emphasizing therapist techniques and another stressing the relationship between therapist and client. These attitudes, though not unrelated to the different theoretical orientations, are not iso-morphic with them. Wilson and Evans (1977) note the early recognition of interpersonal relationships in the behavior therapy literature, which had been criticized for neglecting them. Wolff (1956) and Sundland and Barker (1962), in examining traditional therapies, distinguish between psychotherapies follow-ing formulas of interpretation from those basing themselves on evolving inter-personal relationships. The distinction between technique and relationships is not always clear-cut. As a referent, *technique* appears usually to carry conno-tations of specificity and of the therapist acting on the client, while *relationship* has generally a broader, less particular quality and an implicit emphasis on client motivational attributes.

PSYCHOTHERAPY AS TECHNIQUE

There is an intuitive appeal to the view of psychotherapy as the application of techniques based on psychological principles for the amelioration of partic-ular psychological dysfunctions and distresses. It implies that there are discri-minable and specifiable technical manipulations that, if mastered and applied appropriately, will provide the desired therapeutic results. It also implies that there are discriminable and specifiable dysfunctions and distresses, the correct assessment of which determines the technique or techniques to be applied. The art of clinical practice then resides in the ability to classify the client's problem correctly and to be informed and skillful in the necessary technical actions for its modification. This view is well expressed in the exhortation to researchers to assess "what therapist behaviors are more effective with which type of patient producing which kind of patient changes" (Kiesler, 1971, p. 40). The problem, of course, is in categorizing the dysfunctions and in specifying the techniques. Leaving aside for the moment the issues involved in determining the salient dimensions on which clients and changes should be differentiated, let us take therapist behaviors as our focus.

The desirability of maximum specification of treatment procedures, both for research and clinical purposes, has led to the development of treatment manuals. Such manuals specify the techniques involved in a particular method of treatment and are used both to guide training of therapists and as a basis for evaluating whether alternative therapies do or do not involve similar components or attributes (Luborsky, Woody, McLellan, O'Brien, & Rosenzweig, 1982). Although the early manuals of this kind were developed for the behavior therapies, other manuals can and have been developed for therapies stemming from different theoretical persuasions. That nonbehavioral treatment techniques can be delineated so as to be reliably distinguished and even quantified was evident in many of the early process studies of psychotherapy (Kiesler, 1973), which examined such variables as interpretation, reflection, dynamic focus, and so on. Although a technique-oriented approach to psychotherapy is perhaps most compatible with a behavioral orientation, it is not specific to that orientation.

The predominance of the specific-technique view at the present time is reflected in its utilization in the Psychotherapy of Depression Collaborative Research Program developed by the Psychotherapy and Behavioral Intervention Section, Clinical Research Branch, National Institute of Mental Health (Parloff & Waskow, 1980). The study is examining the relative "effectiveness and safety of two *well defined* [italics added] forms of psychotherapy applied to a specified class of depressive disorders" (p. 2). One of three criteria for the selection of the two psychotherapeutic approaches to be studied (cognitive behavior therapy and interpersonal psychotherapy) was that "the therapy had been standardized in such a fashion that it could be transmitted to other clinicians and researchers (specifically, a manual should be available for the instruction of therapists in the method)" (p. 4). The study includes several "comparison/control conditions," two of which represent specifically delineated procedures; a pharmacotherapy condition (imipramine and clinical management); and a placebo condition (pill–placebo and clinical management).

Research on psychotherapy, as exemplified in the NIMH study, is frequently conceptualized and spoken of in terms analogous to pharmacological treatment. Various treatment approaches, techniques, or modules are applied either to heterogeneous populations, such as outpatient clients, or to more homogeneous groupings, such as phobics or depressives. In the latter case, the specification is in terms of a syndrome or a symptom or, if these terms seem too medically oriented, in terms of a specific dysfunctional behavior or class of such behavior. The search is for the active ingredient (component) or ingredients of the treatment, and there is concern with control for placebo or for nonspecific factors.

Analogizing psychologic and pharmacologic treatments leads to comparative analyses and research designs that may or may not apply with equal logic to both forms of treatment. The analogy can, if it is made explicit, also clarify differences between them. For example, what are called outcome studies in psychotherapy research are structurally similar to clinical trials in drug research: that is, studies in which a particular treatment is applied to a clinical population to test the efficacy of the treatment in a clinical, nonlaboratory setting. This is a relatively late phase in the testing of a new drug and is usually preceded by chemical analyses and animal studies. Although how the drug achieves its effects may not be fully understood, study of the drug, its composition, and actions does not await the proof of its effectiveness in clinical trials. In psychotherapy research, the usual admonition is that the more effective strategy is first to demonstrate the clinical efficacy of a treatment as an indicator that further study is warranted. This places the psychotherapeutic equivalent of clinical trial close to the beginning of the research sequence rather than toward its later stages. Over the past two decades outcome studies increasingly have been given higher priority than process studies, following admonitions regarding the functional autonomy of the practice of psychotherapy (Astin, 1961) and arguments for proving a treatment's effectiveness before exploring its structure or its mode of operation. Psychotherapy outcome studies at the present time, despite the impact of the individual studies, thus have an exploratory quality about them. This contrasts with the more or less finalizing or late-stage testing of pharmacologic clinical trials.

The influence of the drug analogy is perhaps most apparent in the concept of the active ingredient in psychotherapy. It posits that some kind of substance (in the case of psychotherapy some action or manipulation by the therapist) is applied, and it is this action *per se* that results in the alleviation or remediation of the dysfunction or problem. Presumably from this perspective, our ultimate goal would be something equivalent to the *Physicians' Desk Reference* (1982), in which psychotherapy practitioners could look up a specific technique and find information as to its composition, its action or uses, the ways in which it can be administered, dosage indications, possible side effects (untoward consequences), contraindications, precautions concerning its use, and so on. The image may have a certain appeal for some and may appall others. In conception, it is consistent with the repeated emphasis in the literature on the need for greater specificity both in the delineation of techniques and the problems for which they are most useful.

Such specificity lends itself readily to traditional concepts of experimental methodology. The treatment package can be readily replicated and its effects on specified outcome variables can be examined, as can the influence of organismic characteristics on the independent–dependent variable relationship. A number of experimental manipulations are relatively easy to implement. Each of the components of the package can be viewed as discrete independent variables and, by the use of "dismantling treatment strategy," their necessary inclusion or relative contribution to the efficacy of a multifaceted treatment package can be determined. Variations in the form of individual components (their intensity, timing, sequencing, etc.) can also be examined to see how they affect therapeutic outcome. This type of research has been categorized as the "parametric treatment strategy" (Kazdin & Wison, 1978).

The essential emphasis in this perspective is on the identification of the "active" therapeutic ingredient or ingredients. So formulated, the distinction between clinical trials and laboratory research is to be minimized in that the "treatments applied in laboratory projects are the real thing rather than partial analogues" (Bandura, 1978, p. 81). Similarly, a pharmacotherapeutic substance does not change its composition or its physiological action when applied to a subject in the laboratory or to a client who seeks treatment in a clinical context. Nevertheless, questions may be raised as to whether "the move from the laboratory to applied conditions may introduce several variables that attenuate, moderate or entirely eliminate the influence of variables considered to be central in highly controlled laboratory research" (Kazdin & Rogers, 1978, p. 116). Of equal import is whether, in the applied clinical setting, the *fact of applying* the designated treatment technique rather than the technique itself may account for a significant portion of observed therapeutic effects. Thus, Frank (1978) maintains that it is the mobilization of the client's expectations that he or she will be helped that is central to effective treatment—"that simply by offering any form of therapy the therapist conveys the expectation that it [the therapy] will work, thus creating favorable expectations in the patient" (p. 30). The success of treatment is seen as based on a confiding, emotionally charged relationship, a setting containing the symbols of healing, a rationale that prescribes certain activities, therapeutic zeal, and the commitment of the clinician (Frank, 1972). In drug studies, these factors are examined by the use of a placebo, an inert or inactive substance, in the context of a double-blind study (a study in which neither client nor clinician knows whether the active or inactive ingredient is being administered). In pharmacologic studies the placebo represents a control for psychological effects in contradistinction to the chemical effects of the drug. It has been noted that the placebo concept "is not a useful or meaningful construct in relation to psychotherapeutic procedures" (Kirsch, 1978, p. 262), as these procedures are themselves concerned with psychological effects. Shapiro (1978) suggests that the active ingredient–placebo distinction can be maintained with respect to psychological treatments in that in psychotherapy research the placebo has become synonymous with the search for the appropriate control group against which therapies are compared. In its most direct and obvious sense, in this framework the requirements for placebo control may be met to the extent that it is possible for control subjects and their therapists to believe that they are receiving active treatment when their treatment does not include the ingredient that is presumed necessary for therapeutic action.

The issue here not only relates to whether, in evaluating a psychotherapeutic technique, a placebo control design is practical. Certainly, the double-blind study design so central to effective placebo control is not. And one can question whether a non-double-blind placebo study is worthwhile. The point is that logically the placebo-control concept is consistent with the specific-technique specific-effect conceptualization of psychotherapy treatment packages.

ANALYSIS OF TECHNIQUES

What do we mean when we speak of a therapeutic technique? From a behavioral perspective, such procedures as systematic desensitization, symbolic modeling, covert sensitization, response inhibition, reinforcement schedules, and so on can be cited as examples. Techniques may be simple or complex (i.e., made up of one or many components). Thus, systematic desensitization involves training in relaxation, construction of graded anxiety hierarchies, and induction of relaxation and imagined hierarchy items in an appropriate sequence. If we separate the therapist's activities from those of the client, and if we consider treatment technique as activities or manipulations by the therapist that are designed to or have an impact on the client, then among the techniques involved are instruction in relaxation, induction of relaxation, induction of imaginal processes, interruption by suggestion of the imagery when the client signals anxiety, and instructions or suggestions to enhance relaxation. Therefore the therapist's behavioral skills involve verbal behaviors that are explanatory, suggestive, and facilitative of client cooperation. Such activities occur in a relatively defined sequence that is modulated by the pattern of client response. If we observe a therapist in action and wish to check if he or she is implementing the technique, we might use such categories as explanatory statements, motor suggestions ("Rest your right arm on the arm of the chair," "Bend your hand back at the wrist, hold it there, let your hand fall forward"), experiental suggestions ("Notice the pull, the tension in your forearm, feel the relaxation"), motor and/or experiental suggestions ("Let it relax more and more, feel it spread to other body parts"), and questions ("How much did it raise your anxiety level?"). The category of suggestion might be further subdivided into direct suggestions and indirect suggestions. We might provide subcategories to distinguish between direct suggestions and other comments that have a suggestive or inductive impact, even though the statements themselves take a different form. For example, statements could describe the details of an imagined scene ("It is nighttime," Wolpe, 1973, p. 150) or self-referential statements ("I know that when I go to the barber, it is an opportunity for me to relax and, frequently, I almost doze off," p. 153). Another set of categories might be devised for classification of content and might distinguish references to relaxation, anxiety, anxiety-arousing stimuli, statements referring to the client, therapists' self-references, and so on. Such lists are potentially endless; the choice is influenced by theoretical and practical considerations. Interestingly, this kind of analysis of therapist activity is not engaged in when a technique such as systematic desensitization is dismantled in a search for necessary and/or sufficient components. Rather, the analysis is made in terms of products rather than techniques. Thus, systematic desensitization is broken down into relaxation, greater anxiety arousal, or imaging activity. It is not the therapist's technique that is examined but the occurrence in the client of one or another of the phenomena that the therapist's technical operations are designed to induce.

The distinction is important, because discussions of treatment techniques at times classify as techniques both the therapist actions designed to affect the client and the effect of that impact (i.e., the client's response). Thus, Marks (1978) suggests that behavioral treatments of adult neurotics occur within a "dual-process framework." The two processes include motivation to seek and complete treatment and effective execution by the client of the essential therapeutic actions. The first, seen as necessary but insufficient, accommodates expectancies, relationship variables, and alterations in strength of self-efficacy. The second is illustrated in the principle of exposure (which Marks suggests is central in most behavioral treatments of phobias and obsessive-compulsive syndromes) or by the client's rehearsal of a skill that he or she had observed in a modeling context. The second phase requires client execution of the required behaviors. These actions executed by the client are not readily distinguished from behavioral criteria of successful outcome. Examples include exposing oneself to a previously avoided phobic stimulus, inhibiting a previously uncontrollable compulsive act, or engaging in skilled actions that previously were not part of the individual's repertoire. Organization of timing and degree of exposure by the therapist, arrangement of modeling and practice situations, instructions as to how to proceed, and the setting of successive partial goals are part of the therapist's technical armamentarium.

The degrees of effective execution by the client of the essential therapeutic actions then can be conceived as outcome variables, as indicators of improvement, or as signs of partial attainment of treatment goals rather than as process variables. Recast in this way, they become dependent variables. The therapist's contribution to the attainment of treatment's ultimate goals centers on those *therapist* actions that enable the client to move along the desired change dimensions. This holds for dimensions as diverse as willingness to expose oneself to anxiety-arousing stimuli, increasing tolerance for such exposure, modifying cognitions of one's abilities in a more accurate and positively valued direction, engaging with expanded effectiveness in social interactions, allowing greater and more appropriate experiencing and expression of feelings, developing a clearer and more stable awareness of oneself as an independent and worthwhile individual, or becoming more decisive with respect to one's goals and purposes in life. The partial accomplishments, of course, not only reflect therapeutic progress but, in most instances, also contribute to further client improvement. In this respect, differentiation between treatment process and treatment outcome might seem less clear. Nevertheless, to conceptualize partial goal attainments as techniques introduces a tautological element to the degree that treatment and outcome are confounded.

If one takes treatment techniques to be synonymous with therapist behaviors, one is immediately struck with the diversity of operations that have been proposed. Kiesler (1973) summarized about 95% of the systems used in research, analyzing the process of "insight oriented individual psychotherapy of late adolescents or adults who come to various inpatient clinics or agencies" (p. xix). One of his book's appendices provides a summary listing of therapist variables measured by one or another of the 17 major process systems that he reviewed. Even a superficial perusal of that summary reveals a striking diversity of behaviors that have been examined. A truncated and arbitrary selection will illustrate the point: freshness of words and combinations used by the therapist, voice quality, labeling rather than exploring feelings, drive- or demand-eliciting statements, grammatical form of propositions, affective content of propositions, amount of structure of information contained in messages, mean speech duration, mean speech latency, percentage of interruptions, degree of probing, redirecting responses, type of interpretation, depth of interpretation, level of specificity, initiative, and so on. Some of these headings refer to a single or limited behavioral phenomenon that is quantifiable. Others, such as type of interpretation, refer to a broad class of behaviors that are further categorized as to the form in which they are manifest—for example, connecting two aspects of the content of previous client statements, commenting on the client's bodily or facial expression as manifestations of the client's feelings, and so on (Harway, Dittmann, Raush, Bordin, & Rigler, 1955). Some of these actions are not meant, in and of themselves, as the totality of the therapeutic effort. Rather, they may be presumed to have impact or consequence for the client as a function of when and where they occur in relation to his or her communications and behavior in the interview. But even taken in isolation, they imply that the vocal and verbal activity of the therapist (in its qualitative, quantitative, syntactic, semantic, organizational, and contextual aspects) may individually, cumulatively, or interactively (forming a complex pattern of therapist expression) play a role and can be manipulated in a systematic fashion to achieve desired effects. It is true that process studies are generally concerned with intrainterview effects on client behavior; nevertheless, those effects are directed toward the achievement of ultimate treatment goals. Although many of the variables investigated were chosen on the basis of particular theories of psychotherapy, certainly a number of these variables, by themselves, appear to have an atheoretical quality insofar as traditional therapeutic theories are concerned.

RELATIONSHIP AND MOTIVATION

Motivational factors are a function not only of characteristics of the client and of the situational factors that lead him or her to seek therapy, but also of those therapist techniques that facilitate the client's continuance in treatment and enhance the likelihood that he or she will engage in the necessary activities. The therapist's contribution to the necessary motivational development and enhancement is then the essential component of his or her contribution to psychotherapy. The therapeutic techniques are those actions that enable the client to achieve progressively those actions, understandings, feelings, and attitudes whose fuller development is the goal of treatment. The techniques and process are those related to social influence and flow from the interpersonal interaction of client and therapist. Such factors have been posited as common to all effective psychotherapy and not to any specific theory of therapy.

Earlier in this chapter, it was noted that the literature distinguishes two contrasting attitudes as to the nature of therapeutic activity: one emphasizing technique and the other emphasizing that it is "the quality of the patient–therapist interaction [that] represents the fulcrum upon which therapeutic progress turns" (Strupp, 1981, p. 229). The therapist's role, however, in molding, modifying, and modulating the relationship with the client, is effected through actions that can be considered under the rubric of technique. Viewing psychotherapy as a persuasive process, Glaser (1980), applying rhetorical analysis to the content of the therapeutic dialogue, suggests techniques by which therapists can increase client perception of the therapist's trustworthiness, attractiveness, and expertness and can increase the persuasiveness of the therapist's statements. Client satisfaction and consequent continuation in treatment can be furthered by therapist responses that inform the client that his or her communications are understood—responses that contain content that the client will perceive as pertinent to his or her idea of the important issues (Rosen, 1972). Client expectancy can be enhanced by the therapist's structuring of the treatment, by discussions of the utility and proven efficacy of the procedures, and by the provision of a persuasive rationale that explains the development of the client's problems (Wilson & Evans, 1977). Client perception of therapist effectiveness can be enhanced when therapists refer to their specific areas of expertise (Strong & Matross, 1973) and focus on the here-and-now as well as on the client's resources (Mitchell & Berenson, 1970).

The therapist's contributions to the relationship that develops between the participants may be analyzed in terms of component actions. This does not, however, imply a behaviorally simplified view of the process. The influence of a particular act is not independent of the meaning imposed upon it by the context and the client. The technique–relationship distinction would seem more to concern abstractive level than antithetical constructions. The early psychotherapy process studies were instigated within a Rogerian framework, yet they frequently utilized relatively specific qualitative and quantifiable indices of the therapist's participation in the process.

A detailed examination of the interpersonal motivational influences is beyond the purview of this chapter. Unlike some of the techniques discussed earlier, however, the therapist's influencing techniques are generally more clearly separable from client response. The client's state may be part of the description of an interpersonal technique—for example, empathy defined as the therapist responding accurately to the client's present feelings and communicating awareness of their exact intensity (Truax & Carkhuff, 1967). However, neither the preceding nor consequent client condition or activity need be immediately related to those indices of change or of improvement that are component aspects of ultimate treatment goals.

ASSESSMENT AND TECHNIQUE

What are the issues that face the therapy practitioner? Clearly, a first step is to assess the client's problem and to develop a strategy for enabling him or her to achieve desired changes. Assume that the presenting problem is a phobia and, therefore, that exposure to the anxiety-arousing stimuli is a desired goal. In order to achieve that goal, however, it is necessary that the client continue in treatment and be influenced by therapist suggestions, information, instructions, interpretations, and encouragements so that he or she will engage in the steps that lead to exposure to the previously avoided situation. At this point, then, assessment is needed as to what actions by the therapist, if any, are needed to facilitate client continuation and progress in treatment. This assessment proceeds along different dimensions than those involved, for example, in distinguishing phobia from depression. The therapist has to decide how best to involve the client, which strategies will enable the client to achieve the initial requisite level of comfort in the situation, and what level of abstraction or specificity in communication will most effectively and informatively structure therapy in a way that is meaningful and convincing, so that expectations of success are aroused in the particular client. Does the therapist need to engage in actions that increase his or her credibility, or is the fact of his or her socially defined professional status already sufficient to provide credibility in this instance? Will this client be more responsive to therapist self-revelations or to a more authoritarian and self-assured tone? As treatment proceeds, the immediate demands of the therapeutic process may change and the therapist's interventions (actions, techniques) will be altered in response to these requirements.

Most of the earlier forms of psychotherapy prescribed a single treatment package that was seen as suitable to a rather broad and heterogeneous range of disturbance. As new methods were proposed, they did not so much present themselves as treatments for previously untreatable dysfunctions but represented different theoretical viewpoints challenging the conceptual framework and even value orientations of their predecessors (in essence suggesting that the new methods replace the earlier ones). Therapies were divided into "schools" rather than variant treatments for different problems. This pattern continues into the present, though in the last four decades there has been an increase within several schools of suggestions for modifying treatment procedures with different client subgroups without altering the basic structure of the particular theory. Within psychoanalysis, one can point to the principle of flexibility proposed by Alexander and French in 1946 and later psychodynamic short-term therapies suggested as suitable for certain types of clients (Malan, 1963; Sifneos, 1981). The pattern is most apparent in developments within the behavior therapies and in the profusion of methods, many designed not as general treatments but for specific symptoms or syndromes. Here too, the earlier broad-gauge therapy view still recurs, as a review of a recent volume on the treatment of phobias notes that its significance is "in its willingness to recast most everyone's neurotic problems as phobias, thereby rendering them appropriate for the book's simple, straightforward and effective cure" (Lundy, 1982, p. 101).

The most ready specification of client characteristics has been in terms of clinical manifestation, and treatment techniques are chosen as a function of the predominant clinical syndrome. Research evaluating the effectiveness of different treatment approaches compares a treatment with a control group or compares one treatment with another and focuses on effectiveness of the treatment for the group as a whole rather than for specific clients. The investigators provide information about the relative effectiveness of the treatment for the average manifestation of the dysfunction. What emerges from this is the matching of client and treatment on the basis of symptoms. Discussing the treatment of depression, Craighead (1979) questions whether outpatient unipolar or neurotic depression represents a homogeneous entity; he proposes that treatment interventions need to be selected in relation to the particular problem identified in specific spheres: overt motor, cognitive, and physiological responses. In a similar vein, Liberman (1979) suggests a modular treatment strategy in which the therapist selects, from the full array of problem-specific treatment techniques available, only those called for by the client's specific problems.

> Full appreciation of the diverse potentialities for the development, manifestation, and course of depressive phenomena—including any interactions between the biological, behavioral, and environmental influences—suggests that therapeutic intervention must be flexibly responsive to the varied needs of the individual depressed person. (p. 12)

These analyses speak to the heterogeneity within clinical categories and direct our attention to the need for an idiographic analysis of the syndrome as it takes form in the client being treated. Complementary analyses are needed that include a consideration of the client's perception of the nature and meaning of the therapist's actions (or lack of thereof), that is, how the client construes therapy and the relationship with the therapist, and, beyond that, the client's motivational characteristics, cognitive style, intellect, interpersonal attitudes and values, and other personal attributes as may bear on how he or she is likely to attend, appraise, and respond to different interventions by the therapist. At present, a framework for an integrated systematic assessment of these areas is, if anything, more rudimentary than our conceptualization of dysfunctional syndromes.

The distinctions between therapist and client actions, between therapist actions focused primarily toward the client's dysfunction and those that enhance the client's responsivity to the dysfunction focused interventions, between in-therapy client responses that involve outcome criterion behaviors and those that are necessary to the treatment process although they do not fall on the criterion dimension reemphasize the complexity of psychotherapy. The component actions of the participants are specifiable, as are, potentially, the consequences of those actions. The substantiation of the principles underlying these action–consequence sequences (the organismic and contextual variables that modulate them) comes from the laboratory. Application of these principles in psychotherapy is a technical enterprise; in its etymological derivation, *technique* refers to the "manner of artistic execution" (Onions, 1966).

Psychotherapy is not a unitary or a bidimensional process based on one or two active ingredients. As is true in the development of most human interactions, it is a process that involves a multiplicity of multifaceted interchanges between the participants. In dealing with the idiographic context of clinical practice, the clinician needs to draw upon a body of information that goes beyond that which the literature subsumes under the headings of "psychotherapy" and "psychotherapy research." One suspects that the eclecticism subscribed to by so many practitioners reflects their awareness of and openness to a broader range of issues and findings in psychology—issues and findings that, though they are not specific to psychotherapy, contribute ultimately to its practice.

REFERENCES

Alexander, F., & French. T. M. *Psychoanalytic therapy.* New York: Ronald Press, 1946.

Astin, A. W. The functional autonomy of psychotherapy. *American Psychologist,* 1961, *16,* 75–78.

Bandura, A. On paradigms and recycled ideologies. *Cognitive Therapy and Research,* 1978, *2,* 79–103.

Bordin, E. S. *Research strategies in psychotherapy.* New York: Wiley, 1974.

Craighead, W. E. *Behavior therapy for depression: Issues resulting from treatment studies.* Paper presented at NIMH Conference, Pittsburgh, Pa., April 1979.

Frank, J. D. The bewildering world of psychotherapy. *Journal of Social Issues,* 1972, *28,* 27–43.

Frank, J. D. Expectation and therapeutic outcome—the placebo effect and the role induction interview. In J. D. Frank, R. Hoehn-Saric, S. D. Imber, B. L. Liberman, & A. R. Stone (Eds.), *Effective ingredients of successful psychotherapy.* New York: Brunner/Mazel, 1978.

Glaser, S. R. Rhetoric and psychotherapy. In M. J. Mahoney (Ed.), *Psychotherapy process.* New York: Plenum Press, 1980.

Harway, N. I., Dittman, A. T., Raush, H. L., Bordin, E. S., & Rigler, D. The measurement of depth of interpretation. *Journal of Consulting Psychology,* 1955, *19,* 247–253.

Kazdin, A. E., & Rogers, T. On paradigms and recycled ideologies: Analogue research revisited. *Cognitive Therapy and Research,* 1978, *2,* 105–117.

Kazdin, A. E., & Wilson, G. T. *Evaluation of behavior therapy.* Cambridge, Mass.: Ballinger, 1978.

Kiesler, D. J. Experimental designs in psychotherapy research. In A. E. Bergin & S. L. Garfield (Eds.), *Handbook of psychotherapy and behavior change.* New York: Wiley, 1971.

Kiesler, D. J. *The process of psychotherapy.* Chicago: Aldine, 1973.

Kirsch, I. The placebo effect and the cognitive-behavioral revolution. *Cognitive Therapy and Research,* 1978, *2,* 255–264.

Liberman, R. P. *To each his own: Individualizing treatment strategies for depressed persons.* Paper presented at NIMH Conference, Pittsburgh, Pa., April 1979.

Luborsky, L., Woody, G. E., McLellan, A. T., O'Brien, C. P., & Rosenzweig, J. Can independent judges recognize different psychotherapies? An experience with manual-guided therapies. *Journal of Consulting and Clinical Psychology,* 1982, *50,* 49–62.

Lundy, R. M. Expose yourself. *Contemporary Psychology,* 1982, *27,* 101–102.

Malan, D. H. *A study of brief psychotherapy.* New York: Plenum Press, 1963.

Marks, I. Behavioral psychotherapy of adult neurosis. In S. L. Garfield & A. E. Bergin (Eds.), *Handbook of psychotherapy and behavior change* (2nd ed.). New York: Wiley, 1978.

Mitchell, K. M., & Berenson, B. G. Differential use of confrontation by high and low facilitative therapists. *Journal of Nervous and Mental Disease,* 1970, *151,* 303–309.

Onions, C. T. (Ed.). *The Oxford dictionary of English etymology.* New York: Oxford University Press, 1966.

Parloff, M. B., & Waskow, I. E. *Psychotherapy of depression collaborative research program (Revised research plan).* Clinical Research Branch, NIMH, January 1980.

Physicians' Desk Reference (36th ed.). Oradell, N.J.: Medical Economics, 1982.

Rosen, A. The treatment relationship: A conceptualization. *Journal of Consulting and Clinical Psychology,* 1972, *38,* 329–337.

Shapiro, A. K. Placebo effects in medical and psychological therapies. In S. L. Garfield & A. E. Bergin (Eds.), *Handbook of psychotherapy and behavior change* (2nd ed.). New York: Wiley, 1978.

Sifneos, P. E. Short-term anxiety provoking psychotherapy: Its history, technique, outcome, and instruction. In S. H. Budman (Ed.), *Forms of brief therapy.* New York: Guilford, 1981.

Strong, S. R., & Matross, R. P. Change processes in counseling and psychotherapy. *Journal of Counseling Psychology,* 1973, *20,* 25–37.

Strupp, H. H. (Ed.). *A bibliography of research in psychotherapy.* Chapel Hill, N.C.: University of North Carolina, July 1964.

Strupp, H. H. Toward the refinement of time-limited dynamic psychotherapy. In S. H. Budman (Ed.), *Forms of brief therapy.* New York: Guilford, 1981.

Strupp, H. H., & Bergin, A. E. *Research in individual psychotherapy* (NIMH, U.S. Public Health Service Publication No. 1944). Washington, D.C.: U.S. Government Printing Office, 1968.

Sundland, D. M. Theoretical orientations of psychotherapists. In A. S. Gurman & A. M. Razin (Eds.), *Effective psychotherapy.* New York: Pergamon, 1977.

Sundland, D. M., & Barker, E. N. The orientations of psychotherapists. *Journal of Consulting Psychology,* 1962, *26,* 201–212.

Truax, C. G., & Carkhuff, R. R. *Toward effective counseling and psychotherapy: Training and practice.* Chicago: Aldine, 1967.

Wilson, G. T., & Evans, I. M. The therapist–client relationship in behavior therapy. In A. S. Gurman & A. M. Razin (Eds.), *Effective psychotherapy: A handbook of research.* New York: Pergamon, 1977.

Wolff, W. *Contemporary psychotherapists examine themselves.* Springfield, Ill.: Charles C Thomas, 1956.

Wolpe, J. *The practice of behavior therapy* (2nd ed.). New York: Pergamon, 1973.

Index